Algorithms for Next-Generation Sequencing Data

Mourad Elloumi
Editor

Algorithms for Next-Generation Sequencing Data

Techniques, Approaches, and Applications

Editor
Mourad Elloumi
LaTICE
Tunis, Tunisia

University of Tunis-El Manar
Tunis, Tunisia

ISBN 978-3-319-86710-6 ISBN 978-3-319-59826-0 (eBook)
DOI 10.1007/978-3-319-59826-0

Softcover reprint of the hardcover 1st edition 2017

Printed on acid-free paper

This Springer imprint is published by Springer Nature
The registered company is Springer International Publishing AG
The registered company address is: Gewerbestrasse 11, 6330 Cham, Switzerland

To my parents and my children.

Preface

A *deoxyribonucleic acid* (DNA) macromolecule can be coded by a sequence over a four-letter alphabet. These letters are A, C, G, and T, and they code respectively the bases *Adenine*, *Cytosine*, *Guanine* and *Thymine*. DNA *sequencing* consists then in determining the exact order of these bases in a DNA macromolecule. As a matter of fact, DNA sequencing technology is playing a key role in the advancement of molecular biology. Compared to previous sequencing machines, *Next-Generation Sequencing* (NGS) machines function much faster, with significantly lower production costs and much higher throughput in the form of short *reads*, i.e., short sequences coding portions of DNA macromolecules.

As a result of the extended spread of NGS machines, we are witnessing an exponential growth in the number of newly available short reads. Hence, we are facing the challenge of storing them to analyze huge numbers of reads representing sets of portions of genomes, or even whole genomes. The analysis of this huge number of reads will help, among others, to decode life's mysteries, detect pathogens, make better crops, and improve quality of life. This is a difficult task, and it is made even more difficult not only by the short lengths of the reads and the huge number of these reads but also by the presence of high similarity between the concerned portions of genomes, or whole genomes, and by the presence of many repetitive structures in these genomes, or whole genomes. Such a task requires the development of fast algorithms with low memory requirements and high performance.

This book surveys the most recent developments on algorithms for NGS data, offering enough fundamental and technical information on these algorithms and the related problems, without overcrowding the reader's head. It presents the results of the latest investigations in the field of NGS data analysis. The algorithms presented in this book deal with the most important and/or the newest topics encountered in this field. These algorithms are based on new/improved approaches and/or techniques. The few published books on algorithms for NGS data either lack technical depth or focus on specific topics. This book is the first overview on algorithms for NGS data with both a wide coverage of this field and enough depth to be of practical use to working professionals. So, this book tries to find a balance between theoretical and practical coverage of a wide range of issues in the field of NGS data analysis. The techniques and approaches presented in this book combine sound theory with practical applications in life sciences. Certainly, the list of topics covered in this book is not exhaustive, but it is hoped that these topics will get the reader to think of the implications of the presented algorithms on other topics. The chapters presented in this book were carefully selected for quality and relevance. This book also presents experiments that provide qualitative and quantitative insights into the field of NGS data analysis. It is hoped that this book will increase the interest of researchers in studying a wider range of combinatorial problems related to NGS data analysis.

Preferably, the reader of this book should be someone who is familiar with bioinformatics and would like to learn about algorithms that deal with the most important and/or the newest topics encountered in the field of NGS data processing. However, this book could be used by a wider audience such as graduate students, senior undergraduate students, researchers, instructors, and practitioners in bioinformatics, computer science, mathematics, statistics, and life sciences. It will be extremely valuable and fruitful for these people. They will certainly find what they are looking for or, at least, a clue that will help them to make an advance in their research. This book is quite timely since NGS technology is evolving at a breathtaking speed and will certainly point the reader to algorithms for NGS data that may be the key to new and important discoveries in life sciences.

This book is organized into four parts: *Indexing, Compression, and Storage of NGS Data*; *Error Correction in NGS Data*; *Alignment of NGS Data*; and *Assembly of NGS Data*. The 14 chapters were carefully selected to provide a wide scope with minimal overlap between the chapters to reduce duplication. Each contributor was asked to present review material as well as current developments. In addition, the authors were chosen from among the leaders in their respective fields.

Tunis, Tunisia Mourad Elloumi
April 2017

Contents

Contributors

Ziv Bar-Joseph Computational Biology Department and Machine Learning Department, School of Computer Science, Carnegie Mellon University, Pittsburgh, PA, USA

Guillermo Barturen Centre for Genomics and Oncological Research (GENYO), Granada, Spain

Nadia Ben Nsira Laboratory of Technologies of Information and Communication and Electrical Engineering (LaTICE), Tunis, Tunisia

University of Tunis-El Manar, Tunis, Tunisia

The Computer Science, Information Processing and Systems Laboratory (LITIS), EA 4108, University of Rouen-Normandy, Normandy, France

Gaetan Benoit GenScale, Rennes, France

INRIA, Rennes, France

Matteo Comin Department of Information Engineering, University of Padova, Padova, Italy

Lucas Czech Heidelberg Institute for Theoretical Studies, Heidelberg, Germany

Erwan Drezen GenScale, Rennes, France

INRIA, Rennes, France

Mourad Elloumi LaTICE, Tunis, Tunisia

University of Tunis-El Manar, Tunis, Tunisia

Giulia Fiscon Institute for Systems Analysis and Computer Science "Antonio Ruberti" (IASI), National Research Council (CNR), Rome, Italy

Tomáš Flouri Heidelberg Institute for Theoretical Studies, Heidelberg, Germany

Michael Hackenberg Department of Genetics, University of Granada, Granada, Spain

Costas S. Iliopoulos Department of Informatics, King's College London, London, UK

Géraldine Jean Laboratoire d'Informatique de Nantes-Atlantique (LINA), UMR CNRS 6241, Université de Nantes, Nantes, France

Kassian Kobert Laboratoire de Biométrie et Biologie Évolutive (LBBE), Université de Lyon 1, Lyon, France

Dominique Lavenier GenScale, Rennes, France

IRISA-CNRS, Rennes, France

Thierry Lecroq LITIS, EA 4108, University of Rouen-Normandy, Normandy, France

Claire Lemaitre GenScale, Rennes, France

INRIA, Rennes, France

Jimmy Lin Rare Genomics Institute, St. Louis, MO, USA

Yongchao Liu School of Computational Science and Engineering, Georgia Institute of Technology, Atlanta, GA, USA

José L. Oliver Department of Genetics, University of Granada, Granada, Spain

Solon P. Pissis Department of Informatics, King's College London, London, UK

Andreea Radulescu Laboratoire d'Informatique de Nantes-Atlantique (LINA), UMR CNRS 6241, Université de Nantes, Nantes, France

M. Sohel Rahman ALEDA Group, Department of Computer Science and Engineering, Bangladesh University of Engineering and Technology (BUET), Dhaka, Bangladesh

Hugues Richard Analytical Genomics Group, Laboratory of Computational and Quantitative Biology, UMR7238 University Pierre and Marie Curie, Paris, France

Guillaume Rizk GenScale, Rennes, France

IRISA-CNRS, Rennes, France

Irena Rusu Laboratoire d'Informatique de Nantes-Atlantique (LINA), UMR CNRS 6241, Université de Nantes, Nantes, France

Michele Schimd Department of Information Engineering, University of Padova, Padova, Italy

Bertil Schmidt Institut für Informatik, Johannes Gutenberg University, Mainz, Germany

Marcel H. Schulz Excellence Cluster for Multimodal Computing and Interaction, Saarland University, Saarbrücken, Germany

Computational Biology and Applied Algorithms, Max Planck Institute for Informatics, Saarbrücken, Germany

Carol Shen School of Medicine, Washington University in St. Louis, St. Louis, MO, USA

Tony Shen School of Medicine, Washington University in St. Louis, St. Louis, MO, USA

Enrico Siragusa IBM Watson Research, Yorktown Heights, NY, USA

Alexandros Stamatakis Heidelberg Institute for Theoretical Studies, Heidelberg, Germany

Karlsruhe Institute of Technology, Institute for Theoretical Informatics, Karlsruhe, Germany

Evangelos Theodoridis Springer-Nature, London, UK

David Weese SAP Innovation Center, Potsdam, Germany

Emanuel Weitschek Department of Engineering, Uninettuno International University, Rome, Italy

Jiajie Zhang Heidelberg Institute for Theoretical Studies, Heidelberg, Germany

Part I
Indexing, Compression, and Storage of NGS Data

Chapter 1 Algorithms for Indexing Highly Similar DNA Sequences

Nadia Ben Nsira, Thierry Lecroq, and Mourad Elloumi

1.1 Introduction

The availability of numerical data grows from one day to the other in an extraordinary way. This is the case for DNA sequences produced by new technologies of high-throughput *Next Generation Sequencing* (NGS). Hence, it is possible to sequence several genomes of organisms and a project[1] now provides about 2500 individual human genomes (sequences of more than 3 billion characters (A, C, G, T). In fact, we come across a huge, and growing, amount of *highly similar DNA sequences* that are more than 99% identical to a reference genome. These data form a large collection of sequences with great rates of similarity, *highly similar sequences*, where each entry differs from another by a few numbers of differences (often

[1] http://www.1000genomes.org.

N.B. Nsira (✉)
Laboratory of Technologies of Information and Communication and Electrical Engineering (LaTICE), Tunis, Tunisia

University of Tunis-El Manar, Tunis, Tunisia
e-mail: nadia.bennsira@etu.univ-rouen.fr

The Computer Science, Information Processing and Systems Laboratory (LITIS), EA 4108, University of Rouen-Normandy, Normandy, France
e-mail: Thierry.Lecroq@univ-rouen.fr

T. Lecroq
LITIS, EA 4108, University of Rouen-Normandy, Normandy, France
e-mail: mourad.elloumi@gmail.com

M. Elloumi
LaTICE, Tunis, Tunisia

University of Tunis-El Manar, Tunis, Tunisia

M. Elloumi (ed.), *Algorithms for Next-Generation Sequencing Data*,
DOI 10.1007/978-3-319-59826-0_1

involving *Single Nucleotide Variations* (SNVs) of type *substitutions*, *insertions* and *deletions* (indels), *Copy Number Variations* (CNVs) and translocations). Nowadays, the generated sequences are frequently used. Hence, a great need to store and process data has emerged. This is the reason why *indexing* has become one particular area of extreme interest in the last few decades.

Several *data structures* have been designed that enable storing these impressive amounts of data efficiently, while allowing to search quickly among them. Thus, all occurrences of any given pattern can be found without traversing the whole sequences. In short, indexing is profitable and useful if it is used regularly. So, one takes advantage of the index if sequences are available beforehand, very long and do not change periodically. In general, we aim at constructing an index that provides efficient answers to queries with reasonable building and maintenance costs. *Basic*, also called *classical data structures* such as *tries* [10], *suffix trees* [14, 45, 58, 60] and *suffix arrays* [42], *Directed Acyclic Word Graphs* (DAWG) [6] and the compact version *CDAWG* [12] are arguably very interesting data structures for string analysis, especially when searching over large sequence collections. Yet, these structures are *full-text indexes*: they require a large amount of space to represent a sequence.

A recent trend of *compressed* data structures has made much progress in terms of saving space requirements. This new family of data structures develops indexes based on *text compression*, so that their size depends on the compressed text length. For instance, some indexes are based on the *Lempel–Ziv compression* [37]. From this has emerged a new concept of a *self-index* data structures, which are compressed indexes able to replace the original text and to support search operations. The first self-indexes were the *FM-index* proposed by Ferragina and Manzini [16] and the *Compressed Suffix Array* (CSA) by Grossi et al. [23]. Compressed data structures enable compressing sequence collections close to their *high-order entropy*. Yet, it was proven that the entropy is a good measure of compression only when one considers a single repetitive sequence, but when considering a large set of similar sequences the memory requirement is only improved slightly compared to basic structures. Recently, *more advanced* solutions appeared in order to further minimize the cost in space while saving the search time. These solutions seek to represent a collection of similar sequences by storing only a reference sequence with differences from each of the other sequences. At this point, indexes exist that have successfully exploited the advantage of high similarity between DNA sequences coming from the same species.

The classical application for indexes is to solve the *exact pattern matching* problem in linear time. The problem is formalized as follows. Given a text T of length n and a pattern P of length $m \leq n$, a data structure can report an occurrence of P in T in time $O(m)$.

In this chapter, we make a survey of index construction algorithms that we deem to be the most interesting ones. We first give an overview of the *basic* indexes that represent one text or a set of texts considering the *generalized version* of the index. Then, we cover the algorithms for *advanced* structures and explain the relationship between the entropy measure and the index. Finally, we aim at indexes that work for collections of repetitive sequences and focus on how each solution takes advantage of the high rate of similarity between sequences. We explain for each data structure how it supports and solves the search functionality.

1.2 Basic Concepts

An *alphabet* is a finite set of ordered letters, usually called Σ, and its size is denoted by $\sigma = |\Sigma|$. A *string* or a *sequence* T is a sequence of *characters* of Σ. The *length* of a string T is denoted by $|T|$. Note that a string can be empty; in this case we denote it by ε and $|\varepsilon| = 0$. Thus, all possibilities of strings over Σ are elements of Σ^*. Σ^n contains all strings of length n.

The *concatenation* $\cdot$ of two strings T of length n and a string T' of length n' is obtained by appending T' at the end of T. The obtained string is in the form $T \cdot T' = TT'$ of length $n + n'$.

A string T of length n is indexed from 0 to $n - 1$, so that $T[i]$ corresponds to the $i + 1$-th character of T. A *substring* or a *factor* of T is composed from consecutive characters of T. A factor is written $T[i \dots j] = T[i]T[i + 1] \cdots T[j]$ where i is the starting position of the factor in T and j is its ending position. A *prefix* is a factor starting at the first position of T and is denoted by $T[0 \dots j]$. A *suffix* is a factor ending at the last position of T and is of the form $T[i \dots n-1]$. A prefix w or a suffix w of T is *proper* if $w \neq T$. The *mirror* (reverse) of T is the string that corresponds to $T[n - 1]T[n - 2] \cdots T[0]$ and is denoted by $T^{\sim}$.

The *lexicographic order* $\leq$ is an order over strings induced by an order over characters denoted in the same way. Let $u, v \in \Sigma^*$, $u \leq v$ if u is a prefix of v or $u = u'au_1$ and $v = u'bv_1$ and $a < b$ with $u', u_1, v_1 \in \Sigma^*$ and $a, b \in \Sigma$.

All indexes studied in the following enable identifying *exact occurrences* of a given (*relatively short*) *pattern* $P \in \Sigma^*$ of length m in a given sequence T of length n, $m \leq n$. We say that P *occurs* in T at position i if it is equal to $T[i \dots i + m - 1]$.

A *circular shift* (permutation) or *conjugate* of T is a string of length equal to T obtained by concatenating a suffix of T with a prefix of T. The circular permutation starting at position i is said to be on an order of i.

The empirical entropy consists of a common measure of the compressibility of a sequence. The zero-order empirical entropy of a text T of length n, $H_0(T)$ is the average number of bits needed to represent a character of T if each symbol always receives the same code. Formally it is defined as follows[2]:

$$H_0(T) = \sum_{c \in \Sigma} \frac{n_c}{n} \log^2 \frac{n}{n_c}$$

where n_c is the number of occurrences of the symbol c in T.

[2]In this exposition all logarithms are in base 2 unless stated otherwise.

If one wants to achieve better ratios, one can encode each character according to the appearance probability of the k previous characters. Hence, one gets the k-th order empirical entropy defined as follows:

$$H_k(T) = \sum_{s \in \Sigma^k} \frac{|T^s|}{n} H_0(T^s)$$

where T^s is the sequence of characters preceded by the factor s in T.

The *compression* of a text consists of representing the text in a reduced fashion. The main goal of the existing compressed indexes is to efficiently represent information in the most reduced way, without risk of loss, so that one can easily retrieve the original text after decompression. A *compressed index data structure* is an index that requires a space storage proportional to the *entropy* of the text. A *self-index* representation is a compressed index that advantageously replaces the original text and can reproduce any substring, while maintaining the fast and efficient search on the index.

Most compressed data structures support basic queries such as *rank* and *select*. The operation $rank_c(T, i)$ counts the number of a given character c up to position i, and $select_c(T, i)$ finds the position of the i-th occurrence of c in T. Note that those indexes can support other useful queries on characters.

The problems we focus on in this chapter are defined as follows. Given a reference sequence T_0, a large set $\{T_1, \ldots, T_{r-1}\}$ of sequences highly similar to T_0, which contains few differences compared to T_0, and a (relatively short) pattern P of length m, all over an alphabet $\Sigma = \{\mathtt{A}, \mathtt{C}, \mathtt{G}, \mathtt{T}\}$, we have two main problems of interest:

1. efficiently indexing the set of sequences, while exploiting the high rate of similarity between them (by alleviating the index from the redundant information);
2. identifying the occurrences of P in the set in possibly $O(m)$ time by solving efficiently queries such as the **counting** of the number of occurrences, and **locating** their positions.

We denote by N the sum of the length of the all sequences $T_0, \ldots, T_{r-1}$: $N = \sum_{i=0}^{r-1} |T_i|$.

1.3 Basic Data Structures

As a preliminary, classical data structures had birth in the need to study data and to solve problems to facilitate access and analysis in biological sequences. The sequences are usually stored sequentially together in one structure. Although these structures did not show a clever analysis when studying genetic variations between individuals of the same species, they still were versatile and able to answer to queries.

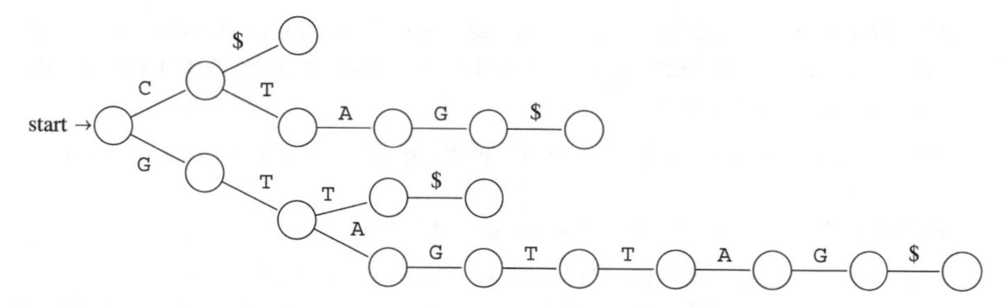

Fig. 1.1 A trie for $S = \{$C$, CTAG$, GTTG$, GTAGTTAG$$\}$. In general the arity of trie nodes may be as large as the alphabet size

1.3.1 Trie

A trie is an edge-labeled tree structure used to represent a set of strings.

Definition 1 A trie of a set of strings $S = \{T_0, T_1, \ldots, T_{r-1}\}$ is a tree where edges are single-labeled (labeled by an individual symbol) and internal nodes represent distinct prefixes of strings of the set. The root node represents the empty string ε, and leaves represent strings of the set. Given a set of strings, its trie is denoted by $Trie(S)$.

An example of a trie is given in Fig. 1.1. When constructing the trie, common prefixes of strings are factorized. Note that to avoid that some strings of S match prefixes of others, a special character called *terminator* (not occurring in the alphabet) is added at the end of the strings, thus avoiding that certain strings end on internal nodes of the trie. This ensures that each leaf represents a distinct string and the trie contains exactly r leaves. In the example in Fig. 1.1, this character is denoted by $.

1.3.1.1 Structure Construction

The algorithm proceeds as follows: given a set of strings $S = \{T_0, T_1, \ldots, T_{r-1}\}$, when adding a new string T_g it identifies the node u representing the longest prefix of T_g already present in the trie; then, the insertion of the remaining suffix of T_g is done from node u. The algorithm runs in $O(N)$, N being the sum of lengths of all strings in the set T, when assuming that the size of the alphabet σ is constant or $O(N \log \sigma)$ for general alphabet (see [10]).

1.3.1.2 Search for a Pattern *P* of Length *m*

The search for a pattern P of length m is performed in $O(m)$ or in $O(m\sigma)$ time (depending on the size of the alphabet) [10]. The search aims to identify the longest prefix of P corresponding to a path in the trie. Indeed, the search consists of finding the path from the root labeled by the symbols of the pattern in the trie. If such a path exists, there are two cases:

1. the path ends in an internal node, then P is a prefix of some strings represented by the trie;
2. the path ends in a leaf, then P is a string of the input set.

If such a path does not exist, P is not a prefix of any string in the set.

In practice, the dependence on the $\log \sigma$ of the above complexities can be reduced to $O(1)$ by using a table of length σ space in each node. Thus, to find the successor of a given node a direct access to the table is performed. Thus, the total size is $O(\sigma)$ times the number of nodes.

Another alternative is to use hashing techniques for the children of each node. This amortizes the complexities in time and space to $O(1)$ per edge. But the construction requires quadratic space.

A trie can be used as an index for an input text T; thereby a data structure called *suffix trie* is used and defined as follows.

Definition 2 A suffix trie of a text T is a trie that represents all suffixes of T.

Being an index, the suffix trie can be exploited to support tasks such as locating a given pattern P of length m in the indexed text T. When one feeds the suffix trie with the characters of the pattern there are two main situations:

1. all the characters of the pattern can be read and a node is reached, and then one can obtain all suffixes of T prefixed by P (positions of the corresponding suffixes are stored in leaves contained in the subtree rooted by the reached node). It should be noted that the number of occurrences of P corresponds to the number of leaves in the subtree;
2. all the characters of the pattern cannot be read and then there is no suffix of T that is prefixed by P. Thus, P does not occur in T.

The search algorithm is simple and reports all occurrences of P in $O(m + occ)$ time where occ is the number of occurrences of P in T.

The suffix trie has $\Theta(n^2)$ nodes. It can be compacted to allow linear construction.

Definition 3 A compact trie is obtained from the original trie by removing internal nodes with exactly one successor and concatenating edge labels between remaining nodes.

Morrison in 1968 [46] designed a type of compact trie called *PATRICIA Tree* that allows linear construction complexity in time and space in the worst case.

1.3.2 Suffix Tree

Another variant that represents the same information as suffix trie by allowing linear construction in time and space in the worst case is known as *suffix tree*.

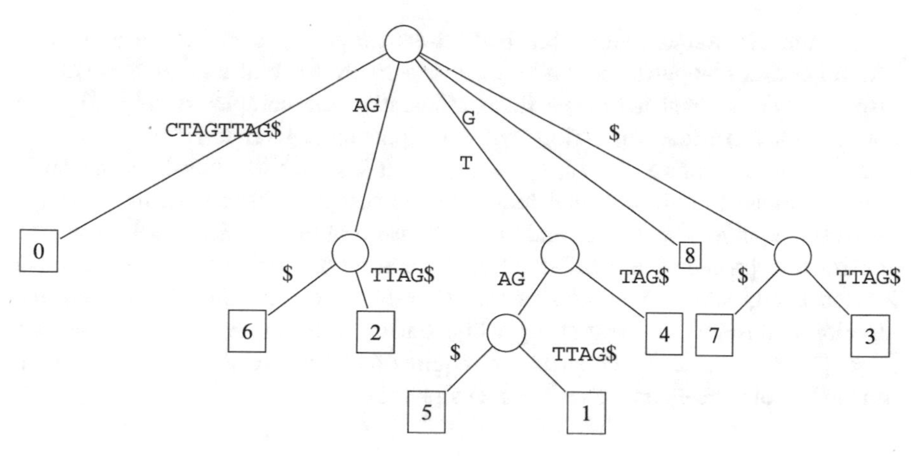

Fig. 1.2 Suffix tree for T = CTAGTTAG$. *Circles* represent internal nodes, while *squares* represent leaves and are numbered by the starting position of the suffixes they are associated with

Definition 4 Given a string T of length n ended by a terminator \$ ($T[n-1] = \$$). The suffix tree $ST(T)$ is a compact trie built over all the suffixes of T. It satisfies the following properties:

1. It has exactly n leaves numbered from 0 to $n-1$ (starting positions of the suffixes).
2. Every internal node has at least two children.
3. Each edge is labeled with a non-empty substring of T.
4. Outgoing edges of a node have string labels beginning with different characters.
5. The string obtained by concatenating all the string labels on the path from the root to leaf i spells out the suffix $T[i \ldots n-1]$, for $i = 0, \ldots, n-1$.

An example of a suffix tree is given in Fig. 1.2.

1.3.2.1 Structure Construction

Weiner was the first to introduce a linear-time algorithm to design a suffix tree. It is constructed in an incremental manner: it scans the input string from right to left and successively inserts suffixes one by one from the shortest to the longest [60]. The algorithm runs in $O(n)$ time but requires quadratic space $O(n^2)$. A short time later, McCreight [45] proposed a novel algorithm allowing the reduction of the memory requirement of the Weiner solution. Moreover, it becomes possible to speed up the insertion of suffixes by the use of *suffix links* [45, 58]. These links connect the node representing the factor $a \cdot u$ with the node representing the factor u, $a \in \Sigma$ and $u \in \Sigma^*$. Furthermore, Ukkonen's algorithm [58] works on-line; this is a strong property that makes the algorithm very useful [23]. Hence, when constructing the suffix tree, there is no need to read the whole string Beforehand, and the tree increases progressively when adding new symbols. Farach [14] presented a truly linear time construction algorithm that works for strings over integer alphabets. The algorithm computes the suffix trees separately for odd and even positions, respectively, and then it merges them efficiently. The complexity of suffix tree construction algorithms for various types of alphabets is given in [15].

For a set of strings $S = \{T_0, T_1, \ldots, T_{r-1}\}$, it is possible to construct the suffix tree in a linear time in the total length of the strings. Such a structure is called *Generalized Suffix Tree* (GST) and is an extension of the classical suffix tree. The construction procedure is similar: at each iteration one string from the input set is considered; when the suffixes of the string have been completely inserted, the algorithm considers the next string while starting from the root. For a given set $S = \{T_0, T_1, \ldots, T_{r-1}\}$, the structure requires $O(N)$ memory space and running time to be constructed on a constant size alphabet.

1.3.2.2 Search for a Pattern *P* of Length *m*

The search algorithm is similar to a search in a trie. Starting from the root of the tree, the search algorithm is as follows: for a query pattern P, the algorithm follows the path such that the starting edge is labeled with the starting character of P. This process is repeated until one of two cases arises: either it is no longer possible to go down in the tree with the current character in P, which means that there is no occurrence of P in the given string T, or all characters of P are recognized. Hence, the number of occurrences is equal to the number of leaves in the subtree deriving from the node located at the end of the reached edge, and typically each leaf in the subtree stores the starting position of an occurrence of P. If a leaf is reached then only one occurrence of P, whose starting position is stored in the leaf, is reported.

Undoubtedly, suffix trees are very efficient data structures and are mentioned as versatile structures [3] allowing supporting many problems for stringology such as basic search operations of a pattern in a string. Nevertheless, this solution is memory costly. As mentioned above, construction algorithms of a suffix tree associated to a string require time and space proportional to the size of the input string [45, 58, 60]. For instance, the memory requirement is about $10n$ bytes and can reach $20n$ in the worst case (n being the length of the input string) [34]. This constraint prevents managing huge sets of data such as NGS data. For instance, the human genome contains about $3 \cdot 10^9$ nucleotides, and its suffix tree requires 45 GB [34].

1.3.3 DAWG

The number of the nodes of the suffix trie of a text T can be reduced by applying a minimization algorithm that gives a structure known as *Direct Acyclic Word Graph* defined as follows:

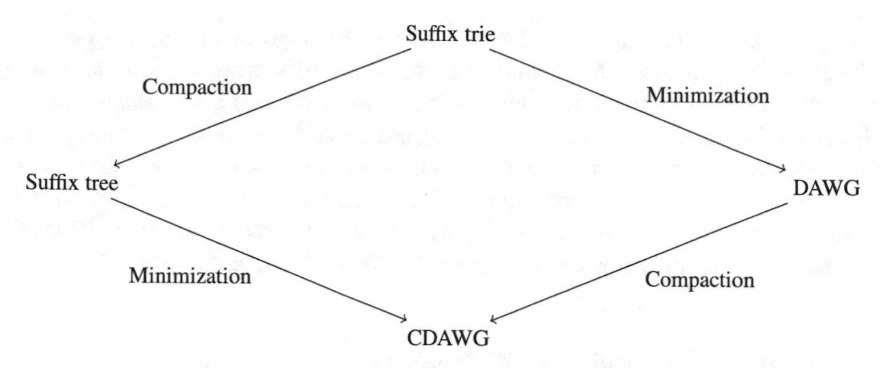

Fig. 1.3 Relationship between suffix trie, suffix tree, DAWG, and CDAWG

Definition 5 The Direct Acyclic Word Graph (DAWG) is the smallest deterministic automaton that accepts all suffixes of a string.

A DAWG can be compacted to produce a *Compact Direct Acyclic Word Graph* (CDAWG) in a similar way as a suffix trie can be compacted to produce a suffix tree. Crochemore and Vérin explain the relationship among suffix tries, suffix trees, DAWGs and CDAWGs [12]. Actually, the CDAWG can be obtained by either compacting the corresponding DAWG or minimizing the corresponding suffix tree, as shown in Fig. 1.3.

1.3.3.1 Structure Construction

The linearity of the DAWG data structure was discovered by Blumer et al. [6], who published a linear construction algorithm. The minimality of the structure as an automaton is from Crochemore [12], who showed how to build the factor automaton of a text with the same complexity. The advantage with the CDAWG is that it preserves some features of both the suffix tree and the DAWG and requires less space than both of them [7, 12]. Several construction algorithms of the CDAWG were proposed. In their paper, Blumer et al. [7] gave a construction based on compaction of the associated DAWG as follows: delete all non-final nodes of out-degree one and their corresponding edges. Although this solution is linear in time (proportional to the size of the input string), it seems not to be very efficient because it needs to calculate the DAWG structure before the compact one and typically it is costly in memory space.

An algorithm that directly constructs CDAWG in linear time and space was first introduced by Crochemore and Vérin [12], based on McCreight's algorithm [45] for constructing suffix trees. It lacks the on-line property: the proposed algorithm requires the whole input string to be known beforehand. Then, the CDAWG has to be rebuilt when a new character is added to the input string. An on-line linear-time algorithm that builds the CDAWG for a single string as well as for a set of strings, inspired from Ukkonen's on-line algorithm [58] for constructing suffix trees, was proposed by Inenaga et al. [28]: it takes advantage of the on-line property. The algorithm processes the characters of the input string from left to right, one by one, with no need to know the whole string beforehand, and it was shown that the algorithm can be easily applied for building the CDAWG for a set of strings [28]. It runs in linear time with respect to the total length of the strings in the input set [28].

As well as other data structures, DAWGs and CDAWGs can solve several combinatorial problems on strings [7, 11, 26, 59]. It was shown that CDAWGs require less memory space than the associated DAWGs and suffix trees [7].

1.3.3.2 Search for a Pattern *P* of Length *m*

The searching algorithm with the DAWG is similar to searching with the suffix trie. The searching algorithm with the CDAWG is similar to searching with the suffix tree.

1.3.4 Suffix Array

The structures mentioned above are automata-oriented. It is obvious that these solutions have proved effective in a field of combinatorial problems on strings. Yet, in practice, storing nodes requires a lot of memory. An alternative consists of using data structures based on arrays to represent the same information. The first array of this kind is the suffix array published in [42].

The suffix array for a text is basically a sorted array of all the suffixes of the text, which can be built in two main steps:

1. Sort the set of suffixes of the text lexicographically.
2. Store sorted suffix positions in the array.

Let us give a formal definition of the suffix array.

Definition 6 Let π_T be a permutation over positions $\{0, 1, \ldots, n-1\}$ defined according to the lexicographic order of suffixes $T[i \ldots n-1], 0 \leq i \leq n-1$ of a string T of length n ending with a terminator $\$ < a$, $\forall a \in \Sigma$. Then

$$T[\pi_T(0) \ldots n-1] < T[\pi_T(1) \ldots n-1] < \cdots < T[\pi_T(n-1) \ldots n-1].$$

The suffix array SA_T is a one-dimensional array such that:

$$SA_T[i] = \pi_T(i)$$

$\forall i, 0 \leq i \leq n-1$.

Figure 1.4 shows an example of a suffix array.

Suffixes	i	$SA_T[i]$	Sorted suffixes
CTAGTTAG$	0	8	$
TAGTTAG$	1	6	AG$
AGTTAG$	2	2	AGTTAG$
GTTAG$	3	0	CTAGTTAG$
TTAG$	4	7	G$
TAG$	5	3	GTTAG$
AG$	6	5	TAG$
G$	7	1	TAGTTAG$
$	8	4	TTAG$

Fig. 1.4 Suffix array of $T =$ CTAGTTAG$. Since we assume that $T[n-1] = \$$ is the lowest character, we always have $SA_T[0] = n - 1$

1.3.4.1 Structure Construction

The suffix array of a string can be obtained in linear time from its suffix tree by performing a lexicographic traversal of the suffix tree; assuming that the children of the nodes of the tree are lexicographically ordered by the edge labels left to right, it is possible to collect leaves in this order. Yet, this construction is not optimal. It is more convenient to adopt a method that allows directly constructing the suffix array. In general, the suffix array can be built using lexicographic sorting of the set of suffixes.

The first $O(n \log n)$ construction algorithm was described in [42]. The algorithm uses a doubling technique; it performs a first sort by the first two characters, and then uses results to perform sorting by the first four characters, and then the first eight and so on. The disadvantage of this construction is that it requires a time longer than that needed to construct a suffix tree because of suffix pre-sorting.

More sophisticated algorithms appeared simultaneously [30, 32, 33], requiring linear time. These algorithms are recursive; they divide suffixes into different classes and perform a sort to each class.

In practice, the recursivity of these solutions will significantly slow down their efficiency compared to algorithms with worse theoretical complexities (not linear time) [29, 36, 43, 44, 56], which are faster and need less memory space [53].

1.3.4.2 Search for a Pattern P of Length m

Since suffixes are sorted, a binary search can be performed on the suffix array. The search outputs an interval $SA_T[s, e]$ containing all the suffixes that start with the query P. Hence, the number of occurrences corresponds to $occ = e - s + 1$, located in positions $SA_T[s], SA_T[s+1], \ldots, SA_T[e]$. The binary search performs $O(\log n)$ steps with m comparisons in the worst case at each step. The final complexity to locate the given query is thus $O(m \log n)$. This complexity can be reduced to $O(m + \log n)$ with the use of additional information. An array called LCP can be used to store the length of the longest common prefixes between two consecutive suffixes [42]. Hence, although this technique needs more memory space, it allows avoiding redundant comparisons especially if there are long repeated substrings within the string.

In addition to the suffix array, some algorithms may consider the *inverse suffix array* denoted by ISA_T. This corresponds to the inverse permutation of SA_T; thus, $ISA_T[j]$ stores the lexicographical *rank* of the j-th suffix. For example, the array of Fig. 1.4 gives $ISA_T[0] = 3$.

All the data structures mentioned above serve as full indexes on the string and are called *basic indexes*. Indeed, they allow direct access to the original input text (from its suffixes), solve problems in stringology, such as locating a given pattern, and give all its positions in the text. Though such data structures have proved their strength and efficiency, the space/time trade-offs that they offer can still be improved even more by the use of indexes called *compressed data structure*. In this survey, we focus on compressed data structures based mainly on suffix arrays.

1.4 Advanced Data Structures

Several data structures are based on the *compression* of the suffix array. Such data structures always allow retrieving $SA_T[i]$, for any i. As well as with the classical suffix array, a range $SA_T[s, e]$ can be computed when searching a pattern P. The search is solved with one of two searches, *forward search* or *backward search*, which are based on the ψ and LF functions (we will give details in the sequel).

In this section, we consider the *FM-index* [18] and the *compressed suffix array* [23, 55].

1.4.1 FM Index

The *FM index* was introduced by Ferragina and Manzini [16, 18]. It is based on the *Burrows–Wheeler Transform* (BWT) [8]. It is composed of a compressed representation of a text from BWT and of a backward search.

1.4.1.1 Burrows–Wheeler Transform

The BWT is a method that reorganizes an input string T of length n. The output, denoted by T^{bwt}, is a permutation of T, which generally compresses more than the original input string T.

Let us consider a $n \times n$ (conceptual) matrix M_T, wose rows are circular shifts (also called *conjugates*) of lexicographically sorted T. The first column of M is denoted by F, and the last one is denoted by L; this latter is actually equal to T^{bwt}. Hence, T^{bwt} is the concatenation of the last character of each sorted circular shift. Because of the presence of the symbol \$ with the lowest order, the matrix M_T is sorted in

Fig. 1.5 BWT of T = CTAGTTAG$: (**a**) circular permutations; (**b**) sorted permutations. The BWT of T corresponds to the last column L of the sorted permutations

```
C T A G T T A G $
T A G T T A G $ C
A G T T A G $ C T
G T T A G $ C T A
T T A G $ C T A G
T A G $ C T A G T
A G $ C T A G T T
G $ C T A G T T A
$ C T A G T T A G
```

(a)

```
F               L
$ C T A G T T A G
A G $ C T A G T T
A G T T A G $ C T
C T A G T T A G $
G $ C T A G T T A
G T T A G $ C T A
T A G $ C T A G T
T A G T T A G $ C
T T A G $ C T A G
```

(b)

right-to-left lexicographic order. Hence, when sorting rows of M_T, one essentially sorts suffixes of the input. This establishes the close relationship between M_T and SA_T. Then, T^{bwt} can be viewed as the concatenation of characters that precede each suffix of SA_T. More explicitly, $SA_T[i]$ corresponds to the suffix of T, which is a prefix of the $(i+1)$-th row of M_T. Formally, $T^{bwt}[i] = T[SA_T[i] - 1]$ if $SA_T[i] > 1$, and $T^{bwt}[i] = T[n-1] = \$$ otherwise.

The BWT is reversible, and it is possible to restore the original string T from its transformation T^{bwt}. This is done by a reversible function called *LF-mapping*. The LF mapping is used to map symbols in the column L to symbols in the column F. The function is defined as follows [8, 18]:

Definition 7 Let $c = T^{bwt}[i]$ then $LF(i) = Count[c] + rank_c(c, i)$, where $Count[c]$ is the total number of symbols smaller than c and $rank_c(T^{bwt}, i)$ is the total number of c in L up to position i.

Note that the LF function gives information about the occurrence of each character stored in column L in column F: $c = T^{bwt}[i]$ is located in F at position $LF(i)$.

An example is given in Fig. 1.5.

1.4.1.2 Backward Search

Ferragina et al. proposed a new binary search that performs a search for a pattern from its last character to its first one, called *backward search*. The crucial property is that when operating, the process does not require the original text and the search is done using only the BWT.

The initial step of the search considers $P[m-1]$ and calculates the range interval $SA_T[s_{m-1}, e_{m-1}]$ of suffixes in SA_T prefixed by $P[m-1]$. The next step computes $SA_T[s_{m-2}, e_{m-2}]$ of suffixes prefixed by $P[m-2 \ldots m-1]$. The process is renewed inductively, such that at each step i $SA_T[s_i, e_i]$ is computed for $P[i \ldots m-1]$ from the previous interval $SA_T[s_{i+1}, e_{i+1}]$ computed for $P[i+1 \ldots m-1]$. The process is completed until either all symbols of the query have been considered and $[s_0, e_0]$ gives all suffixes prefixed by P or an empty interval is returned and there is no suffix that contains P as a prefix. One can notice that the search can be performed by exploiting the relation between SA_T and M_T. The crux is to use the LF function to update bounds of the interval at each step. Consider the range $[s_{i+1}, e_{i+1}]$ in M_T of rows starting with $P[i+1 \ldots m-1]$ and one wants to compute those preceded by $P[i]$. Indeed, all the occurrences of $P[i]$ in $T^{bwt}[s_{i+1}, e_{i+1}]$ appear contiguously in the first column F of M_T; moving backward with LF, we obtain the range $[s_i, e_i]$.

The basic idea to retrieve the positions of suffixes is to store a sample of values of the suffix array every k positions and to use LF to compute unsampled positions. Assume that one wants to retrieve $SA_T[i]$. Since $SA_T[i] = SA_T[LF(i)] + 1$, then eventually after $1 \leq j < k$ backward steps the original position $SA_T[i] = SA_T[LF^k[i]] + k$ is obtained [50].

Count is a small table of $\sigma \log n$ bits. The run time of the backward algorithm depends mainly on the run time of $rank_c$. The classic way to solve $rank_c$ is to perform counting on each query. This requires $O(mn)$ time since T is scanned at each step. Hence, different alternative solutions of the backward search aim to implement $rank_c$ in $O(1)$ time and therefore solve the search in $O(m)$ time only.

Since the suffix array is not explicitly stored, all implementations of the FM index use regular text sampled positions to retrieve the positions covered by the resulting range. Hence, samples of the M rows are marked, and their corresponding values in the suffix array are stored in an array SA'_T.

The original FM index was first described by Ferragina and Manzini in 2000 [16]. They showed that their *opportunistic data structure* compresses T^{bwt} in at most $5nH_k(T) + o(n)$ bits, where H_k is the k-th order empirical entropy of the sequence, for sufficiently low fixed $k \geq 0$, and the search of occurrences is answered in $O(m + \log^{\epsilon} n)$ for a constant $\epsilon > 0$. Note that in practice Ferragina et al. slightly modify the theoretical implementation to achieve more space effectiveness [17]. A very effective alternative that uses *wavelet trees* [50] was later proposed by [19, 23]. The theoretical space is $nH_0 + O(n \log \sigma)$ bits. The use of the *Huffman-shaped wavelet trees* [41] leads to $n(H_0 + 1) + o(n \log \sigma)$ bits and $O(H_0(T) + 1)$ average searching time in practice. Ferragina et al. went further and gave a surprising version of the FM index [20]. The new index achieves $nH_k + o(n \log \sigma)$ bits of space; this holds where $k \leq \alpha \log_\sigma n$, for a constant $0 < \alpha < 1$, and $\sigma = O(\text{polylog}(n))$.

1.4.2 Compressed Suffix Array

The *Compressed Suffix Array* (CSA) represents the suffix array and the text by a sequence of numbers defined by a function ψ.

Definition 8 The function ψ is defined for all $0 \leq i \leq n-1$, such that $\psi(i) = j$ where $SA_T[j] = SA_T[i] + 1 \bmod n$

Consider two suffixes $T[SA_T[i] \ldots n-1]$ and $T[SA_T[i+1] \ldots n-1]$. If they share the same first character, and since entries of SA_T are stored in the increasing order, it holds that $T[SA_T[i] \ldots n-1] < T[SA_T[i+1] \ldots n-1]$, and they differ in their successors pointed to by ψ: $T[SA_T[i] + 1 \ldots n - 1] = T[SA_T[\psi(i) \ldots n - 1]] < T[SA_T[i]1 \ldots n - 1] = T[SA_T[\psi(i + 1)] \ldots n - 1]$. This gives the following lemma.

Lemma 1 *For* $0 \leq i < n - 1$, $\psi(i) < \psi(i + 1)$ *if* $T[SA_T[i]] = T[SA_T[i + 1]]$.

This property was first established by Grossi and Vitter in [22]. In addition, there are long runs where $\psi(i + 1) = \psi(i) + 1$ [41]. These properties permit a compact representation of ψ more effectively than SA_T and allow its fast access.

The first breakthrough in compressed suffix arrays was the idea of Grossi and Vitter [22]. The solution has appeared simultaneously and independently with the *compact suffix array* of Mäkinen [38]. Then, the Grossi solution was quickly improved by Sadakane et al. [55]. This was followed by the Grossi, Gupta and Vitter solution [23, 24] and by the *Compressed Compact Suffix Array* proposed by Mäkinen and Navarro [39].

1.4.2.1 Grossi-CSA

Grossi et al. achieved a hierarchical decomposition in which they consider samples of SA_T based on ψ (recall that the storage of ψ is more efficient than storing SA_T and ψ allows retrieving any value of SA_T). In their decomposition, they consider SA_k the suffix array at the level $k \geq 0$ ($SA_0 = SA_T$), ψ_k computed for odd values of SA_k and a bit vector $B_k[0, n_k]$ where $B_k[i] = 1$ if $SA_k[i]$ is even, 0 otherwise and $n_k = \lceil n/2^k \rceil$. From there, they compute the suffix array SA_{k+1} obtained by the even values of SA_k divided by 2. The procedure is repeated until obtaining the smallest SA_ℓ kept together with all ψ_k and B_k for $0 \leq k \leq \ell - 1$.

To retrieve $SA_k[i]$, one needs to check whether $B_k[i] = 1$. If it is, then $SA_k[i]$ is even and sampled in SA_{k+1}, and its exact position is given by the number of 1s in B_k up to i. Hence, $SA_k[i] = 2SA_{k+1}[rank_1(B_k, i) - 1]$. Otherwise, if $B_k[i] = 0$, then $SA_k[i]$ is computed by $\psi(i)$ (since we know that $\psi(i)$ is even and sampled in SA_{k+1}). Hence, one computes even $SA_k[\psi(i)]$ as mentioned above and simply obtains $SA_k[i] = SA_k[\psi(i)] - 1$. For the examples in Fig. 1.6, to obtain $SA_0[6]$, one

	0	1	2	3	4	5	6	7	8
SA_0	8	6	2	0	7	3	5	1	4
B_0	1	1	1	1	0	0	0	0	1
					0	1	2	3	
ψ_0					0	8	1	2	

	0	1	2	3	4
SA_1	4	3	1	0	2
B_1	1	0	0	1	1
		0	1		
ψ_1		0	4		

	0	1	2
$SA_2[i]$	2	0	1

Fig. 1.6 The hierarchical recursive data structure of Grossi et al. for CTAGTTAG\$: the original suffix SA_0 and data structures ψ_0, SA_1, ψ_1 and SA_2. Recall that the structure consists only of the last (smallest one) SA_ℓ (SA_2 in this example) and all the ψ_k and B_k arrays for $0 \leq k < \ell$, instead of SA_0

can see that $B_0[6] = 0$. Thus, $\psi_0(6) = 1$, and one must compute $SA_0[\psi(6)] = SA_0[1]$. Then, $SA_0[1] = 2SA_1[rank_1(B_0, 1) - 1] = 2SA_1[1] = 2 \cdot 3 = 6$. Finally, one can easily obtain $SA_0[6] = SA_0[\psi(6) = 1] - 1 = 6 - 1 = 5$.

1.4.2.2 Sadakane-CSA

Sadakane et al. give a self-index solution inspired by the previous hierarchical structure of Grossi et al. [21, 22]. Their index, as well as the FM index, allows direct access to any prefix $T[SA_T[i] \ldots n-1]$ rather than access to $SA_T[i]$. This is useful for the search phase.

Sadakane et al. show how to exploit the property of Lemma 1 to give an elegant alternative to the Grossi solution. Essentially, they encode ψ differentially $\psi(i) - \psi(i-1)$, run length encode the long runs of 1s occurring over those differences, and for the rest use an encoding favoring small numbers. Absolute samples are stored at regular intervals to permit the efficient decoding of any $\psi(i)$. This allows access to the entries of these compressed ψ_ℓ arrays in constant time [55].

To speed up the binary search, they use a bit-vector D of size n defined as follows: for $0 \leq i \leq n-1$, $D[i] = 1$ if and only if $i = 0$ or $T[SA_T[i]] \neq T[SA_T[i-1]]$ and a string C, of length at most σ, containing all distinct characters of T in lexicographical order. As mentioned above, one has direct access to any prefix $T[SA_T[i] \ldots n-1]$ of T. Thus, when searching a given pattern P, one compares it against $T[SA_T[i] \ldots n-1]$ and Extracts the needed characters. Then, if one wants the first character of some suffix pointed by $SA_T[i]$, one just computes $c = T[SA_T[i]] = C[rank_1(D, i)]$. Once computed, one needs to determine the next character. Recall that the next character is given by $SA_T[i] + 1 = SA_T[\psi(i)]$. So one just needs to move to $\psi(i)$ and to repeat the same process to obtain $T[SA_T[i+1]]$. The process is repeated as needed. This enables computing the range $SA_T[s, e]$ of starting positions of suffixes of the text T prefixed by the pattern P in $O(m \log n)$ worst-case time.

The access to $SA_T[i]$ is done via the inverse suffix table ISA_T. Sadakane et al. were inspired by the hierarchical structure introduced by Grossi et al. to compute $ISA_T[i]$. It suffices to store the inverse permutation of the last level SA_ℓ explicitly. Then, it is possible to retrieve the value going back through the levels. This requires $O(\log^\epsilon n)$ time for a constant ϵ.

Note that in practice Sadakane et al. do not implement the inverse suffix array to retrieve $SA_T[i]$. They sample SA_T at regular intervals of a fixed length. Then, they use ψ repeatedly until retrieving the required value. This implementation in practice requires $nH_0(T) + O(n \log\log\sigma)$ bits of space. The space and search time trade-off was later improved in [39, 40] by achieving $nH_k(T) + O(n \log\log\sigma)$ bits of space for $k \leq \alpha \log_\sigma n$ while maintaining $O(m \log n)$ search time.

1.4.2.3 Grossi-Gupta-CSA

The previous solutions were improved by Grossi, Gupta et al. [23]. Their implementation reduces further the representation of ψ to nH_k bits rather than nH_0.

The solution uses σ lists one for each character in Σ. A position i is in the *list* of the character a if the suffix starting at a position $SA_T[i]$ is preceded by a.

A *context* is defined as the prefix of length k starting at $\psi(i)$. In brief, each value $\psi(i)$ belongs to the list of $T[SA_T[\psi(i)] - 1]$ and to the context $T[SA_T[\psi(i)] \ldots SA_T[\psi(i)]+k-1]$. Values of ψ are increasing along the same context (sorted by $T[SA_T[i] \ldots SA_T[i]+k]$, in increasing order of i), so that the representation verifies Lemma 1.

Let $n_s = |T^s|$ be the number of characters (lists) that precedes a given context, then the encoding of the context needs $n_s H_0(T^s)$ bits and the summing up over all contexts requires $nH_k(T)$ bits.

To retrieve $\psi(i)$, one needs to answer in which *list* and *context* it is stored and find the right position. First, one needs to compute $c = T[SA_T[\psi(i)] - 1]$ by using either a bit-vector G_k where $c = rank_1(G_k, i)$ or bit arrays L_y^k and b_y^k computed for each list such that the first one is aligned to values of ψ owned by the list c and the second one contains a bit to indicate whether a context is non-empty in the list c. G_k requires space in memory that sums over $nH_0(G_k)$ bits, L_y^k over $n_c H_0(L_y^k)$ where n_c is the number of values in the list c and b_y^k over $\sigma^k \log(n_c)$ bits. Both techniques answer to queries in constant time, and it is possible to have information about values of the suffix array (the range) stored at each context. For this a bit array, F_k that indicates if the context changes when crossing SA_T is used, and its space requirement is $O(n)$ bits. Finally, to determine the needed value of SA_T, a bit array Z is used that for each couple (context, list) indicates which values stored at F_k belong to the given list.

The representation of F_k over all contexts and lists adds up to nH_k bits. It should be noted that this representation works for all levels of ψ. Recall that $O(\log \sigma)$ time is needed to identify a level h, and then $O(\log\log n + \log \sigma)$ time is needed to retrieve $SA_T[i]$.

1.4.3 Lempel–Ziv Indexes

A new trend of indexes based on a Lempel–Ziv compression has arisen lately. It gives new alternatives of self-indexes that allow achieving a considerable space/time trade-off. Let us first define the family of the Lempel–Ziv algorithms and then illustrate the indexes based therein.

1.4.3.1 LZ78 Algorithm

The Lempel–Ziv algorithms were first introduced by Lempel and Ziv [37] in 1976. They are based on the concept of encoding future segments of the text T via the maximum length encoding recent past segments. Thus, the encoding is performed progressively when scanning the text, taking advantage of repetitions (repetitions are replaced by pointers to their occurrences in T). More precisely, the algorithm factorizes the text from left to right into *phrases* (that are different, non-overlapping and the longest possible ones): let w_i be a string, and w_{-1} corresponds to ε. Then, $T = w_0 \ldots w_{n'}$, where (1) $w_i \neq w_j$, for $0 \leq i < r$ and $-1 \leq j < i$, and (2) for $1 \leq i \leq n'$ every proper prefix of w_i is equal to some w_j, for $0 \leq j < i$.

One can take advantage of the properties of the LZ78 parsing to represent the text using the least possible space. It is based on the idea that all phrases are prefix-closed such that each $w_i = w_j \cdot c$ for some $c \in \Sigma$. Then, the compressor outputs the couple (j, c). A trie can be built with these phrases, called *LZ78-trie*, which consists of n' nodes, where n' is the total number of phrases. According to [5] $n' \log n' = O(n \log \sigma)$. Furthermore, the size of the output text (slowly) converges to the entropy of the original text [9], and [18] shows that the number n' is related to the empirical entropy $H_k(T)$. Thus, the trie can be stored in $O(nH_k)$ bits.

The LZ index was originally thc purpose of Kärkkäinen and Ukkonen in 1996 [31]. Then, it was improved in 2002 by Navarro [49]. In 2005, Ferragina and Manzini [18] proposed an efficient solution based on the LZ78 parsing. The LZ index of Arroyuelo [4] permits restricting the space requirement and the search time. This latter will be detailed in this section.

1.4.3.2 LZ Index

The LZ index of T of [4] consists of four components:

1. **LZ-trie**: trie over all phrases $\{w_0, \ldots, w_{n'}\}$, based on the LZ78 parsing of T;
2. **RevTrie**: trie over the reverse of all phrases $(\{\widetilde{w_0}, \ldots, \widetilde{w_{n'}}\})$;
3. **Node**: mapping from phrase identifiers to their node in LZ trie;
4. **Range**: a 2D $[0, n'] \times [0, n']$ data structure that stores the value k at $(revpos(\widetilde{w_k}), pos(w_{k+1}))$ for $k \in \{0, \ldots, n'\}$, where *revpos* is the lexicographic position in reversed phrases and *pos* is the lexicographic position in all phrases.

Each of these four structures requires $n \log n(1 + o(1))$ bits when using the representation of [4]; it permits answering queries in $O((occ + 1) \log n)$ time. The overall complexity for the LZ index is $4nH_k(T) + o(n \log \sigma)$ bits for $k = o(\log_\sigma n)$. This is true because we have $n \log n = nH_k(T) + O(kn \log \sigma) \leq n \log \sigma$, for any k [4].

Occurrences of P are divided into three cases:

1. occurrences fully inside one phrase;
2. occurrences that cross two phrases;
3. occurrences that span more than three phrases.

We will now detail the three types of occurrences:

1. P is contained inside a phrase w_k, such that P is not a suffix of w_k. This means that some w_ℓ exists where $w_k = w_\ell \cdot c$ (remember that by the property of LZ78 all phrases are prefix-closed). Thus, the crux is to find the shortest possible w_ℓ ended by P. This is done recursively by going up from w_k. The phrase can be easily found by searching for $P^\sim$ in RevTrie and reporting the corresponding node. This can be done in $O(m^2 \log \sigma)$ time. When such a node is found, by the property of the LZ78, each node descending from it corresponds to a phrase ended by P. One must then restart the process on all such descendants by reporting all subtries corresponding to each one. All type (1) occurrences can be reported in $O(m^2 \log \sigma + occ_1)$ time where occ_1 is the number of type (1) occurrences.
2. In this case, P crosses the boundary between two consecutive phrases w_k and w_{k+1}: $P[0 \dots i]$ matches a suffix of w_k and $P[i+1 \dots m-1]$ matches a prefix of w_{k+1}. Hence, $(P[0 \dots i])^\sim$ is first searched using the RevTrie and the remainder of P is searched using the LZ trie. When one gets the range for occurrences in each case, the 2D structure range is used to find those matching pairs. The overall complexity for type (2) occurrences requires $O(m^3 \log \sigma + (m + occ_2) \log n)$ time where occ_2 is the number of type (1) occurrences.
3. To report type (3) occurrences, the reasoning is as follows. In a first step, one needs to identify the maximal concatenation of phrases $w_k \cdots w_\ell$ that totally overlaps a factor $P[i \dots j]$ and then to determine whether w_{k-1} ends with $P[0 \dots i-1]$ and $w_{\ell+1}$ starts with $P[j+1 \dots m-1]$. Since several concatenations of phrases can match a factor of P, one seeks to expand the phrases on the right to obtain the maximum matching with the longest factor of P. This ensures that factors that match are unique for each choice of i and j. One uses the LZ trie to extract all $w_i \cdot w_{i+1}$ as being factors of P in $O(m^2 \log \sigma)$ time. The $O(m^2)$ maximal concatenations of phrases are obtained in the $O(m^2 \log m)$ worst case time and $O(m^2)$ time on average. Each of the obtained concatenations is verified in $O(m \log \sigma)$ time, and all occurrences in this case require $O(m^3 \log \sigma)$ time.

The search for a pattern in the LZ index requires $O(m^3 \log \sigma + (m + occ) \log n)$ worst case time where occ is the total number of occurrences of the pattern. This is due to the fact that one considers all prefixes, suffixes and factors of the pattern when exploring the three types of occurrences.

1.5 More Advanced Data Structures

In this section, we present *more advanced* solutions that minimize further the cost in space while saving the search time. These solutions seek to represent a collection of similar sequences by storing only a reference sequence with differences from each of the other sequences.

1.5.1 Common and Differentiating Segments

Huang et al. [27] give an efficient solution to index highly similar sequences that exploits the BWT and the suffix array data structures. Their index works for two different representations (models) of data. Let T_0 be a reference sequence (arbitrary selected).

Model 1 Let $T_0, T_1, \ldots, T_{r-1}$ be r strings not necessarily of the same length. Each T_i is of the form $R_{i,0}C_0R_{i,1}C_1 \cdots C_{k-1}R_{i,k}$, where $R_{i,j}$ $(0 \leq j \leq k)$ are called *differentiating segments* (or R segments) and vary from the corresponding segments in T_0 and C_j $(0 \leq j \leq k-1)$ are *common segments*. Note that $R_{i,j}$ may be empty for any $i \neq i'$ and that $R_{i,j}$ and $R_{i',j}$ may not have the same length.

Model 2 Let $T_0, T_1, \ldots, T_{r-1}$ be r strings all of the same length n. Each T_i $(i \neq 0)$ differs from T_0 in x_i positions, where $x_i << n$. Note that these positions may not be the same for T_i and $T_{i'}$, with $i \neq i'$.

1.5.1.1 Model 1

Description Let C be the concatenation of the common parts separated by an end marker \$ smaller than all the other symbols: $C = C_0\$C_1\$\cdots\$C_{k-1}\$$ and $|C| = n' + k$ where $n' = \sum_0^{k-1} |C_j|$. Let $Start_C$ be an array of k elements such that $Start_C[j]$ is equal to the starting position of C_j in C. Let R be the concatenation of the R segments of all the sequences T_is: $R = R_{0,0} \cdots R_{0,k}R_{1,0} \cdots R_{1,k} \cdots R_{r-1,0} \ldots R_{r-1,k}$, and let N' denote the length of R. A segment number is assigned to each $R_{i,j}$, which is its rank in R. Let $Start_R$ be an array of $r(k+1)$ elements such that $Start_R[j]$ stores the starting position of the R segment j in R. A *differentiating suffix* is a suffix starting inside an R segment. Each differentiating suffix is referred to by a pair (w, s) where w is the number of the R segment and s is the starting position of thc suffix inside the R segment.

For Model 1, common parts and R segments are indexed separately. The index for common segments consists of the BWT for $C^{\sim}$, the mirror of C and the array $Start_C$. The latter is used to recover the text of any C_j.

For R segments, the index includes: a suffix array SAR of size N' that stores the pairs associated with the differentiating suffixes sorted in lexicographic order, an array SAR_0 that stores entries of SAR where the starting position is 0 (differentiating suffixes starting from the first character of an R segment) and the array $Start_R$. Furthermore, SAR_0 stores a field *c-rank*: let w be the number of the R segment $R_{i,j}$, then its *c-rank* is equal to the rank of the suffix $\$C^{\sim}_{j-1}\$\cdots\$C^{\sim}_0$ among all suffixes of $C^{\sim}$ (*c-rank* $= -1$ if $j = 0$). This is useful in the search phase when one needs to know whether the pattern crosses The boundary between a common segment and a R segment of some T_i. The overall space required by these data structures is $O(n' + N' \log rk + rk(\log n' + \log N'))$ bits.

An example is given in Fig. 1.7.

T_0 = AACGCGCCGG
T_1 = CACGAGCCGG
T_2 = TACGATCCGC

C_0 = ACG
C_1 = CCG

$R_{0,0}$ = A
$R_{0,1}$ = CG
⋮
$R_{2,2}$ = C

(a)

$R =$ ACGGCAGGTATC

$Start_R$:

0	1	3	4	5	6	7	8	9
0	1	3	4	5	7	8	9	11

$C =$ ACG\$CCG\$

$Start_C$:

0	1
0	4

$C^{\sim} =$ \$GCC\$GCA

(b)

order	suffix	*SAR*
0	AACGCGCCGG	(0,0)
1	AGCCGG	(4,0)
2	ATCCGC	(7,0)
3	C	(8,0)
4	CACGAGCCGG	(3,0)
5	CGCCGG	(1,0)
6	G	(2,0)
7	G	(5,0)
8	GCCGG	(1,1)
9	GCCGG	(4,1)
10	TACGATCCGC	(6,0)
11	TCCGC	(7,1)

SAR_0	*c-rank*
(0,0)	-1
(4,0)	0
(7,0)	0
(8,0)	1
(3,0)	-1
(1,0)	0
(2,0)	1
(5,0)	1
(6,0)	-1

(c)

Fig. 1.7 An example for the data structure. (**a**) Input set of sequences, common parts and R segments (*underlined*); (**b**) R, C and arrays $Start_C$ and $Start_R$; (**c**) all differentiating suffixes in lexicographical order and arrays *SAR*, SAR_0 and *c-rank*

Search for a Pattern *P* of Length *m* Occurrences of a given pattern P of length m with this Model 1 are divided into three types:

1. P appears completely inside a common segment;
2. P appears as a prefix of a differentiating suffix of some T_i;
3. P spans the boundary between a common segment and a R segment.

The procedure for searching for the first two types is simple. For type (1) occurrences, the BWT of $C^{\sim}$ is used for searching $P^{\sim}$. The answer to this type takes $O(m + occ_1)$ time where occ_1 is the number of type (1) occurrences of P in the T_is. The type (2) occurrences are found by using *SAR*, and this takes $O(m \log N' + occ_2)$ time where occ_2 is the number of type (2) occurrences of P in the T_is.

Type (3) occurrences are non-trivial, and the reasoning is as follows. P is divided into non-empty substrings $P_1 \cdot P_2$ where P_1 is a suffix of some C_{j-1} and P_2 is a prefix of the following $R_{i,j}$ in some T_i. $(P_1\$)^{\sim}$ is searched using the BWT of $C^{\sim}$ to identify the common segments suffixed by P_1. This yields to a range LR_1. Then, P_2 is searched using SAR_0 (looking for occurrences of P_2 starting at the first character of $R_{i,j}$), and a range LR_2 is obtained. The idea is to check for each value in LR_1 that refers to C_{j-1} if there is a value in LR_2 that points to the following segment $R_{i,j}$ in T_i. This is done by computing *c-rank* in *SAR* for each value in LR_2. Recall that *c-rank* is the rank of $\$C_{j-1}^{\sim}\$ \cdots \$C_0^{\sim}$ with respect to all suffixes in $C^{\sim}$. Thus, if *c-rank* is within LR_1, P_1 appears as a suffix of C_{j-1}.

The above checking can be made in constant time. The array $Start_C$ is used for searching the suffix arrays *SAR* and SAR_0. The BWT index supports the backward search, so that the range of $P[0], P[0\dots1]^{\sim}, \dots, P[0\dots m-1]^{\sim}$ can be computed incrementally in $O(m)$ time. Thus, the range of $(P[0\dots i]\$)^{\sim}$ can be obtained in $O(1)$ time from the range of $(P[0\dots i])^{\sim}$.

The search of type (3) occurrences requires $O(m + PSC(P)(m\log(rk) + rk) + occ_3)$ time, where $PSC(P)$ is the number of prefixes of P that are suffixes of some common segments, and where occ_3 is the number of type (3) occurrences of P in the T_is. It is possible to avoid the reliance on the size of LR_2. This is worth noting since it is possible that no entries in LR_2 could validate an occurrence of P. A 2*D*-range structure [51] is used to check whether any *c-rank* stored in SAR_0 (specified by LR_2) falls in the range LR_1. The range search index requires $O(rk\log n)$ bits. The check is done in $O(\log n)$ time, and each occurrence can be retrieved in $O(\log n)$ time. Thus, the complexity becomes $O(m + PSC(P)(m\log(rk)) + occ_3\log n)$ time. The overall complexity requires $O(n + N'\log rk + rk(\log n + \log N'))$ bits.

1.5.1.2 Model 2

The index for Model 2 is similar to that for Model 1.

Description In this model, a differentiating suffix is a suffix that starts at one of x_i positions. An R segment is the maximal region of characters in T_i that differs from the corresponding characters in T_0. In similar way as in Model 1, the index exploits the BWT index for $T_0^{\sim}$, *SAR* and SAR_0 for all differentiating suffixes. Let s be the number of R segments. All R segments are concatenated in a sequence $R = R_0R_1\cdots R_{s-1}$. So, each differentiating suffix in R is recovered by the use of $Start_R$, which stores the starting position of each R segment in R and its starting position in its original sequence T_i. The array B is such that $B[v] = v'$, where v is the rank in SAR_0 of the differentiating suffix $T_i[j\dots n-1]$ and v' is the rank of $T_0[0\dots j-1]^{\sim}$ with respect to $T_0^{\sim}$.

Search for a Pattern *P* of Length *m* Given a pattern P of length m, the same three types of occurrences are considered as in Model 1. The BWT for $T_0^{\sim}$ allows locating all occurrences that are in T_0 and then checking for each of these occurrences if it also occurs in T_i at the same position, for all $i > 0$. The array *SAR* is used for occurrences completely inside a differentiating suffix. The last kind of occurrence is processed similarly. The pattern is split into $P_1 = P[0\dots j]$ with $j < m-2$ and

$P_2 = P[j+1 \ldots m-1]$. P_1 is searched using the BWT of $T_0^{\sim}$, and a range LR_1 is obtained. Then, P_2 is searched using SAR_0, and a range LR_2 is obtained. Now it remains to check for each $v \in LR_2$ if $B[v] \in LR_1$. If it is true, an occurrence of P has been found. It requires $O(n + N' \log N' + s(\log n + \log N'))$ bits and $O(m + m \log N' + PSC'(P)(m \log s + s) + occ)$ time where $N' = |R|$ and $PSC'(P)$ are the number of prefixes of P that occur in T_0.

1.5.1.3 A Second Solution

Adopting a similar view of similar sequences as blocks of common segments with differences between then, Alatabbi et al. [2] proposed an index that represents each sequence as a concatenation of *permutations* of *identical blocks* together with bounded length *differences* between them. Formally, each T_i is of the form $T_i = R_{i_1} C_{\sigma_1} R_{i_2} C_{\sigma_2} \cdots R_{i_k} C_{\sigma_k}$ where $|R_j| \leq \omega/2$ and ω denotes the length of the computer word. Hence, one needs to store: (1) all common sequences $C = C_1 C_2 \cdots C_k$, for each T_i (2) $\tau_i = \sigma_{i_1} \sigma_{i_2} \cdots \sigma_{i_k}$ and (3) $R_i = R_{i_1} \cdots R_{i_k}$.

Occurrences of P belong essentially to one of four classes:

Simple matches P occurs entirely within a common region C_p or within a differentiating region R_v.

Border matches P is either of the form $C'_{p_{j-1}} R_j C''_{p_j}$ where $C'_{p_{j-1}}$ is a suffix of some $C_{p_{j-1}}$ and where C''_{p_j} is a prefix of some C_{p_j} or of the form $C'_{p_{j-1}} R''_j$ where $C'_{p_{j-1}}$ is a suffix of $C_{p_{j-1}}$ and R''_j is a prefix of R_j.

Complex matches P, is of the form $R'_1 C_{p_1} R'_2 C_{p_2} \cdots R'_n C_{p_n}$, where R'_j is a suffix of R_j, for $1 \leq j \leq n$.

Complex border matches P is of the form $P = C'_{p_0} R_1 C_{p_1} \cdots R_v C_{p_v} R''_{v+1}$, where C'_{p_0} is a suffix of C_{p_0} and R''_{v+1} is a prefix of R_{v+1}.

Simple and border matches can be found using the suffix array SA_C of the common regions.

Complex matches are found in four steps. The first step consists of computing all the valid factorizations of the pattern P. A valid factorization of the pattern is of the form $P = C'_{p_0} R_1 C_{p_1} R_2 \cdots R_v C_{p_v} R''_{v+1}$ such that $|R_j| \leq \omega/2$ where C'_{p_0} is a suffix of a common region and R''_{v+1} is a prefix of a differentiating region. Valid factorizations of P are found with an Aho-Corasick automaton [1] of the common regions $C_1, \ldots, C_k$. Then, to each valid factorization corresponds a permutation $\tau_p = p_1 \cdots p_v$, which is the order in which the common regions occur in P. In a second step, the sequences τ_i containing these τ_p are found using another Aho-Corasick automaton of the valid factorizations fed by the τ_is. In a third step, each occurrences of τ_p is checked, and then in a fourth step the leading and trailing parts are checked.

Since each differentiating region has a length smaller than $\omega/2$, then by encoding nucleotides on 2 bits, the differentiating regions can be represented by integers. Most of the checking parts can then be realized in constant time.

The occurrences of a pattern P of length m can be found in $O(m + vk \log k + occ_v(m/\omega + v) + (PSC(p)mr)/\omega + \log n)$ time using $O(n' \log n' + kh + vh \log vh)$ bits where v is the number of valid factorizations each of length h, k is the number of differentiating regions, occ_v is the number of occurrences of valid factorizations, $PSC(p)$ is the number of prefixes of P that occur as suffixes of common regions, and n' is the total length of common regions.

1.5.2 Fast Relative Lempel–Ziv Self-index

Compressed indexes based on factorization (or parsing) of texts using the Lempel–Ziv compression are not suitable to index highly similar texts in theory [54] and in practice [57]. Do et al. [13] proposed a solution based on a new alternative of factorization called *relative Lempel–Ziv* (RLZ) scheme [35]. Experiments show that this method gives good results for repetitive sequences.

Description The RLZ compression is defined as follows:

Definition 9 The RLZ parsing LZ(S/T_0) of a set $S = \{T_1, \ldots, T_{r-1}\}$ constitutes a way to represent S by a sequence of phrases that are substrings (factors) of a reference sequence T_0 of length n such that each $T_i = w_0^i w_1^i \cdots w_{c_i}^i$ is built as follows: The first phrase is $w_{-1}^i = \varepsilon$. For all $j > 0$, assume that w_{j-1} has already been processed, then w_j^i is the longest prefix of $T_i[(|w_{-1}^i \ldots w_{j-1}^i| + 1) \ldots |T_i| - 1]$ that occurs in T_0.

An example is displayed Fig. 1.8.

The output of an RLZ compressor is a pair (s_i, ℓ_i), where p_i is the starting position of the phrase w_i in T_0 and ℓ_i is its length. Note two facts: (1) all the phrases in an RLZ parsing are not necessarily different from each other; (2) the factorization guarantees that no phrase can be extended any further to the right. This gives an important property of the RLZ parsing:

T_0 = ACGTGACATAGT
T_1 = GATAGAC = GA, TAG, AC
T_2 = TGCA = TG, CA
T_3 = TGACGT = TGAC, GT

$T_F[id]$	Factor	Pos. in T_0
0	AC	0..1
1	CA	6..7
2	GA	4..5
3	GT	2..3
4	TAG	8..10
5	TG	3..4
6	TGAC	3..6

Fig. 1.8 The RLZ compression of the set of strings $S = \{T_1, T_2, T_3\}$ with respect to T_0 (*top*). The distinct factors (longest ones) are stored in lexicographical order in the array T_F (*bottom*) as pairs of starting and ending positions in T_0

Lemma 2 *$LZ(S/T_0)$ represents S using the smallest possible number of factors of T_0.*

Let T_F be an array of size f of distinct factors that appear in the factorization of S sorted in lexicographical order. Then, $T_F[j] = (s_j, e_j)$, where $T_0[s_j \dots e_j]$, is the factor of rank j in the lexicographic order in T_F. Note that $f \leq \min\{n^2, n_F\}$, where n is the length of T_0 and n_F is the minimum number of factors needed to compress S. The encoding of S requires $O(n_F \log f) = O(n_F \log n)$ bits.

Search for a Pattern P of Length m Occurrences of P belong to two main types whether (type 1) it appears completely inside one factor w_p^i or (type 2) it is of the form $X_{p-1}^i w_p^i \cdots w_q^i Y_{q+1}^i$, where X_{p-1}^i is a suffix of w_{p-1}^i and Y_{q+1}^i is a prefix of w_{q+1}^i.

Type 1 For the type 1 occurrences, the searching algorithm locates first occurrences of P in T_0 and then finds factors in T_F that contain entirely P. Recall, as mentioned above, that each factor $T_F[j]$ is represented as an interval of positions on the reference sequence T_0. Hence the searching algorithm maps the problem as a cover problem such that after locating P in T_0 using SA_{T_0}, and obtaining a range $[s_P, e_P]$, it reports all factors that cover positions in the obtained range.

Before going into the details of the data structure one needs some further definitions. Then, $T_F[j] = (s_j, e_j)$ covers a position p if $s_j \leq p \leq e_j$. A factor $T_F[j]$ is at the left of the factor $T_F[j']$ if $s_j \leq s_{j'}$ and $e_j < e_{j'}$. Let G be the array of indices where $G[i] = j$ if $T_F[j]$ is the i-th leftmost factor of T_F. When $G[i] = j$, we define $I_s[j] = s_{G[i]}$ and $I_e[j] = e_{G[i]}$. Naturally, $I_s[0]$ stores the starting position of the leftmost factor, and values in I_s are non-decreasing.

Consider an array D such that $D[p]$ stores the distance between the position p and the rightmost ending position of all factors that covers p. Thus, formally, $D[p] = \max_{j=0..f-1}\{I_e[j] - p + 1 \mid I_s[j] \leq p\}$, for $p \in \{0, 1, \dots, n-1\}$. Since occurrences of P are sorted in SA_{T_0}, then one needs an additional array that remembers the $D[p]$ order. Let D' be an array of n elements where $D'[p] = D[SA_{T_0}[p]]$. Hence, each entry of D' gives information about the length of the longest interval whose starting position is equal to $SA_{T_0}[p]$.

One can check if a substring of T_0 is covered by at least one factor according to the following lemma:

Lemma 3 *For any index p and length ℓ, a factor $T[j]$ exists that covers positions $SA_{T_0}[p], \dots, SA_{T_0}[p] + \ell - 1$ in T_0 if and only if $D'[p] \geq \ell$.*

In summary the search algorithm considers the range $[s_P, e_P]$ as being the range of P in SA_{T_0}. The following procedure is performed:

1. Find an index q such that $s_P \leq q \leq e_P$ where $D[SA_{T_0}[q]]$ has the largest value and check that q verifies the condition of Lemma 3.
2. From a given position p of P in T_0 and q, report the set $\{T_F[G[i]] \mid I_s[i] \leq p \text{ and } q \leq I_e[i]\}$.

The data structures for this case use $2n + o(n) + O(s \log n)$ bits. Note that the arrays D, D', I_s, I_e are not explicitly stored. For any position $p \in \{0, \dots, n-1\}$, one needs $O(1)$ time to compute $D[p]$ and $O(\log n)$ time to compute $D'[p]$. Also, $I_s[i]$

and $I_e[i]$ can be computed in $O(1)$ time given $G[i]$ and information about factors. To report all occurrences of P inside factors in $T_F[0\ldots f-1]$, one needs $O(occ_1(\log^{\epsilon} n + \log\sigma))$ time, where occ_1 is the number of type 1 occurrences and ϵ is a constant.

Type 2 For the type 2 occurrences, every occurrence of P is divided into two subcases:

1. **the first subcase**: locate every occurrence of a prefix of P that equals a suffix X^i_{p-1} of a factor w^i_{p-1} of S.
2. **the second subcase**: find all occurrences of suffixes of P that are equal to a prefix of a suffix starting with a factor in S, i.e., of the form $w^i_p \cdots w^i_q Y^i_{q+1}$.

The result of the first subcase is the set of factors in T_F having a prefix of P as a suffix. First note that every non-empty prefix P' of P is considered as a separate query pattern. Second, the searching algorithm searches P' in the reference sequence using SA_{T_0} and obtains its suffix range $[s_{P'}, e_{P'}]$. Since the array T_F stores the f distinct factors of the form $w^i_j \in S$ sorted lexicographically, the searching algorithm can report all occurrences of P by the following:

Theorem 1 *For any suffix range $[s_P, e_P]$ in SA_{T_0} of a query pattern P, the data structure can report the maximal range $[p, q]$ such that P is a prefix of every element $T_F[p] \ldots T_F[q]$.*

Some properties are required to check whether the query P is a prefix of any factor:

Lemma 4 *Let $[s_P, e_P]$ and $[s_j, e_j]$ be the suffix ranges respectively of P and the factor $T_F[j]$ in SA_{T_0}. Then, P is a prefix of $T_F[j]$ if and only if either: (1) $s_P < s_j \le e_P$; or (2) $s_P = s_j$ and $|T_F[j]| > m$.*

One can observe from the above Theorem 1 a mapping between the suffix array SA_{T_0} and the sorted array of factors T_F to determine whether a pattern is a prefix of any factor. For every $i = 0, \ldots, n-1$ let $\Gamma(i) = \{|T[j]| \mid s_j = i$ and $[s_j, e_j]$ is the suffix range of $T_F[j]$ in $SA_{T_0}\}$; in other words, $\Gamma(i)$ stores the set of lengths of factors whose suffix ranges start at i in SA_{T_0}. $\Gamma(i)$ is used to compute the mapping from a suffix range in SA_{T_0} to a range of factors in T_F as follows:

Lemma 5 *Suppose $[s_P, e_P]$ is the suffix range of P in SA_{T_0}. Then, $[p, q]$ is the range in $T_F[0\ldots f-1]$ such that P is a prefix of all $T_F[j]$ where $p \le j \le q$ where $p = 1 + \sum_{i=0}^{s_P - 1} |\Gamma(i)| + |\{x \in \Gamma(ST_P) \mid x < m\}|$ and $q = \sum_{i=0}^{e_P} |\Gamma(i)|$.*

The data structure uses $O(f \log n) + o(n)$ bits. It can report the maximal range $[p, q]$ such that P is the prefix of all $T_F[j]$, where $p \le j \le q$, in $O(\log\log n)$ time.

Let us now turn our interest to the second subcase. Each suffix P' of P can be factorized using the reference sequence LZ(P'/T_0). Then, all the factorizations are matched with the sequences in one scan from right to left using dynamic programming.

Let F be the lexicographically sorted array of all non-empty suffixes in S that start with a factor, i.e., each element in F is of the form $w^i_p w^i_{p+1} \cdots w^i_{c_i}$ and it is called *factor suffix*.

The result of the second subcase is the set of factor suffixes in F such that a suffix $P[i \ldots m-1]$ of P, for $0 \leq i \leq m-1$, is a prefix of these factor suffixes. Then, the algorithm reports, for an index p in F, the unique location in S where the factor suffix $F[p]$ occurs. To resolve the problem, the algorithm performs two main phases:

1. Compute the relative locations in S of a factor suffix in F, given an index p of P in F, and return i and j so that $F[p]$ starts at w_j^i in S.
2. Convert the relative locations in S to the exact locations in S, given i, j and return $1 + \sum_{q=1}^{j-1} |w_q^i|$; this computes the locations of w_j^i in the input string $S_i \in S$.

The built data structure, for a given query P, computes the range $[p, q]$ of $P[i \ldots m-1]$, for $0 \leq i \leq m-1$ in F, where $P[i \ldots m-1]$ is a prefix of each $F[p], \ldots, F[q]$. Let $Q[i]$ denote the range for each i.

The head of $F[i]$ is the first factor of $F[i]$. Let $\Im$ be the concatenation of the factor representations of all strings in S and β be a general BWT index of $\Im$. Define $A[i] = P[i \ldots j]$, where j is the largest index such that $P[i \ldots j]$ is a factor of S, if one exists, and *nil* otherwise. Let $Y[i]$ be the range $[p, q]$ in F such that $P[i \ldots m-1]$ is the prefix of all the heads of factor suffixes $F[p], \ldots, F[q]$, if one exists, *nil* otherwise. Then, Q is computed as follows:

$$Q[i] = \begin{cases} Y[i] & \text{if } Y[i] \neq nil \\ \text{BackwardSearch}(A[i], Q[i + |A[i]|]) & \text{if } Y[i] = nil \text{ and } A[i] \neq nil \\ nil & \text{otherwise.} \end{cases}$$

The last step consists of encoding combinations of X_{p-1}^i and $w_p^i \cdots w_q^i Y_{q+1}^i$, which are adjacent in some $T_i \in S$. For that, one needs to construct a $(f \times n_F)$ matrix to find the needed combinations. If this matrix is stored in $O(n_F \log f)$ bits, the task can be answered in $O(\log f / \log\log f + occ \cdot \log^{\epsilon} f)$ time, where occ is the number of answers.

The data structures for type 2 occurrences use $O(n) + (2 + 1/\epsilon) n H_k(T_0) + o(n \log \sigma) + O(n_F \log n)$ bits. It reports all suffix ranges of F that match some suffix of P of length m', where $1 \leq m' \leq m-1$, in $O(m'(\log \sigma / \log\log n + \log\log n))$ time.

The whole data structure for this solution requires $2nH_k(T_0) + 5.55n + O(n_F \log n \log\log n)$ bits and $O(m(\log \sigma + \log\log n) + occ \cdot (\log n \frac{\log n_F}{\log n}))$ query time, where occ is the number of occurrences of a given P.

1.5.3 BIO-FMI Index

The authors of [52] presented another idea of combining indexing similar DNA sequences and compression. They proposed a self-index called *BIO-FMI* based on known *alignments* between every single sequence with a chosen reference by using the FM index.

1.5.3.1 Description

Data Model Given a set of similar sequences $\{T_0, \ldots, T_{r-1}\}$, let T_0 be the reference sequence. All other sequences T_i, for $1 \leq i \leq r-1$, are represented as: $T_i = R_{i,0}C_{i,0}R_{i,1}C_{i,1} \cdots C_{i,k_i-1}R_{i,ki}$, where $C_{i,j}$ $(0 \leq j < k_i)$ are segments common to T_i and T_0 and $R_{i,j}$ $(0 \leq j \leq k_i)$ are distinct segments.

Let d be a string that stores the concatenation of all the variations in every T_i of the form: $C''_{i,j-1}R_{i,j}C'_{i,j}$ where $C''_{i,j-1}$ is a suffix of $C_{i,j-1}$ and $C'_{i,j}$ is a prefix of $C_{i,j}$ of length $\ell_c - 1$ (ℓ_c being a parameter). Such a suffix and a prefix are respectively called left and right *context* of $R_{i,j}$. Note that $R_{i,j}$ may be an empty string (if deletion) or a single symbol (if SNV).

The index will be able to store quadruples (Δ, p, ℓ, o), where Δ is the type of variation (indel, SNV), p is the position in T_0, ℓ is the length of the variation, and o is the offset between T_i and T_0, i.e., $o = q - p$,where q is the position of the variation in T_i. This defines the essential idea behind the proposed solution that consists of *tracking the variations* (alignments) between every T_i and T_0. The approach contains two FM indexes: the first one built over T_0 and the second one over the string d.

Let I_0 be the index for T_0. A wavelet tree over T_0^{bwt} is constructed (this implies that $SA(T_0)$ and T_0^{bwt} are precomputed). Also consider a bit vector *IsLoc* over the built wavelet tree. The process is to use the regular text sampled position over T_0 with $\log n$ rate. Hence, each $\log n$-th symbol is marked using *IsLoc*. By applying *rank* operations over $IsLoc_0$, one can obtain a pointer to the *Loc* array that contains the base position of the corresponding symbol in T_0.

The data structure for I_0 consumes $n + n + n\log\sigma + \log n + 4n\log\log n$ bits. This is because Loc_0 and $IsLoc_0$ each requires n bits. The $n\log\sigma$ is the space for the wavelet tree. The number of samples is $\frac{n}{\log^2 n} + \frac{2n}{\log n}$, and they are recorded using simple byte codewords of $\log n$ and $\log\log^2 n$ bit length, respectively .

For the string d, the data structure I_d includes a wavelet tree built in the same way as for T_0, $IsLoc_d$, which recognizes positions of variations in T_i and an array Loc_d. The process for the array Loc_d is different: it will be able to extract each T_i containing a variation and to give information for quadruples (Δ, p, ℓ, o). Hence, it points to an array *aIndex* that stores start positions of variations in sequences and stores the ranks of its variations for each T_i. For every T_i, the variations are sorted in increasing order and then stored in additional arrays, *aBasePos*, that contain the start position of a variation in T_0, *aOffset* that stores the offset of a variation and *aOp* that gives information about the type and the length of the variation. An auxiliary array is used to accelerate the extraction of the i–th sequence called *SampleStart* that stores pointers to the start positions of a single sequence T_i.

Let $N' = O(\Sigma_{i=1}^{i<r}(k_i + 1))$ and $N = \Sigma_{i=1}^{i<r}\Sigma_{j=0}^{j<k_i}|R_{i,j}|$ be respectively the number and the total length of all distinct segments. For the data structure I_d, the wavelet tree uses $O(N\log\sigma)$, and the bit vector $IsLoc_d$ uses N bits in the same way as for I_0.

The numbers of bits used by the other arrays are as follows: $O(N'(\log r + \log \frac{N'}{r}))$ for Loc_d, $O(r\log' N)$ for *aIndex*, $O(N'\log n)$ for *aBasePos*, $O(N'\log n)$ for *aOffset*, $O(N')$ for *aOp* and $O(r\log N')$ for *SampleStart*. The overall complexity gives $O(N+$

$N'(\log r + \log \frac{N'}{r}) + N \log \sigma + \frac{N}{\log N} + 4N\frac{\log\log N}{\log N} + r \log N' + N' \log n + N' \log n + N' + r \log N)$ memory space.

Search for a Pattern *P* of Length *m* First, the pattern P is divided into chunks of length ℓ_c as follows: $P = P_{1,\ell_c} \cdots P_{|\frac{m}{\ell_c}|\ell_c,m}$, and then each chunk is searched separately. Each single chunk is first searched for using the index I_d and for each occurrence, occ_j, its position $aBasePos_j + aOffset_j$ is stored in a hash table. Next the chunks are searched again using the index I_0, and each base occurrence in T_0 is stored in the hash table using the $aBasePos_j$ position. Then, the algorithm evaluates whether the occurrences in T_0 can be propagated to all the sequences. For each occurrence occ_j in T_0 at position p_j, for each T_i it looks for the distinct segment just preceding position p_j to see whether the occurrence can be validated. At a final step, the algorithm reports all positions with complete occurrence of the pattern.

The algorithm takes $O(\log \sigma)$ time for the wavelet tree in I_0 and I_d. It needs $O(s \log \sigma)$ time for each chunk to locate its occurrences in T_0 and T_i, where s is the length of a sample with a localization pointer. To find occurrences in all sequences T_i, it needs $O(r \log \frac{N'}{r})$ time.

1.5.4 *Suffix Tree of an Alignment*

The *suffix tree of an alignment* of Na et al. [47] is an improvement over the classical generalized suffix tree based on a new representation of the alignment of sequences. The space usage is strongly reduced without compromising the linear time of the search.

The alignment is a way to map between two given strings so that it can retrieve one string from another. It is obtained by replacing substrings of one string into those of the other one. Let give the formal definition.

Definition 10 Given $T_0 = \alpha_1\beta_1 \cdots \beta_k\alpha_{k+1}$ and $T_1 = \alpha_1\delta_1 \cdots \delta_k\alpha_{k+1}$ two strings for some $k \geq 1$, the alignment of T_0 and T_1 consists of replacing each β_i with δ_i, and it is denoted by $\alpha_1(\beta_1/\delta_1) \cdots \alpha_k(\beta_k/\delta_k)\alpha_{k+1}$.

An important property of the alignment is that it satisfies the following conditions: α_i cannot be the empty string (except for α_1); either β_i or δ_i can be empty; the first characters of $\beta_i\alpha_{i+1}$ and $\delta_i\alpha_{i+1}$ are distinct. Note that a special character \$ is inserted at the end of α_{k+1} to ensure that it is never empty.

for simplicity, assume that T_0 and T_1 are of the following form: $T_0 = \alpha\beta\gamma$ and $T_1 = \alpha\delta\gamma$ and the alignment is of the form $\alpha(\beta/\delta)\gamma$. Let α^i be the longest suffix of α, which occurs at least twice in T_i with $0 \leq i \leq 1$, and let α^* be the longest of α^0 and α^1. *Alignment-suffixes* (also called *a-suffixes*) are defined as follows:

1. suffixes of γ;
2. suffixes of $\alpha^*\beta\gamma$ longer than γ;
3. suffixes of $\alpha^*\delta\gamma$ longer than γ;

4. suffixes of form $\alpha'(\beta/\delta)\gamma$, where α' is a suffix of α longer than α^* (this represents two usual suffixes derived from T_0 and T_1).

Definition 11 The suffix tree of alignment $\alpha(\beta/\delta)\gamma$ is a compacted trie representing all a-suffixes.

Common suffixes of T_0 and T_1 (a-suffixes of type 1) are represented by an identical leaf. A-suffixes of type 4 are connected with one leaf, so the leaf is labeled with the alignment. A-suffixes of type 2 and 3 are dealt with separately. Let an a-suffix be of the form $\alpha''(\beta/\delta)\gamma$, where α'' is a suffix of α^*. From the definition of α^*, α'' appears twice in T_0 or in T_1, and then (β/δ) may not be present in a single arc. So, a-suffixes $\alpha''\beta\gamma$ of type 2 and $\alpha''\delta\gamma$ of type 3 are represented by two leaves.

1.5.4.1 Structure Construction

First, the suffix tree $ST(T_0)$ of T_0 is constructed. Second, suffixes of T_1 are mapped into a-suffixes according to the three categories described above (types 1, 3 and 4). Then, the obtained a-suffixes are inserted into $ST(T_0)$. For a-suffixes of type 1, there is nothing particular to do; they already exist in the tree. The insertion of a-suffixes longer than γ is done in three steps. The first and the second steps aim to insert *explicitly* a-suffixes of type 3 (which are shorter than or equal to $\alpha^*\delta\gamma$). The last step handles *implicitly* the a-suffixes of type 4. The three steps consist of:

1. finding α^0 and inserting suffixes of $\alpha^0\delta\gamma$ longer than γ;
2. finding α^* and inserting suffixes of $\alpha^*\delta\gamma$ longer than $\alpha^0\delta\gamma$;
3. inserting suffixes of $\alpha\delta\gamma$ longer than $\alpha^*\delta\gamma$.

For step 1, $ST(T_0)$ is used to find α^0, by checking for some suffixes of α, whether or not two leaves exist to represent them. Let $\alpha^{(i)}$ be the suffix of α of length i. The method consists of checking whether suffix $\alpha^{(i)}$ is represented by at least two leaves for $i = 1, 2, 4, 8, \ldots$ using a doubling technique described in [45]. Let h be the smallest i in the previous set such that $\alpha^{(h)}$ is represented by only one leaf. Let $\alpha^{(k)} = \alpha^0$ be the longest suffix of α that is represented by two leaves, and then $h/2 \leq |\alpha^{(k)}| < h$. Then, $\alpha^{(k)}$ is searched by scanning $\alpha^{(h-1)}, \alpha^{(h-2)}, \ldots$ using suffix links. After finding α^0, suffixes from $\alpha^0\delta\gamma$ to $d\gamma$, where d is the last character of δ, are inserted is $ST(T_0)$ with the classical method resulting in tree ST'.

Note that finding α^* for step 2 is not obvious, since ST' is not complete:

- suffixes of T_1 longer than $\alpha^0\delta\gamma$ are not contained in ST';
- all suffixes of T_0 and some suffixes of T_1 are represented in ST';
- suffixes of type 1 share leaves with suffixes of T_0 but some suffixes of type 3 do not.

It is easy to demonstrate that ST' can provide efficient information to retrieve α^*. From the definition of α^* it appears at least twice in T_0 or in T_1. Then, it can be found by checking for some suffix α' longer than α^0 in ST' if at least two leaves exist to represent it as for finding α^0 is $ST(T_0)$. After finding α^*, the end of step 2 consists of inserting suffixes of type 3 in ST'. For the last step, a-suffixes of type 4 of the form $\alpha'\delta\gamma$ have to be inserted where α' is longer than α^*. Recall that these suffixes share a common leaf with the a-suffix $\alpha'\beta\gamma$. These a-suffixes are implicitly inserted by just replacing every $\beta\gamma$ by $(\beta/\delta)\gamma$.

The first step requires $O(|\alpha^0|)$ time to find α^0 and $O(|\alpha^0\delta\hat{\gamma}|)$ time for the insertion where $\hat{\gamma}$ is the longest prefix of γ such that $d\hat{\gamma}$ occurs at least twice in T_0 and T_1 where d is the character preceding γ. $O(\alpha^*)$ time is spent to find α^* and $O(|\alpha^*\delta\hat{\gamma}|)$ time to perform the insertions. The last step is implicitly done. Then, the overall time for the construction of the suffix tree of alignment $\alpha(\beta/\delta)\gamma$ is $O(|\alpha|+|\beta|+|\delta|+|\gamma|)$.

This method can be easily extended to represent general alignments of form $\alpha_1(\beta_1/\delta_1)\cdots\alpha_k(\beta_k\delta_k)\alpha_{k+1}$. Furthermore, the previous representation can be extended into alignments of more than two strings.

This method has been extended to construct the suffix array of an alignment [48].

1.5.4.2 Search for a Pattern *P* of Length *m*

The searching algorithm with the suffix tree of alignment is similar to searching with the usual suffix tree.

1.5.5 *Suffix Array of an Alignment*

Since suffix trees are costly in practice, the *suffix array of an alignment* (SAA) of Na et al. [48] was proposed as an economical version of the suffix tree of alignment [47].

The alignment is defined as in [47] (see previous subsection).

Assume that T_0, T_1 and T_2 are of the following form: $T_0 = \alpha\beta\gamma$, $T_1 = \alpha\delta\gamma$ and and $T_2 = \alpha\vartheta\gamma$. Then, the alignment is of the form $\alpha(\beta/\delta/\vartheta)\gamma$. Let α^i be the longest suffix of α that occurs at least twice in T_i with $0 \leq i \leq 2$, and let α^* be the longest of α^0, α^1 and α^2. Then, a-suffixes are classified into five categories as follows:

1. suffixes of γ;
2. suffixes $\omega^0\gamma$, where ω^0 is a non-empty suffix of $\alpha^*\beta$;
3. suffixes $\omega^1\gamma$, where ω^1 is a non-empty suffix of $\alpha^*\gamma$;
4. suffixes $\omega^2\gamma$, where ω^2 is a non-empty suffix of $\alpha^*\vartheta$;
5. suffixes of the form $\alpha'(\beta/\delta/\vartheta)\gamma$, where α' is a suffix of α longer than α^* (this represents three usual suffixes derived from T_0, T_1 and T_2).

Definition 12 The suffix array of alignment $\alpha(\beta/\delta/\vartheta)\gamma$ is a lexicographically sorted list of all the a-suffixes.

The sorted order is well defined for a-suffixes of types 1-4, since an a-suffix of these types represents one string. Let an a-suffix of type 5 of the form $\alpha'(\beta/\delta/\vartheta)\gamma$ be used. We know that $|\alpha'| > |\alpha^*|$ and that it appears as a prefix only once in each string $\alpha'\beta\gamma$, $\alpha'\delta\gamma$ and $\alpha'\vartheta\gamma$, i.e., substrings of T_0, T_1 and T_2, respectively.

Then, the order of $\alpha'(\beta/\delta/\vartheta)\gamma$ is well defined by α'. The lexicographically sorted order between the a-suffixes is well defined. Hence, the longest common prefix *LCP* between a-suffixes of $\alpha(\beta/\delta/\vartheta)\gamma$ is well defined.

1.5.5.1 Structure Construction

The generalized suffix array (*GSA*) of all a-suffixes of the alignment is constructed. A-suffixes of type 1 are suffixes of γ in T_0. The suffixes of T_0 longer than $\alpha^*\beta\gamma$ are converted *implicitly* to the a-suffixes of type 5. The insertion of a-suffixes of type 3 and 4 needs a particular reasoning.

Assume γ^0, γ^1 and γ^2 are the longest prefix of γ, appearing at least twice in T_0, T_1 and T_2, respectively. Let γ^* be the longest of γ^0, γ^1 and γ^2. The construction of the $SA_{\alpha(\beta/\delta/\vartheta)\gamma}$ is done in three steps that consist of:

1. find $|\alpha^*|$ and $|\gamma^*|$;
2. construct the *GSA* for T_0, $\alpha^*\delta\gamma^*d$ and $\alpha^*\vartheta\gamma^*d$, where d is the symbol following γ^* in γ;
3. delete suffixes of γ^*d derived from $\alpha^*\delta\gamma^*d$ and $\alpha^*\vartheta\gamma^*d$.

From the definition of α^*, it appears at least twice in T_0, T_1 or T_2. Then, it can be found by searching for the longest suffix α' of α. First, α^0 is computed. It can be found by using $SA_{T_0^{\sim}}$. Once found, the next step consists of finding the longest suffix $\alpha^{(k)}$, $k = 1, 2$ longer than α^0. For computing $\alpha^{(k)}$, one can omit the method followed for α^0, for the simple reason of avoiding complexities proportional to the total length of input strings. Since $\alpha^{(k)}$ is longer than α^0, it is easy to demonstrate that it occurs in some suffix prefixed by α^0 in T_k. Hence, the suffix arrays of the substrings $\alpha^0\delta\gamma^0$ and $\alpha^0\vartheta\gamma^0$ are sufficient to deduce it.

Then, γ^* is found symmetrically using the suffix arrays of T_0, $\alpha^0\delta\gamma^0$ and $\alpha^0\vartheta\gamma^0$ to obtain γ^0, γ^1 and γ^2, respectively, and retain the maximal one. Note that the suffix array constructed for each substring is needed only in the corresponding substep.

The step 2 constructs the *GSA* of T_0, $\alpha^*\beta\gamma^*d$, $\alpha^*\vartheta\gamma^*d$, where d is the symbol that follows γ^* in γ. It is not obvious to determine the order of a-suffixes of form $\omega^k\gamma$ (type 3, 4), since γ^*d occurs only once in each string. Hence, the order is determined by $\omega\gamma^*d$.

In step 3, the *GSA* is scanned to delete suffixes of γ^*d in $\alpha^*\delta^*d$ and $\alpha^*\vartheta\gamma^*d$ that are redundant with a-suffixes of type 1.

The step for finding α^* and γ^* requires $O(|T_0| + |\alpha^*\delta\gamma| + |\alpha^*\vartheta\gamma|)$ time and $O(|T_0|)$ space. Both steps 2 and 3 spend $O(|T_0| + |\alpha^*\delta\gamma| + |\alpha^*\vartheta\gamma|)$ in time and space. For a given alignment, its suffix array of alignment is constructed in $O(|T_0| + |\alpha^*\delta\gamma^*| + |\alpha^*\vartheta\gamma^*|)$ time and working space.

This data structure can be generalized to consider alignments with multiple non-common regions of the form $\alpha_1(\beta_1/\delta_1\vartheta_1)\cdots\alpha_k(\beta_k/\delta_k\vartheta_k)\alpha_{k+1}$. A non-common region is of the form $\beta_{i-1}\alpha_i\beta_i$. Hence, the *GSA* is extended to represent $2k + 1$ strings, and it requires at $O(|T_0| + \Sigma_{i=1}^{k}(2|\alpha_i^*| + |\delta_i| + |\vartheta_i|))$ space.

1.5.5.2 Search for a Pattern P of Length m

The searching algorithm with the suffix array of alignment is similar to the searching with the usual suffix array.

1.6 Conclusion

In this chapter, we made a summary of the state of the art to the most known data structures for DNA sequences and gave a generic view on each family of indexes. We presented basic indexes and showed that although they proved a strong ability to solve problems in the area of stringology, these indexes are memory consumers. Advanced families that are based on the essential ideas related to text compressibility permit to improving the memory usage while allowing an efficient search over the index. Recent efforts yield to more advanced data structures directly related to highly similar sequences. We focused on the essential ideas that take advantage from the high rate of similarity in a given set of sequences.

Instead of classical indexes that do not exploit the property of the high similarity between sequences and that yield to a redundant representation, novel solutions offer data structures devoted to highly similar data, which store only a reference sequence with differences with other sequences. This yields to an efficient *compact* index while providing fast searching over the sequences.

In other words, the more advanced indexes that we have reviewed take space close to the length of one sequence with the number of differences.

In Table 1.1, we gave a brief summary of complexities of data structures we have presented. We presented only the most recent and useful indexes with the best results for each category.

Table 1.1 Table depicting some complexities (memory consumptions and search times): n is the length of the input sequences, n' and N' denote respectively the number of common and differentiating segments, and N_{psc} is the number of prefixes of a given pattern that are suffixes of some common parts (Huang et al.), N_F is the number of factors of a reference sequence used to compress the other sequences of the input set (RLZ)

Structure	Space	Search time
$ST(T)$	$O(n)$	$O(m)$
$SA(T)$	$O(n)$	$O(m + \log n)$
FM index	$nH_k + o(n \log \sigma)$	$O(m)$
CSA	$nH_k(T)$	$O(\log \log n + \log \sigma)$
LZ index	$4nH_k(T) + o(n \log \sigma)$	$O((occ + 1) \log n)$
Huang et al.	$O(n' + N' \log rk + rk(\log n' + \log N'))$	$O(m + N_{psc}(m \log(rk)) + occ_3 \log n)$
RLZ	$2nH_k(T_0) + 5.55n +$ $O(N_F \log n \log \log n)$	$O\big(m(\log \sigma + \log \log n) + occ \cdot$ $\big(\log n \frac{\log N_F}{\log n}\big)\big)$
SAA	$O(n + \lvert\alpha^* \delta \gamma^*\rvert + \lvert\alpha^* \vartheta \gamma^*\rvert)$	$O(m + occ)$

The number of occurences of a pattern is denoted by *occ*

References

1. Aho, A.V., Corasick, M.J.: Efficient string matching: an aid to bibliographic search. Commun. ACM **18**(6), 333–340 (1975)
2. Alatabbi, A., Barton, C., Iliopoulos, C.S., Mouchard, L.: Querying highly similar structured sequences via binary encoding and word level operations. In: Iliadis, L.S., Maglogiannis, I., Papadopoulos, H., Karatzas, K., Sioutas, S. (eds.) Proceedings of the International Workshop on Artificial Intelligence Applications and Innovations, AIAI 2012, Part II. IFIP Advances in Information and Communication Technology, vol. 382, pp. 584–592. Springer, Cham (2012)
3. Apostolico, A.: The myriad virtues of subword trees. In: Apostolico, A., Galil, Z. (eds.) Combinatorial Algorithms on Words. NATO Advance Science Institute Series, vol. 12, pp. 85–96. Springer, Berlin (1985)
4. Arroyuelo, D., Navarro, G., Sadakane, K.: Reducing the space requirement of LZ-index. In: Lewenstein, M., Valiente, G. (eds.) Proceedings of the 17th Annual Symposium on Combinatorial Pattern Matching, CPM 2006, Barcelona. Lecture Notes in Computer Science, vol. 4009, pp. 318–329. Springer, Berlin (2006)
5. Bell, T., Cleary, J.G., Witten, I.H.: Text Compression. Prentice Hall, Upper Saddle River (1990)
6. Blumer, A., Blumer, J., Haussler, D., Ehrenfeucht, A., Chen, M.-T., Seiferas, J.: The smallest automation recognizing the subwords of a text. Theor. Comput. Sci. **40**, 31–55 (1985)
7. Blumer, A., Blumer, J., Haussler, D., McConnell, R., Ehrenfeucht, A.: Complete inverted files for efficient text retrieval and analysis. J. ACM **34**(3), 578–595 (1987)
8. Burrows, M., Wheeler, D.J.: A block-sorting lossless data compression algorithm. Technical Report 124, Digital SRC Research (1994)
9. Cover, T.M., Thomas, J.A.: Elements of Information Theory. Wiley, Hoboken (2012)
10. Crochemore, M., Lecroq, T.: Trie. In: Liu, L., Özsu, M.T. (eds.) Encyclopedia of Database Systems, pp. 3179–3182. Springer, Heidelberg (2009)
11. Crochemore, M., Rytter, W.: Text Algorithms. Oxford University Press, Oxford (1994)
12. Crochemore, M., Vérin, R.: On compact directed acyclic word graphs. In: Mycielski, J., Rozenberg, G., Salomaa, A. (eds.) Structures in Logic and Computer Science. A Selection of Essays in Honor of Andrzej Ehrenfeucht. Lecture Notes in Computer Science, vol. 1261, pp. 192–211. Springer, Berlin (1997)
13. Do, H.H., Jansson, J., Sadakane, K., Sung, W.-K.: Fast relative Lempel-Ziv self-index for similar sequences. In: Snoeyink, J., Lu, P., Su, K., Wang, L. (eds.) Proceedings of the Joint International Conference on Frontiers in Algorithmics and Algorithmic Aspects in Information and Management, FAW-AAIM 2012, Beijing. Lecture Notes in Computer Science, vol. 7285, pp. 291–302. Springer, Berlin (2012)
14. Farach, M.: Optimal suffix tree construction with large alphabets. In: Proceedings of the 38th Annual Symposium on Foundations of Computer Science, FOCS 1997, Miami Beach, FL, pp. 137–143 (1997)
15. Farach-Colton, M., Ferragina, P., Muthukrishnan, S.: On the sorting-complexity of suffix tree construction. J. ACM **47**(6), 987–1011 (2000)
16. Ferragina, P., Manzini, G.: Opportunistic data structures with applications. In: 41st Annual Symposium on Foundations of Computer Science, FOCS 2000, Redondo Beach, CA, pp. 390–398 (2000)
17. Ferragina, P., Manzini, G.: An experimental study of an opportunistic index. In: Proceedings of the 12th Annual ACM-SIAM Symposium on Discrete Algorithms, SODA 2001, Washington, DC, pp. 269–278. Society for Industrial and Applied Mathematics, Philadelphia (2001)
18. Ferragina, P., Manzini, G.: Indexing compressed text. J. ACM **52**(4), 552–581 (2005)
19. Ferragina, P., Manzini, G., Veli, M., Navarro, G.: An alphabet-friendly fm-index. In: Apostolico, A., Melucci, M. (eds.) Proceedings of the 11th International Conference on String Processing and Information Retrieval, SPIRE 2004, Padova. Lecture Notes in Computer Science, vol. 3246, pp. 150–160. Springer, Berlin (2004)

20. Ferragina, P., Manzini, G., Mäkinen, V., Navarro, G.: Compressed representations of sequences and full-text indexes. ACM Trans. Algorithms **3**(2), 20 (2007)
21. Grossi, R., Vitter, J.S.: Compressed suffix arrays and suffix trees with applications to text indexing and string matching (extended abstract). In: Yao, F.F., Luks, E.M. (eds.) Proceedings of the 32nd Annual ACM Symposium on Theory of Computing, STOC 2000, Portland, OR, pp. 397–406 (2000)
22. Grossi, R., Vitter, J.S.: Compressed suffix arrays and suffix trees with applications to text indexing and string matching. SIAM J. Comput. **35**(2), 378–407 (2005)
23. Grossi, R., Gupta, A., Vitter, J.S.: High-order entropy-compressed text indexes. In: Proceedings of the 14th Annual ACM-SIAM Symposium on Discrete Algorithms, SODA 2003, Baltimore, MD, pp. 841–850 (2003)
24. Grossi, R., Gupta, A., Vitter, J.S.: When indexing equals compression: experiments with compressing suffix arrays and applications. In: Proceedings of the 15th Annual ACM-SIAM Symposium on Discrete Algorithms, SODA 2004, New Orleans, LA, pp. 636–645. Society for Industrial and Applied Mathematics, Philadelphia (2004)
25. Gusfield, D.: Algorithms on Strings, Trees and Sequences: Computer Science and Computational Biology. Cambridge University Press, Cambridge (1997)
26. Holub, J., Crochemore, M.: On the implementation of compact DAWG's. In: Champarnaud, J.-M., Maurel, D. (eds.) Proceedings of the 7th International Conference on Implementation and Application of Automata, CIAA 2002, Revised Papers, Tours. Lecture Notes in Computer Science, vol. 2608, pp. 289–294. Springer, Berlin (2003)
27. Huang, S., Lam, T.W., Sung, W.-K., Tam, S.-L., Yiu, S.-M.: Indexing similar DNA sequences. In: Chen, B. (ed.) Proceedings of the 6th International Conference on Algorithmic Aspects in Information and Management, AAIM 2010, Weihai. Lecture Notes in Computer Science, vol. 6124, pp. 180–190. Springer, Berlin (2010)
28. Inenaga, S., Hoshino, H., Shinohara, A., Takeda, M., Arikawa, S., Mauri, G., Pavesi, G.: On-line construction of compact directed acyclic word graphs. In: Lewenstein, M., Valiente, G. (eds.) Proceedings of the 17th Annual Symposium on Combinatorial Pattern Matching, CPM 2006, Barcelona. Lecture Notes in Computer Science, vol. 4009, pp. 169–180. Springer, Berlin (2006)
29. Itoh, H., Tanaka, H.: An efficient method for in memory construction of suffix arrays. In: Proceedings of String Processing and Information Retrieval Symposium, 1999 and International Workshop on Groupware, pp. 81–88 (1999)
30. Kärkkäinen, J., Sanders, P.: Simple linear work suffix array construction. In: Baeten, J.C.M., Lenstra, J.K., Parrow, J., Woeginger, G.J. (eds.) Proceedings of the 30th International Colloquium on Automata, Languages and Programming, ICALP 2003, Eindhoven. Lecture Notes in Computer Science, vol. 2719, pp. 943–955. Springer, Berlin (2003)
31. Kärkkäinen, J., Ukkonen, E.: Lempel-Ziv parsing and sublinear-size index structures for string matching. In: Proceedings of the 3rd South American Workshop on String Processing (WSP). Citeseer (1996)
32. Kim, D.K., Sim, J.S., Park, H., Park, K.: Linear-time construction of suffix arrays. In: Baeza-Yates, R.A., Chávez, E., Crochemore, M. (eds.) Proceedings of the 14th Annual Symposium on Combinatorial Pattern Matching, CPM 2003, Morelia, Michocán. Lecture Notes in Computer Science, vol. 2676, pp. 186–199. Springer, Berlin (2003)
33. Ko, P., Aluru, S.: Space efficient linear time construction of suffix arrays. In: Baeza-Yates, R.A., Chávez, E., Crochemore, M. (eds.) Proceedings of the 14th Annual Symposium on Combinatorial Pattern Matching, CPM 2003, Morelia, Michocán. Lecture Notes in Computer Science, vol. 2676, pp. 200–210. Springer, Berlin (2003)
34. Kurtz, S.: Reducing the space requirement of suffix trees. Softw.-Pract. Exper. **29**(13), 1149–1171 (1999)
35. Kuruppu, S., Puglisi, S.J., Zobel, J.: Relative Lempel-Ziv compression of genomes for large-scale storage and retrieval. In: Chávez, E., Lonardi, S. (eds.) Proceedings of the 17th International Symposium on String Processing and Information Retrieval, SPIRE 2010, Los Cabos. Lecture Notes in Computer Science, vol. 6393, pp. 201–206. Springer, Berlin (2010)

36. Larsson, N.J., Sadakane, K.: Faster suffix sorting. Theor. Comput. Sci. **387**(3), 258–272 (2007)
37. Lempel, A., Ziv, J.: On the complexity of finite sequences. IEEE Trans. Inf. Theory **22**(1), 75–81 (1976)
38. Mäkinen, V.: Compact suffix array-a space-efficient full-text index. Fundam. Inform. **56**(1–2), 191–210 (2003)
39. Mäkinen, V., Navarro, G.: Compressed compact suffix arrays. In: Sahinalp, S.C., Muthukrishnan, S., Dogrusöz, U. (eds.) Proceedings of the 15th Annual Symposium on Combinatorial Pattern Matching, CPM 2004, Istanbul. Lecture Notes in Computer Science, vol. 3109, pp. 420–433. Springer, Berlin (2004)
40. Mäkinen, V., Navarro, G.: New search algorithms and time/space tradeoffs for succinct suffix arrays. Technical Report C-2004-20, University of Helsinki (2004)
41. Mäkinen, V., Navarro, G.: Succinct suffix arrays based on run-length encoding. Nordic J. Comput. **12**(1), 40–66 (2005)
42. Manber, U., Myers, G.: Suffix arrays: a new method for on-line string searches. SIAM J. Comput. **22**(5), 935–948 (1993)
43. Maniscalco, M.A., Puglisi, S.J.: Faster lightweight suffix array construction. In: Proceedings of the 17th Australasian Workshop on Combinatorial Algorithms, Ayers Rock, Uluru, pp. 16–29 (2006)
44. Manzini, G., Ferragina, P.: Engineering a lightweight suffix array construction algorithm. Algorithmica **40**(1), 33–50 (2004)
45. McCreight, E.D.: A space-economical suffix tree construction algorithm. J. ACM **23**(2), 262–272 (1976)
46. Morrison, D.: Patricia-practical algorithm to retrieve information coded in alphanumeric. J. ACM **15**(4), 514–534 (1968)
47. Na, J.C., Park, H., Crochemore, M., Holub, J., Iliopoulos, C.S., Mouchard, L., Park, K.: Suffix tree of alignment: an efficient index for similar data. In: Lecroq, T., Mouchard, L. (eds.) Proceedings of the 24th International Workshop on Combinatorial Algorithms, IWOCA 2013, Rouen. Lecture Notes in Computer Science, vol. 8288. Springer, Berlin (2013)
48. Na, J.C., Park, H., Lee, S., Hong, M., Lecroq, T., Mouchard, L., Park, K.: Suffix array of alignment: a practical index for similar data. In: Oren Kurland, M.L., Porat, E. (eds.) Proceedings of the 20th International Symposium on String Processing and Information Retrieval, SPIRE 2013, Jerusalem. Lecture Notes in Computer Science, vol. 8214, pp. 243–254. Springer, Berlin (2013)
49. Navarro, G.: Indexing text using the Ziv-Lempel trie. In: Laender, A.H.F., Oliveira, A.L. (eds.) Proceedings of the 9th International Symposium on String Processing and Information Retrieval, SPIRE 2002, Lisbon. Lecture Notes in Computer Science, vol. 2476, pp. 325–336. Springer, Berlin (2002)
50. Navarro, G., Mäkinen, V.: Compressed full-text indexes. ACM Comput. Surv. **39**(1), 2 (2007)
51. Nekrich, Y.: Orthogonal range searching in linear and almost-linear space. In: Dehne, F.K.H.A., Sack, J.-R., Zeh, N. (eds.) Proceedings of the 10th International Workshop on Algorithms and Data Structures, WADS 2007, Halifax. Lecture Notes in Computer Science, vol. 4619, pp. 15–26. Springer, Berlin (2007)
52. Procházka, P., Holub, J.: Compressing similar biological sequences using FM-index. In: Bilgin, A., Marcellin, M.W., Serra-Sagristà, J., Storer, J.A. (eds.) Data Compression Conference, DCC 2014, Snowbird, UT, 26–28 March 2014, pp. 312–321. IEEE, New York (2014)
53. Puglisi, S.J., Smyth, W.F., Turpin, A.H.: A taxonomy of suffix array construction algorithms. ACM Comput. Surv. **39**(2), 4 (2007)
54. Rytter, W.: Application of Lempel-Ziv factorization to the approximation of grammar-based compression. Theor. Comput. Sci. **302**(1), 211–222 (2003)
55. Sadakane, K.: New text indexing functionalities of the compressed suffix arrays. J. Algorithms **48**(2), 294–313 (2003)
56. Schürmann, K.-B., Stoye, J.: An incomplex algorithm for fast suffix array construction. In: Demetrescu, C., Sedgewick, R., Tamassia, R. (eds.) Proceedings of the 7th Workshop on Algorithm Engineering and Experiments and the Second Workshop on Analytic Algorithmics and

Combinatorics, ALENEX/ANALCO 2005, Vancouver, BC, pp. 77–85. SIAM, Philadelphia (2005)
57. Sirén, J., Välimäki, N., Mäkinen, V., Navarro, G.: Run-length compressed indexes are superior for highly repetitive sequence collections. In: Amir, A., Turpin, A., Moffat, A. (eds.) Proceedings of the 15th International Symposium on String Processing and Information Retrieval, SPIRE 2008, Melbourne. Lecture Notes in Computer Science, vol. 5280, pp. 164–175. Springer, Berlin (2008)
58. Ukkonen, E.: On-line construction of suffix trees. Algorithmica **14**(3), 249–260 (1995)
59. Ukkonen, E., Wood, D.: Approximate string matching with suffix automata. Algorithmica **10**(5), 353–364 (1993)
60. Weiner, P.: Linear pattern matching algorithms. In: Proceedings of the 14th Annual Symposium on Switching and Automata Theory, SWAT (FOCS), Iowa City, IA, vol. 1873, pp. 1–11. IEEE Computer Society, Washington (1973)

Chapter 2
Full-Text Indexes for High-Throughput Sequencing

David Weese and Enrico Siragusa

2.1 Introduction

Recent advances in *High-Throughput Sequencing* (HTS) demand for novel algorithms working on efficient data structures specifically designed for the analysis of large volumes of sequence data. This chapter describes such data structures, called *full-text indexes*, to represent all substrings (or substrings up to a certain length) contained in a given text (or text collection).

Full-text indexes are a fundamental component of almost all HTS applications. For instance, they are used in read error correction [48], de novo [9, 50] or reference-guided assembly [46], short DNA read mapping [32, 33, 52], short RNA read mapping [13], structural variation detection [14], and fast local alignment [31]. Besides HTS, the applicability of full-text indexes in repeating a search in multiple large sequences allows finding identical, conserved regions which are later extended to multiple genome alignments [45]. The tree structure exposed by some full-text indexes allows to efficiently count frequencies in multiple databases for frequency-based string mining [57] or to construct variable order Markov chains [47].

This chapter covers three classes of full-text indexes: *suffix tries*, *suffix trees*, and *q-gram indexes*. It describes practical index realizations that have been implemented in *SeqAn* [12], the open-source software library for sequence analysis, and used in the aforementioned bioinformatics applications. It omits other compressed full-text indexes, e.g., compressed suffix arrays [23] that have not been applied yet to bioinformatics, even though efficient implementations are provided by other open-source libraries, such as SDSL [22], libcds2, and strmat [25]. For a comprehensive description of compressed full-text indexes, we refer the reader to [43].

D. Weese (✉) • E. Siragusa
Department of Mathematics and Computer Science, Freie Universität Berlin, Berlin, Germany
e-mail: david.weese@fu-berlin.de; enrico.siragusa@fu-berlin.de

M. Elloumi (ed.), *Algorithms for Next-Generation Sequencing Data*,
DOI 10.1007/978-3-319-59826-0_2

The rest of the chapter is organized as follows: Sects. 2.1.1 and 2.1.2 give fundamental definitions required throughout the following sections. Sections 2.2.1 and 2.2.2 introduce the *suffix trie*, a fundamental index of all substrings of a single string or multiple strings, and the *suffix tree*, a more compact representation of the suffix trie. The *Burrows-Wheeler Transform*, described in Sect. 2.2.4, is a transformation of the text applicable to text compression and full-text indexing, e.g., for the FM-index explained in Sect. 2.3.2. Section 2.3.1 introduces the *suffix array* and how it can be used like a suffix trie. Sections 2.4.1 and 2.4.2 cover two suffix tree indexes, the theoretical optimal *enhanced suffix array* and the dynamically constructed *lazy suffix tree*. Section 2.5 describes the *q-gram index*, a simple index of all substrings of length q. Finally, Sects. 2.6.1, 2.6.2, and 2.6.3 show how the proposed indexes can be applied to practical HTS problems, e.g., exact and approximate string matching, and how they perform on real data.

2.1.1 Notations

An *alphabet* $\Sigma = \{c_0, c_1, \ldots, c_{\sigma-1}\}$ is a finite ordered set of characters c_i. A *string* S over Σ is a finite sequence of characters from Σ. We denote with $S[i]$ the ith character counting from 0, with $S[i,j]$ the *concatenated* characters $S[i] \ldots S[j]$, and with $|S| = n$ the length of S. Σ^n denotes the set of all strings of length n over Σ where $\Sigma^0 = \{\varepsilon\}$ contains only the *empty string*; the set of all strings is denoted with $\Sigma^* = \cup_{n=0}^{\infty} \Sigma^n$. Moreover, we denote the concatenation of two strings S and T by ST or $S \cdot T$; if for strings $S, T, U, V \in \Sigma^*$ holds $S = T \cdot U \cdot V$, we call T prefix, U infix or substring, and V suffix of S. For convenience, we denote the suffix of S beginning at position i simply by S_i. In the following, all intervals are integer intervals, i.e., $[i,j] = \{i, i+1, \ldots, j\}$ and $[i,j) = [i, j-1]$.

Definition 1 (Structure) A structure is a collection of variables. By $x = \{a, b, c\}$, we denote a structure x consisting of variables a, b, c. We refer to these variables as $x.a$, $x.b$, $x.c$.

Definition 2 (String Collection) A *string collection* is an ordered multiset $\mathbb{S} = \{S^0, S^1, \ldots, S^{\kappa-1}\}$ of not necessarily distinct strings over Σ. We denote the total length of the string collection with $\|\mathbb{S}\| = \sum_{i \in [0,\kappa)} |S^i|$. The notations for characters and substrings naturally generalize to string collections, e.g., $\mathbb{S}_{(d,i)}$ denotes the suffix S_i^d.

Definition 3 (Reversal) For a string S, $\overline{S}$ denotes its reversal, i.e., $\overline{S}[i] = S[|S| - 1 - i]$ for $i \in [0, |S|)$. For a string collection $\mathbb{S} = \{S^0, S^1, \ldots, S^{\kappa-1}\}$, $\overline{\mathbb{S}}$ denotes the reversed collection $\{\overline{S^0}, \overline{S^1}, \ldots, \overline{S^{\kappa-1}}\}$.

Definition 4 (Lexicographical Order) The *lexicographical order* $<_{\text{lex}}$ between two non-empty strings S, T is defined as $S <_{\text{lex}} T \iff S[0] < T[0] \vee (S[0] = T[0] \wedge S_1 <_{\text{lex}} T_1)$. We remark that the lexicographical order is a partial order.

We denote by $\min_{\text{lex}}(\Sigma)$ the lexicographically smallest character $c \in \Sigma$ and by $\text{next}_{\text{lex}}(c)$ the lexicographical successor of c in Σ.

2.1.2 *Padding*

We define *padding* of strings and string collections in order to well define full-text indexes. *Padding* is necessary to ensure that no suffix is a prefix of a string in the collection. To this end, we define the special *terminator* character $\$ \notin \Sigma$, such that $\$ < a$ for any $a \in \Sigma$.

Definition 5 (Padded String) For a string S over Σ, we call $S\$$ the *padded string* over $\Sigma_\$$, with $\Sigma_\$ = \Sigma \cup \{\$\}$.

Definition 6 (Padded String Collection) For a string collection $\mathbb{S} = \{S^0, S^1, \ldots, S^{\kappa-1}\}$ over Σ, we call $\{S^0\$^0, S^1\$^1, \ldots, S^{\kappa-1}\$^{\kappa-1}\}$ the *padded string collection* over $\Sigma_\$$, where $\Sigma_\$ = \Sigma \cup \{\$^0, \$^1, \ldots, \$^{\kappa-1}\}$ and $\$^0, \$^1, \ldots, \$^{\kappa-1}$ are individual terminator characters with $\$^i < \$^j \iff i < j$.

We remark that such terminator characters are used only for ease of explanation, e.g.,, to have a one-to-one relationship between suffixes and their start positions, to well define a lexicographical order on them, and to associate the *Burrows-Wheeler Transform* (BWT) [8] with the suffix array. Explicit representation of terminator characters increases the size of the padded alphabet $\Sigma_\$$ and thus complicates the implementation of any full-text index. Thus, almost all careful full-text index implementations avoid representing terminator characters explicitly.

2.2 Background

In the following, we define the concepts of suffix trie, suffix tree, and suffix array full-text indexes in general. Then, we will introduce the Burrows-Wheeler Transform of a text and how it can be inverted.

2.2.1 *Suffix Trie*

Definition 7 (Suffix Trie) The *suffix trie* $\mathcal{T}_{\text{strie}}(S)$ [41] of a padded string S (Definition 5) is a lexicographically ordered tree data structure having one node designated as the root and $|S|$ leaves labeled by $0, 1, \ldots, |S| - 1$, where leaf i refers to suffix S_i, such that:

1. Incoming edges of inner nodes are labeled by characters from Σ, and incoming edges of leaves are labeled by the terminator character.

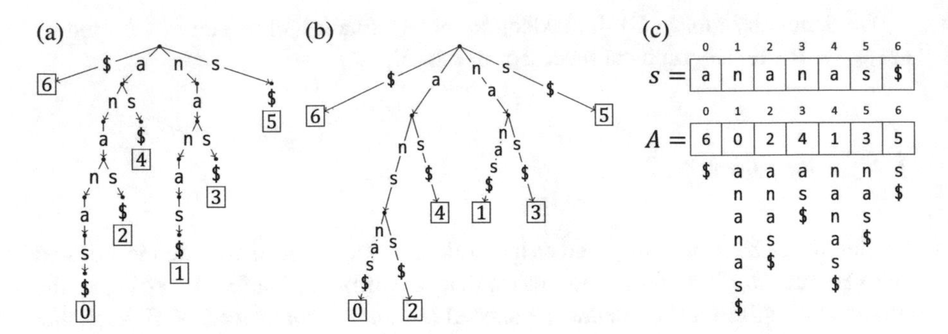

Fig. 2.1 Suffix trie (**a**), suffix tree (**b**), and suffix array (**c**) of the string ananas$

2. The outgoing edges of each inner node are labeled by distinct characters.
3. The path from the root to the leaf i spells suffix $S_i\$$.

An example of a suffix trie is given in Fig. 2.1a.

Definition 8 (Generalized Suffix Trie) The *generalized suffix trie* $\mathscr{T}_{\text{strie}}(\mathbb{S})$ of a padded string collection $\mathbb{S}$ (Definition 6) is a suffix trie having $\|\mathbb{S}\|$ leaves labeled by pairs (i,j) which refer to suffixes $\mathbb{S}_{(i,j)}$, and the path from the root to leaf (i,j) spells suffix $\mathbb{S}_{(i,j)}\$^i$.

In the following, we introduce a set of generic operations to traverse any $\mathscr{T}_{\text{strie}}$ realization in a top-down fashion. Note that the top-down traversal stops when encountering terminator edges and thus does not reach suffix tree leaves. Given a suffix trie $\mathscr{T}_{\text{strie}}(S)$, we define the following operations on the node v pointed during the traversal of $\mathscr{T}_{\text{strie}}(S)$:

- ISROOT(v) returns *true* if and only if the pointed node is the root.
- ISTERMINAL(v) returns *true* if and only if all outgoing edges labels are terminators.
- LABEL(v) returns the character labeling the edge entering v.
- OCCURRENCES(v) returns the list of positions pointed by leaves below v.
- PARENT(v) returns the parent node of v.

Moreover, we define the following operations moving from pointed node v and returning *true* on success and *false* otherwise:

- GODOWN(v) moves to the lexicographically smallest child of v.
- GODOWN(v, c) moves to the child of v whose entering edge is labeled by c.
- GORIGHT(v) moves to the lexicographically next sibling of v.

Time complexities of the above operations depend on the specific realization of the (generalized) suffix trie. For realizations shown in what follows in this chapter, LABEL is $\mathscr{O}(1)$, both variants of GODOWN and GORIGHT range from $\mathscr{O}(1)$ to $\mathscr{O}(\log n)$, and while OCCURRENCES is linear in the number of occurrences, PARENT is $\mathscr{O}(1)$ but requires $\mathscr{O}(n \log n)$ bits to stack all parent nodes. Recursive algorithms implicitly implement PARENT by stacking all parent nodes during recursive calls.

The (generalized) suffix trie contains $\mathcal{O}(n^2)$ nodes, where n is the total text length. However, its practical realizations consume only $\mathcal{O}(n \log n)$ bits as they do not explicitly represent the internal nodes. In the following section, we describe the *(generalized) suffix tree* [25, 58], a data structure that represents a text (collection) in $\mathcal{O}(n)$ nodes but can be traversed like a suffix trie.

2.2.2 *Suffix Tree*

The *suffix tree* [25, 58] arises from the suffix trie by compacting each path where only start and end nodes are branching into a single edge labeled by the concatenation of edge labels. Thus, the suffix tree comes with the restriction that internal nodes must have more than one child and with the property that edges can labeled by strings of arbitrary length.

Definition 9 (Suffix Tree) The *suffix tree* $\mathcal{T}_{\text{stree}}(S)$ of a padded string S is a lexicographically ordered tree data structure having one node designated as the root and $|S|$ leaves labeled by $0, 1, \ldots, |S| - 1$, where leaf i refers to suffix S_i, such that:

1. Edges are labeled by strings over $\Sigma_{\$}$.
2. The outgoing edges of each inner node begin with distinct characters.
3. The path from the root to the leaf i spells suffix $S_i \$^i$.
4. Each internal node is *branching*, i.e., it has at least two children.

An example of a suffix tree is given in Fig. 2.1b.

Definition 10 (Generalized Suffix Tree) The *generalized suffix tree* $\mathcal{T}_{\text{stree}}(\mathbb{S})$ of a given padded string collection $\mathbb{S}$ is a suffix tree having $\|\mathbb{S}\|$ leaves, where leaf (i,j) refers to suffix $\mathbb{S}_{(i,j)}$ and the path from the root to leaf (i,j) spells suffix $\mathbb{S}_{(i,j)}\$^i$.

The (generalized) suffix tree of a string (collection) of length n can be constructed in $\mathcal{O}(n)$ time and $\mathcal{O}(n \log n)$ bits space [15, 40, 55]. The space consumption can be achieved by representing edge labels implicitly, i.e., by storing only their begin and end positions in the underlying input string.

As a consequence of property 2, every tree node v can uniquely be identified by the concatenation of edge labels on the path from the root to v. For a node v, we denote this string by $\textsc{concat}(v)$ and call it the *concatenation string* or *representative* of v. Note that, conversely to tries, $\textsc{label}$ on (generalized) suffix trees returns a substring instead of a character.

2.2.3 *Suffix Array*

The *suffix array* (SA) has been introduced by Manber and Myers [36] as a practical and memory-efficient replacement for the suffix tree in string matching applications.

The SA represents explicitly all leaves of the suffix trie, while it omits internal nodes and outgoing edges. An example of a suffix array is given in Fig. 2.1c.

Definition 11 (Suffix Array) The *suffix array* SA(S) of a string S of length n is an array A containing a permutation of the interval $[0, n)$, such that $S_{A[i-1]} <_{\text{lex}} S_{A[i]}$ for all $i \in [1, n)$.

The first generalization of the SA to string collections can be found in [49]. Recently, several authors have implemented the generalized suffix array, for instance, [4, 35, 52, 53].

Definition 12 (Generalized Suffix Array) The *generalized suffix array* SA($\mathbb{S}$) of a string collection $\mathbb{S}$ is an array A of length $n = \|\mathbb{S}\|$ containing a permutation of all pairs (i, j) where i refers to a string $S^i \in \mathbb{S}$ and j refers to one of the $|S^i|$ suffixes of S^i. Pairs are ordered such that $\dot{\mathbb{S}}_{A[i-1]} <_{\text{lex}} \dot{\mathbb{S}}_{A[i]}$ for all $i \in [1, n)$, where $\dot{\mathbb{S}}$ is the corresponding padded string collection of $\mathbb{S}$.

Various algorithms have been proposed to construct the SA in linear time. An interesting review can be found in [44].

2.2.4 *Burrows-Wheeler Transform*

The *Burrows-Wheeler Transform* (BWT) [8] is a transformation defining a permutation of an input string. The transformed string exposes two important properties: *reversibility* and *compressibility*. The former property allows to reconstruct the original string from its BWT, while the latter property makes the transformed string more amenable to compression [2, 39]. Because of these two properties, the BWT has become a fundamental method for text compression. Only some years after its introduction, the BWT has been proposed by Ferragina and Manzini [17] as a method for full-text indexing.

Let S be a padded string (Definition 5) of length n over an alphabet $\Sigma_\$$. In the following, we consider S to be cyclic and its subscript $S[i]$ to be *modular*, i.e., $S[i] = S[i \bmod n]$ for any $i \in \mathbb{Z}$.

Definition 13 (Burrows-Wheeler Transform) Given a string S and its SA A, the BWT L of S is a string over $\Sigma_\$$ with $L[i] = S[A[i] - 1]$.

To visualize the BWT, consider the square matrix consisting of all cyclic shifts of the string S sorted in lexicographical order, where the ith cyclic shift has the form $S[i, n) \cdot S[0, i)$. The BWT is the string obtained by concatenating the characters in the last column of the sorted cyclic shifts matrix of S. Note how the cyclic shifts matrix is related to the SA A of S: The ith cyclic shift in lexicographical order is $S[A[i], n) \cdot S[0, A[i])$. An example of an SA and the cyclic shifts matrix is given in Fig. 2.2.

i	$A[i]$	$S_{A[i]}$
0	10	`$`
1	9	`a$`
2	2	`atctctta$`
3	4	`ctctta$`
4	6	`ctta$`
5	8	`ta$`
6	1	`tatctctta$`
7	3	`tctctta$`
8	5	`tctta$`
9	7	`tta$`
10	0	`ttatctctta$`

(a) suffix array with suffixes

i	$\Psi(i)$	$LF(i)$	$S[A[i]]$	...	$S[A[i]-1]$
0	10	1	`$`	`ttatctctt`	`a`
1	0	5	`a`	`$ttatctct`	`t`
2	7	6	`a`	`tctctta$t`	`t`
3	8	7	`c`	`tctta$tta`	`t`
4	9	8	`c`	`tta$ttatc`	`t`
5	1	9	`t`	`a$ttatctc`	`t`
6	2	10	`t`	`atctctta$`	`t`
7	3	2	`t`	`ctctta$tt`	`a`
8	4	3	`t`	`ctta$ttat`	`c`
9	5	4	`t`	`ta$ttattc`	`c`
10	6	0	`t`	`tatctctta`	`$`

(b) Ψ, LF, and sorted cyclic shifts with BWT

Fig. 2.2 Suffix array (**a**) and sorted cyclic shifts (**b**) with permutations Ψ and LF of the string $S =$ `ttatctctta$`. The BWT of S corresponds to $S[A[i]-1]$, i.e. the last column of the sorted cyclic shifts matrix

The cyclic shifts matrix is conceptual and does not have to be constructed explicitly to derive the BWT of S. The BWT can be obtained in linear time by scanning the SA A and assigning the character $S[A[i]-1]$ to the ith BWT character. However, constructing the BWT from the SA is not desirable in practice, especially for strings over small alphabets like DNA, as the SA consumes $\Theta(n \log n)$ bits in addition to the $\Theta(n \log \sigma)$ bits of the BWT. Therefore, various direct BWT construction algorithms working within $o(n \log \sigma)$ bits plus constant space have been recently proposed in [5, 11, 18, 27].

2.2.4.1 Inversion

We now describe how to invert the BWT to reconstruct the original string S. For convenience, we denote the first column $S[A[i]]$ of the cyclic shifts matrix by F. BWT inversion relies on two simple observations [8] on the cyclic shifts matrix.

Observation 2.1 *For all $i \in [0, n)$, character $L[i]$ precedes character $F[i]$ in the original string S.*

Observation 2.2 *For all characters $c \in \Sigma_\$$, the ith occurrence of c in F corresponds to the ith occurrence of c in L.*

These observations are evident as F is column $S[A[i]]$ and L is column $S[A[i]-1]$ (see Definition 13 and Fig. 2.2b).

Inverting the BWT means being able to know where any BWT character occurs in the original string. To this intent, we define two permutations, $LF : [0, n) \rightarrow [0, n)$ and $\Psi : [0, n) \rightarrow [0, n)$, with $LF = \Psi^{-1}$; the value of $LF(i)$ gives the position j in F where character $L[i]$ occurs and the value $\Psi(j)$ gives back the position i in L where $F[j]$ occurs. The iterated Ψ is defined as:

$$\begin{aligned} \Psi^0(j) \quad &= j \\ \Psi^{i+1}(j) &= \Psi(\Psi^i(j)) \end{aligned} \tag{2.1}$$

and the iterated LF as:

$$\begin{aligned} LF^0(j) \quad &= j \\ LF^{i+1}(j) &= LF(LF^i(j)). \end{aligned} \tag{2.2}$$

Character $S[i]$ corresponds to $F\left[\Psi^i(j)\right]$, and $\overline{S}[i]$ corresponds to $L\left[LF^i(j)\right]$ with j being the position of \$ in L. Thus, the full string S is recovered by starting in F at the character preceded by \$ and following the cycle defined by the permutation Ψ. Conversely, the reverse string $\overline{S}$ is recovered by starting in L at the position of \$ and following the cycle defined by the permutation LF.

Permutation LF needs not to be stored explicitly as it is computable from the BWT with the help of some additional character counts. Given Observations 2.1, 2.2, permutation LF is computed by:

$$LF(i) = C(L[i]) + Occ(L[i], i) \tag{2.3}$$

where $C : \Sigma_\$ \rightarrow [0, n)$ denotes the total number of occurrences in L of all characters alphabetically smaller than c, while $Occ : \Sigma_\$ \times [0, n] \rightarrow [0, n)$ denotes the number of occurrences of character c in the prefix $L[0, i)$. The key problem of encoding the permutation LF lies in representing function Occ, as function C is easily tabulated by a small array of size $\mathcal{O}(\sigma \log n)$ bits. In Sect. 2.3.2.1, we address the problem of representing function Occ efficiently.

The BWT easily generalizes to string collections. Indeed, Definition 13 still holds for a padded string collection $\mathbb{S}$ (Definition 6) and its cyclic shifts matrix sorted in lexicographical order. Inverting the generalized BWT works in the same way. Indeed, permutations Ψ and LF are composed of κ cycles, where each cycle corresponds to a distinct string S^i in the collection. The string S^i is recovered by starting at the position of $\$^i$ in the BTW and by following the cycle of Ψ (or LF) associated to S^i. See [5, 38] for further details on the generalized BWT.

2.3 Suffix Trie Realizations

In the following, we introduce two different full-text indexes that both can be traversed like a suffix trie. They both emulate the traversal of the potentially quadratic number of suffix trie nodes, either by conducting binary searches or by utilizing a property of the Burrows-Wheeler Transform, and hence have a space consumption linear in the length of the text.

2.3.1 *Suffix Array*

We now describe how to emulate the top-town traversal of a (generalized) suffix trie $\mathcal{T}_{\text{strie}}(\mathbb{S})$ using the SA A. Any node v of $\mathcal{T}_{\text{strie}}(\mathbb{S})$ is univocally identified by an interval of A. Thus, while traversing the trie, we remember the interval $[l, r)$ identifying the current node; the root node is initialized by the interval $[0, n)$ spanning the whole SA. In addition, we remember the right interval *parentR* of the parent node initialized to n for the root node, and the depth d of the current node initialized to 0. The incoming edge label of an internal node at depth d is identified by the $(d-1)$th character within the corresponding suffix $\mathbb{S}_{A[i]}$ for any $i \in [l, r)$ and undefined for the root node. Terminal nodes are defined to have all their outgoing edges labeled by terminator characters. The occurrences below any node correspond by definition to the interval $[l, r)$ of A. Summing up, we represent the current node v by the integers $\{l, r, parentR, d\}$ and define the following operations on it:

- GOROOT(v) initializes v to $\{0, n, n, 0\}$.
- ISTERMINAL(v) returns *true* if and only if $|\mathbb{S}_{A[v.r-1]}| = v.d + 1$.
- LABEL(v) returns $\mathbb{S}[A[v.l] + v.d - 1]$.
- OCCURRENCES(v) returns $A\,[v.l, v.r)$.

Binary search is the key to achieve function GODOWN by character (Algorithm 2). By means of two calls to function B (Algorithm 1), GODOWN by character (Algorithm 2) computes in $\mathcal{O}(\log n)$ binary search steps the left and right SA intervals identifying the child node reached by going down the edge labeled by a given character c. Algorithms 3 and 4 achieve respectively GODOWN and GORIGHT with time complexity independent of the alphabet size σ. As they rely on a single call of function B, their time complexity is also $\mathcal{O}(\log n)$.

2.3.2 *FM-Index*

The FM-index [17] is the most popular *succinct* full-text index for HTS. Over the last years, the FM-index has been widely employed under different re-implementations by many popular bioinformatics tools, e.g., *Bowtie* [32] and

Algorithm 1 $\mathrm{B}(v, c, \lhd)$

Input v : SA iterator
c : character to query
$\lhd$: lexicographic comparator
Output integer denoting the left or right SA interval

1: $(i_1, i_2) \leftarrow (v.l, v.r)$
2: **while** $i_1 < i_2$ **do**
3: $\quad j \leftarrow \lfloor \frac{i_1+i_2}{2} \rfloor$
4: $\quad$ **if** $\mathbb{S}[A[j] + v.d] \lhd c$ **then**
5: $\quad\quad i_1 \leftarrow j + 1$
6: $\quad$ **else**
7: $\quad\quad i_2 \leftarrow j$
8: **return** i_1

Algorithm 2 GODOWN(v, c)

Input v : SA iterator
c : character to query
Output boolean indicating success

1: **if** ISTERMINAL(v) **then**
2: $\quad$ **return false**
3: $l \leftarrow \mathrm{B}(v, c, <)$
4: $r \leftarrow \mathrm{B}(v, c, \leq)$
5: **if** $l < r$ **then**
6: $\quad v \leftarrow \{l, r, v.r, v.d + 1\}$
7: $\quad$ **return true**
8: **else**
9: $\quad$ **return false**

Algorithm 3 GODOWN(v)

Input v : SA iterator
Output boolean indicating success

1: **if** ISTERMINAL(v) **then**
2: $\quad$ **return false**
3: $v.l \leftarrow \mathrm{B}(v, \$, \leq)$
4: $c_l \leftarrow \mathbb{S}[A[v.l] + v.d]$
5: $c_r \leftarrow \mathbb{S}[A[v.r - 1] + v.d]$
6: $v.parentR \leftarrow v.r$
7: **if** $c_l \neq c_r$ **then**
8: $\quad v.r \leftarrow \mathrm{B}(v, c_l, \leq)$
9: $v.d \leftarrow v.d + 1$
10: **return true**

BWA [34], and is now considered a fundamental method for the full-text indexing of genomic sequences.

The FM-index allows to emulate efficiently the top-town traversal of a (generalized) suffix trie. This is possible thanks to the BWT, provided an efficient representation of function *Occ* (see Sect. 2.2.4). Thus, we first show how to represent function *Occ*, and afterwards, we describe the top-town traversal.

Algorithm 4 GORIGHT(v)

Input v : SA iterator
Output boolean indicating success

1: **if** $v.r = v.parentR$ **then**
2:     **return false**
3: $v.l \leftarrow v.r$
4: $c_r \leftarrow \mathbb{S}[A[v.r-1]+v.d]$
5: $c_p \leftarrow \mathbb{S}[A[v.parentR-1]+v.d]$
6: $v.r \leftarrow v.parentR$
7: **if** $c_r \neq c_p$ **then**
8:     $v.r \leftarrow \mathrm{B}(v, c_r, \leq)$
9: **return true**

2.3.2.1 Rank Dictionaries

Function *Occ*, answering the question "*how many times does a given character c occur in the prefix* $L[0,i)$?" has to be done ideally in constant time and succinct space. The general problem on arbitrary strings has been tackled by several studies on the succinct representation of data structures. This specific question takes the name of *rank query*, and a data structure answering rank queries is called *rank dictionary* (RD).

Definition 14 Given a string S over an alphabet Σ and a character $c \in \Sigma$, $\mathrm{RANK}_c(S,i)$ returns the number of occurrences of c in the prefix $S[0,i)$.

The key idea of RDs is to maintain a succinct (or even compressed) representation of the input string and attach a dictionary to it. By doing so, Jacobson shows how to answer rank queries in constant time (on the RAM model) using $o(n)$ additional bits for an input binary string of n bits [29]. Here, we follow the explanation of Navarro and Mäkinen [42], yet we stick only to the most practical RDs for HTS. We first consider RDs for strings over a binary alphabet $\mathbb{B} = \{0,1\}$. We start by describing a simple *one-level* rank dictionary answering rank queries in constant time but consuming $\mathcal{O}(n)$ additional bits. Subsequently, we describe an extended *two-levels* RD consuming only $o(n)$ additional bits. Finally, we generalize these RDs for strings over small alphabets, e.g., DNA.

The one-level binary RD partitions the binary input string $S \in \mathbb{B}^*$ in blocks of b characters and complements it with an array R of length $\lfloor n/b \rfloor + 1$. The jth entry of R provides a summary of the number of occurrences of the bit 1 in S before position jb, i.e., $R[0] = 0$ and $R[j] = \mathrm{RANK}_1(S, jb)$ for any $j > 0$. Note that R summarizes only RANK_1, as $\mathrm{RANK}_0(S,i) = i - \mathrm{RANK}_1(S,i)$. Therefore, the rank query is rewritten by:

$$\mathrm{RANK}_1(S,i) = R[\lfloor i/b \rfloor] + \mathrm{RANK}_1(S_{\lfloor i/b \rfloor \cdot b}, i \bmod b). \tag{2.4}$$

The query is answered in $\mathcal{O}(b)$ time by(i) fetching the rank summary from R in constant time and (ii) counting the number of occurrences of the bit 1 within a block

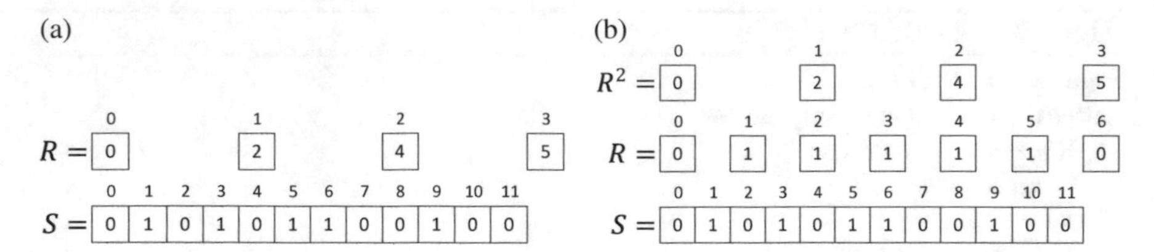

Fig. 2.3 Binary RDs of the string $S = 010101100100$. (**a**) One-level RD with $b = 4$; for instance, $\text{RANK}_1(S, 6) = 3$ is answered as $R[1] + \text{RANK}_1(S_4, 2)$. (**b**) Two-levels RD with $b = 2$; $\text{RANK}_1(S, 6) = 3$ is answered as $R^2[1] + R[2] + \text{RANK}_1(S_4, 2)$

of $\mathcal{O}(b)$ bits. An example of a one-level rank dictionary is given in Fig. 2.3a. [29] answers step (ii) in $\mathcal{O}(1)$ time by posing $b = \lceil \log n \rceil$ and using the four-Russians tabulation technique [3]. However, despite being theoretically optimal, the four-Russians method can be quite slow in practice; a more practical solution consists in using the SSE 4.2 popcnt instruction [28] to implement step (ii) in $\mathcal{O}(b/w)$ time, where w is the SSE register width, e.g., $w = 256$ bits on modern processors. As the array R stores $\mathcal{O}(n/\log n)$ positions and each position in S requires $\lceil \log n \rceil$ bits, R consumes $\mathcal{O}(n)$ bits. Thus, the one-level binary RD adds $\mathcal{O}(n)$ bits of memory.

The *two-levels* binary RD squeezes space consumption down to $o(n)$ bits of additional memory. The idea is to divide S in superblocks of length b^2 and add another array R^2 to store the number of occurrences of the bit 1 in S before each superblock. The initial array R now counts bit 1 only within the overlapping superblock defined by R^2. Accordingly, the rank query becomes:

$$\text{RANK}_1(S, i) = R^2\left[\left\lfloor i/b^2 \right\rfloor\right] + R[\lfloor i/b \rfloor] + \text{RANK}_1(S_{\lfloor i/b \rfloor \cdot b}, i \bmod b). \tag{2.5}$$

Figure 2.3b shows an example of two-levels binary RD. Each value in R is less than b^2 and thus consumes only $\mathcal{O}(\log b^2) = \mathcal{O}(\log b)$ bits. Summing up, this RD consumes n bits for the input string, $\mathcal{O}(n \log n/b^2)$ bits for R^2 and $\mathcal{O}(n \log b/b)$ bits for R. By posing $b = \lceil \log n \rceil$ as in the above one-level RD, it follows $\mathcal{O}(n/\log n)$ bits for R^2 and $\mathcal{O}(n \log\log n/\log n)$ bits for R. Hence, the two-levels binary RD adds $o(n)$ bits of memory.

It is easy to extend the above binary RDs to support strings over arbitrary alphabets; in the following, we show how to extend the one-level RD. Consider an input string S of n characters over Σ, thus consisting of $n \lceil \log \sigma \rceil$ bits. As in the binary case, this one-level RD partitions S in blocks of b bits. It complements the string S with a matrix R_σ of size $\lfloor n/b + 1 \rfloor \times \sigma$, summarizing the number of occurrences for each character in Σ. The rank query is rewritten accordingly:

$$\text{RANK}_c(S, i) = R_\sigma[\lfloor i/b \rfloor, c] + \text{RANK}_c(S_{\lfloor i/b \rfloor \cdot b}, i \bmod b). \tag{2.6}$$

Figure 2.4 shows an example of one-level DNA RD. Answering this query requires counting the number of occurrences of the character c inside a block of b bits.

Fig. 2.4 One-level DNA rank dictionary of the string $S = \text{atttttacc\$}$ with $b = 4$. For instance, $\text{RANK}_\text{t}(S, 6) = 5$ is answered as $R_\sigma[1, \text{t}] + \text{RANK}_1(S_4, 2)$

	0	1	2
$R_\sigma[\cdot,\text{a}] =$	0	1	2
$R_\sigma[\cdot,\text{c}] =$	0	0	0
$R_\sigma[\cdot,\text{g}] =$	0	0	0
$R_\sigma[\cdot,\text{t}] =$	0	3	6

	0	1	2	3	4	5	6	7	8	9	10
$S =$	a	t	t	t	t	t	t	a	c	c	$

In order to answer this query in constant time, we consider blocks of $b = \lfloor \lceil \log n \rceil / \lceil \log \sigma \rceil \rfloor$ characters, i.e., each block consumes not more than $\lceil \log n \rceil$ bits as in the binary RD case. The matrix R_σ has thus σ rows and $\mathcal{O}(n \log \sigma / \log n)$ columns, where each entry consumes $\lceil \log n \rceil$ bits. Thus, this RD requires additional $\mathcal{O}(n\sigma \log \sigma)$ bits of memory.

To reduce space consumption, Grossi et al. propose *wavelet trees* [24]. The wavelet tree of a string of n characters over Σ consumes $\mathcal{O}(n \log^2 \sigma)$ bits of memory, including the input string itself. In practice, wavelet trees give no advantages for DNA strings. Their practical memory footprint is equivalent to those of level RDs, yet they are twice as slow [51]. For wavelet trees, we refer the reader to [24, 42].

2.3.2.2 Top-Down Traversal

We now describe how to emulate the top-town traversal of a (generalized) suffix trie $\mathcal{T}_{\text{strie}}(\mathbb{S})$ using permutation *LF* [see Eq. (2.3)]. The top-down traversal of $\mathcal{T}_{\text{strie}}(\mathbb{S})$ spells all suffixes of $\mathbb{S}$. Therefore, to emulate the traversal, it is sufficient to enumerate all suffixes of $\mathbb{S}$ by inverting the BWT using the counting argument of permutation *LF*. This enumeration provides not only all suffixes of $\mathbb{S}$ but also their corresponding SA intervals, as permutation *LF* and the SA are closely related. However, we have seen that permutation *LF* enumerates suffixes of $\overline{\mathbb{S}}$ in a *backward* direction, while we would rather need permutation Ψ to enumerate suffixes of $\mathbb{S}$ in a *forward* direction. Since $LF = \Psi^{-1}$, we consider the BWT of $\overline{\mathbb{S}}$, so that its permutation *LF* enumerates suffixes of $\mathbb{S}$. *LF* of $\overline{\mathbb{S}}$ computes intervals on the backward SA $\overline{A}$ of $\overline{\mathbb{S}}$, so OCCURRENCES has to reverse the occurrences of $\overline{A}$ back as those of the forward SA A.

We represent the current node v by the elements $\{l, r, e\}$, where $[l, r)$ represents the current SA interval and e is the label of the edge entering the current node. Therefore, we define the following node operations:

- GOROOT(v) initializes v to $\{0, n, \varepsilon\}$.
- ISTERMINAL(v) returns *true* if and only if $Occ(\$, v.r) - Occ(\$, v.l) = v.r - v.l$.
- LABEL(v) returns $v.e$.
- OCCURRENCES returns each value (i, j) of $\overline{A}[l, r)$ as $(i, n^i - j - x.d - 1)$.

The traversal easily goes from the root node to a child node following the edge labeled by character c, as it suffices to determine the interval $[C(c), C(\text{next}_{\text{lex}}(c))$. Now, suppose the traversal is on an arbitrary suffix trie node v of known interval $[v.l, v.r)$ such that the path from the root to v spells the substring CONCAT(v). The traversal goes down to a child node w of unknown interval $[w.l, w.r)$ such that the path from the root to w spells CONCAT(v) $\cdot\, c$ for some $c \in \Sigma$. Thus, the known interval $[v.l, v.r)$ contains all suffixes of $\mathbb{S}$ starting with CONCAT(v), i.e., all prefixes of $\overline{\mathbb{S}}$ ending with CONCAT(v), while the unknown interval $[w.l, w.r)$ contains all suffixes of $\mathbb{S}$ starting with CONCAT(v) $\cdot\, c$, i.e., all prefixes of $\overline{\mathbb{S}}$ ending with $c\,\cdot$ CONCAT(v). All these characters c are in $L[w.l, w.r)$ since $L[i]$ is the character $\mathbb{S}[A[i] - 1]$ preceding the suffix pointed by $A[i]$. Moreover, these characters c are *contiguous* and *in relative order* in F (see Observations 2.1 and 2.2). If $p_0, p_1, \ldots, p_{m-1}$ are all positions in L within $[v.l, v.r)$ such that $L[p_i] = c$ and $v.l \le p_0 < \cdots < p_{m-1} < v.r$, then $w.l = LF(p_0)$ and $w.r = LF(p_{m-1}) + 1$. Therefore, $LF(p_i)$ becomes:

$$\begin{aligned} LF(p_i) &= C(L[p_i]) + Occ(L[p_i], p_i) \\ &= C(c) + Occ(c, p_i) \\ &= C(c) + Occ(c, v.l) + i \end{aligned} \tag{2.7}$$

and analogously $LF(p_{m-1}) + 1$ becomes:

$$\begin{aligned} LF(p_{m-1}) + 1 &= C(L[p_{m-1}]) + Occ(L[p_{m-1}], p_{m-1}) + 1 \\ &= C(c) + Occ(c, p_{m-1}) + 1 \\ &= C(c) + Occ(c, v.r) \end{aligned} \tag{2.8}$$

Algorithm 5 uses the above equations to achieve the operation GODOWN by character. Conversely, Algorithms 6 and 7 realize GODOWN and GORIGHT by enumerating all characters in Σ. Below, we show how function Occ can be stored in succinct space and queried in $\mathcal{O}(1)$ time. Thus, Algorithm 5 runs in $\mathcal{O}(1)$ time, while GODOWN and GORIGHT run in $\mathcal{O}(\sigma)$ time.

Algorithm 5 GODOWN(v, c)

Input v : FM-index iterator
c : char to query
Output boolean indicating success

1: **if** ISTERMINAL(v) **then**
2:     **return false**
3: $x.l \leftarrow C(c) + Occ(c, x.l)$
4: $x.r \leftarrow C(c) + Occ(c, x.r)$
5: $x.e \leftarrow c$
6: **return** $x.l < x.r$

Algorithm 6 GODOWN(v)

Input v : FM-index iterator
Output boolean indicating success

```
1: if not ISTERMINAL(v) then
2:     c ← min_lex(Σ)
3:     do
4:         if GODOWN(v, c) then
5:             return true
6:     while c ← next_lex(Σ)
7: return false
```

Algorithm 7 GORIGHT(v)

Input v : FM-index iterator
Output boolean indicating success

```
1: if not ISROOT(v) then
2:     c ← LABEL(v)
3:     w ← PARENT(v)
4:     while c ← next_lex(c) do
5:         if GODOWN(w, c) then
6:             return true
7: return false
```

2.3.2.3 Sparse Suffix Array

SA A is required to achieve function OCCURRENCES; however, the whole SA would take $\mathcal{O}(n \log n)$ bits of memory. As proposed by Ferragina and Manzini [17], we maintain a *sparse* SA A^{ε} containing suffix positions sampled at regular intervals from the input string. In order to determine if and where any $A[i]$ is sampled in A^{ε}, we employ an auxiliary binary RD I of length n, such that $I[i] = 1$ if $A[i]$ is sampled and stored at $A^{\varepsilon}[\text{RANK}_1(I, i)]$ or $I[i] = 0$ otherwise. Any $A[i]$ is obtained by finding the smallest $j \geq 0$ such that $LF^j(i)$ is in A^{ε}, and then $A[i] = A[LF^j(i)] + j$. By sampling one text position out of $\log^{1+\varepsilon} n$, for some $\varepsilon > 0$, A^{ε} consumes $\mathcal{O}(n/\log^{\varepsilon} n)$ bits of memory and OCCURRENCES returns all o occurrences in $\mathcal{O}(o \cdot \log^{1+\varepsilon} n)$ time [17]. In practice, we sample positions at rates between 2^{-3} and 2^{-5}. The rank dictionary I consumes $n + o(n)$ additional bits of memory independently of the sampling rate.

2.4 Suffix Tree Realizations

In this section, we show two suffix tree indexes. While the first has optimal running time and space consumption, the second might be faster in practice as it doesn't need to be fully constructed in advance but can be constructed dynamically on demand.

2.4.1 Enhanced Suffix Array

The enhanced suffix array (ESA) has been introduced by Abouelhoda et al. [1] as a memory-efficient replacement for suffix trees. In general, the ESA of a string S (or string collection $\mathbb{S}$) consists of the (generalized) suffix array, the longest common prefix (LCP) table, and the child table. Each table is a string of the total text length n that consists of values between -1 and n and thus requires $\mathcal{O}(n \log n)$ bits of memory. In the following, we define the LCP and the child table and their relation to suffix trees. Then, we describe how the ESA can be traversed like a suffix tree. Note that the ESA is generalized to a collection $\mathbb{S}$.

2.4.1.1 LCP Table

The LCP table (also known as *height array*) [37] stores the lengths of the longest common prefix between every consecutive pair of suffixes in the suffix array A.

Definition 15 (LCP Table) Given a string S of length n and its suffix array A, the LCP table H of S is a string of length $n+1$ over the alphabet $[-1, n-1]$. For every $i \in [1, n)$ hold:

$$H[0] = -1 \tag{2.9}$$

$$H[i] = \left|\mathrm{LCP}\left\{S_{A[i-1]}, S_{A[i]}\right\}\right| \tag{2.10}$$

$$H[n] = -1. \tag{2.11}$$

The definition applies as well for string collections with $H[i] = \left|\mathrm{LCP}\left\{\mathbb{S}_{A[i-1]}, \mathbb{S}_{A[i]}\right\}\right|$ in (2.10). For the sake of simplicity, we extended the LCP table by two boundary values (-1) which are implicitly needed by some algorithms if the text is not padded. However, they are not explicitly required in the implementations of these algorithms. We call the table entries *LCP values* and $H[i]$ the LCP value of $S_{A[i]}$ or the ith LCP value.

Manber and Myers [37] introduced the LCP table and how to construct it as a by-product of the $\mathcal{O}(n \log n)$ suffix array construction. The first optimal algorithm, proposed by Kasai et al. [30], constructs the LCP table for a text and a given suffix array in linear time. In [56], Kasai et al.'s algorithm has been extended to construct the generalized LCP table. Other optimal construction algorithms for the generalized LCP table have been proposed in [4, 35].

2.4.1.2 Child Table

Besides the linear-time algorithm to construct H, Kasai et al. [30] proposed a method to traverse the suffix tree $\mathcal{T}_{\text{stree}}(S)$ in a bottom-up fashion by solely scanning H from left to right and updating a stack that represents the path from the traversed node to the root. This method is used in [1] to construct the child table cld, which contains links to the siblings and children of a node and thus represents the structure of the suffix tree. To understand the child table, we first need to introduce LCP intervals.

Definition 16 (LCP Interval) An interval $[i, j) \subseteq [0, n)$ with $i + 1 < j$ is called an *LCP interval* of value ℓ or *ℓ-interval* $[i, j)$ if the following hold:

1. $H[i] < \ell$,
2. $H[j] < \ell$,
3. $\forall_{k \in (i,j)}\ H[k] \geq \ell$,
4. $\exists_{k \in (i,j)}\ H[k] = \ell$.

For completeness, interval $[i, i + 1)$ is defined to be a (singleton) *ℓ-interval* with $\ell = |S_{A[i]}|$ and $i \in [0, n)$.

An ℓ-interval $[i, j)$ is associated to a set of suffixes $S_{A[i]}, S_{A[i+1]}, \ldots, S_{A[j-1]}$. Then, the following hold: (1) The longest common prefix ω of these suffixes has length ℓ by properties 3 and 4; this ℓ-interval is called an *ω-interval.* (2) A non-singleton ℓ-interval $[i, j)$ is maximal by properties 1 and 2, i.e. every extension to the left or right is no longer an ℓ-interval. In Fig. 2.5, $[6, 8)$ is an ℓ-interval or ω-interval with $\ell = 3$ and $\omega =$ `tct`.

Lemma 1 (Node-Interval-Duality) *For every suffix tree node v in $\mathcal{T}_{\text{stree}}(S)$ there is an ω-interval $[i, j)$, and vice versa. If v is an inner node, it holds $\omega = \text{CONCAT}(v)$ and otherwise $\omega\$ = \text{CONCAT}(v)$.*

In the following, we consider the suffix tree $\mathcal{T}_{\text{stree}}(S)$ and for each node v, denote with $S(v)$ the suffixes represented by the leaves below v.

Corollary 1 *Every suffix tree node v is identified by an LCP interval $[i, j)$; both node and interval represent the same set of suffixes $S(v) = \{S_{A[i]}, S_{A[i+1]}, \ldots, S_{A[j-1]}\}$.*

Fig. 2.5 Suffix array and LCP table of `ttatctctta`

	0	1	2	3	4	5	6	7	8	9
$S =$	t	t	a	t	c	t	c	t	t	a

	0	1	2	3	4	5	6	7	8	9	
$A =$	9	2	4	6	8	1	3	5	7	0	
$H =$	-1	1	0	2	0	2	1	3	1	3	-1
	a	atctctta	ctctta	ctta	ta	tatctctta	tctctta	tctta	tta	ttatctctta	

Let v be an inner suffix tree node with children $w_1, \ldots, w_m$. W.l.o.g. let $\text{CONCAT}(w_1) <_{\text{lex}} \text{CONCAT}(w_2) <_{\text{lex}} \cdots <_{\text{lex}} \text{CONCAT}(w_m)$. Obviously, the sets $S(w_1)$, $S(w_2)$, …, $S(w_m)$ form a partition of the set $S(v)$. As a consequence of Corollary 1, the LCP intervals of the children (child intervals) are subintervals that form a partition $[l_0, l_1)$, $[l_1, l_2)$, …, $[l_{m-1}, l_m)$ of the ℓ-interval $[i, j)$ of v, where $l_0 = i$ and $l_m = j$. The length of the longest common prefix of suffixes from different child subtrees is $\ell = |\text{CONCAT}(v)|$, whereas the LCP length of suffixes from the same subtree is greater than ℓ. Thus, for $x \in [i, j)$ it holds $H[x] = \ell$ only if $x \in \{l_1, l_2, \ldots, l_{m-1}\}$ and $H[x] > \ell$ otherwise. The indexes $l_1, \ldots, l_{m-1}$ uniquely define the partition into subintervals and are called *ℓ-indexes* of the LCP interval $[i, j)$. The set $\{l_1, l_2, \ldots, l_{m-1}\}$ is denoted by ℓ-indexes(i, j).

The parent-child relationship of LCP intervals corresponds to the parent-child relationship of suffix tree nodes and constitutes the so-called *LCP interval tree* [1], compare Fig. 2.6a, b. The child table is a linked list of ℓ-indexes and stores for each ℓ-index so-called up, down, and nextℓIndex values, see Fig. 2.6c. It can be represented as three subtables which are strings of length $n + 1$ over the alphabet $[0, n]$ (columns u, d, and n in Fig. 2.6d).

For $l_k \in \ell$-indexes(i, j), nextℓIndex(l_k), if it exists, is the next greater ℓ-index l_{k+1} in the set ℓ-indexes(i, j). up(l_k) and down(l_k), if they exist, are the smallest ℓ-indexes in the sets ℓ-indices(l_{k-1}, l_k) and ℓ-indexes(l_k, l_{k+1}). For an arbitrary ℓ-index i, the values up, down, and nextℓIndex can formally be defined as follows [1]:

$$\mathsf{up}(i) = \min\{q \in [0, i) \mid H[q] > H[i] \wedge \forall_{k \in (q,i)} H[k] \geq H[q]\}, \quad (2.12)$$

$$\mathsf{down}(i) = \max\{q \in (i, n] \mid H[q] > H[i] \wedge \forall_{k \in (i,q)} H[k] > H[q]\}, \quad (2.13)$$

$$\mathsf{next}\ell\mathsf{Index}(i) = \min\{q \in (i, n] \mid H[q] = H[i] \wedge \forall_{k \in (i,q)} H[k] > H[i]\}. \quad (2.14)$$

Abouelhoda et al. proposed an easy and elegant way to reduce the memory consumption of the child table by two-thirds. The authors take advantage of redundancies, e.g., identical down and up values of adjacent ℓ-values or undefined values, to store the three subtables in a single string cld of size n instead of using three separate strings of that size. Figure 2.6d shows where the three values are stored in cld. For more details, we refer the reader to [1].

2.4.1.3 Top-Down Traversal

The ESA, consisting of suffix array, LCP table, and child table, allows to emulate a top-down traversal of the suffix tree by means of a node traversal of the corresponding LCP interval tree. The top-down iterator represents the current node by its LCP interval boundaries l and r. These boundaries and the ℓ-values of the current interval become the boundaries of the children's intervals. The iterator can be moved down to the leftmost child by replacing the value of r with the first ℓ-value. Moving right to the next sibling corresponds to replacing the value of l with the value

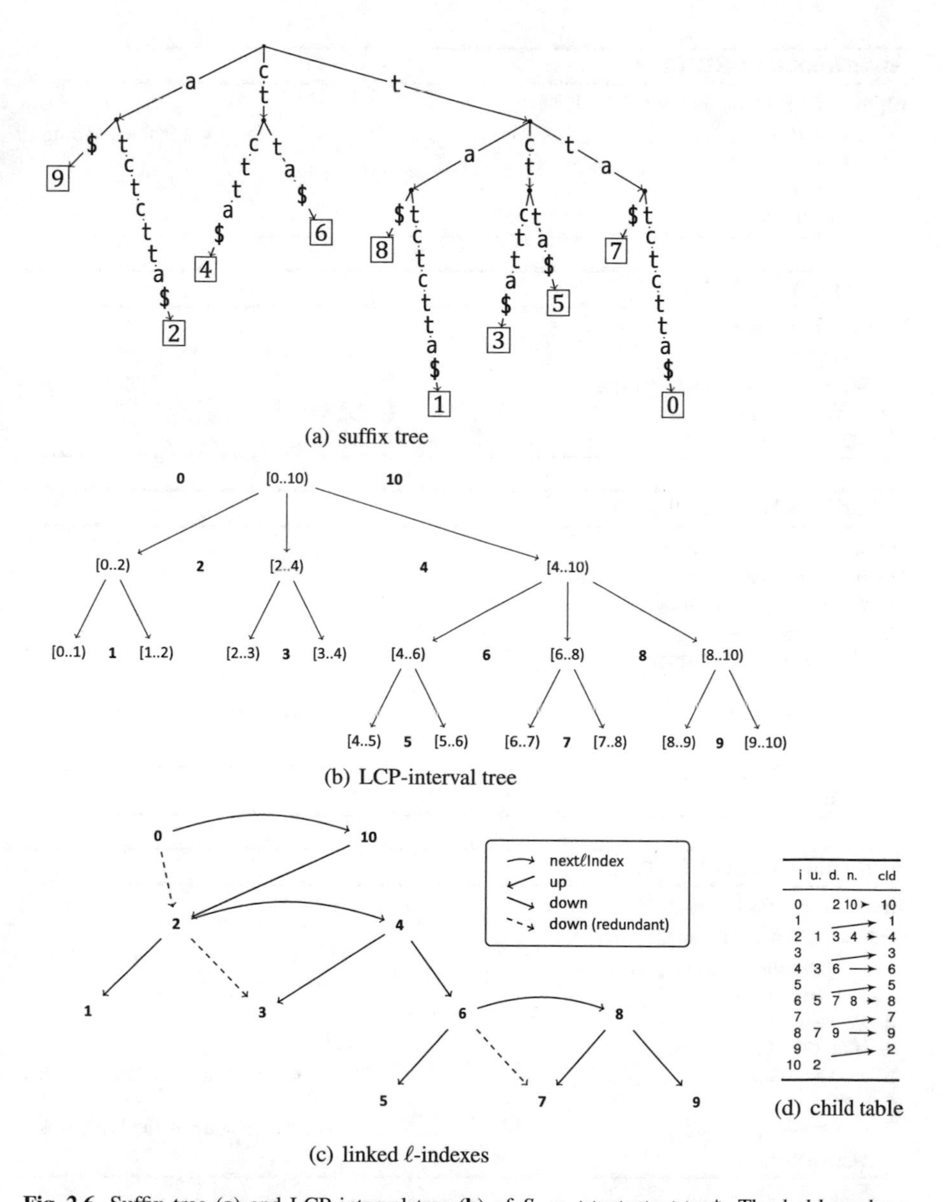

i	u.	d.	n.	cld
0		2	10	10
1				1
2	1	3	4	4
3				3
4	3	6		6
5				5
6	5	7	8	8
7				7
8	7	9		9
9				2
10	2			

Fig. 2.6 Suffix tree (**a**) and LCP interval tree (**b**) of S = `ttatctctta$`. The *bold numbers* between LCP interval nodes (**b**) are ℓ-indexes of the parent interval above. The child table (**d**) stores for every ℓ-index i the up, down, and nextℓIndex values (compare with **c**), which are the first ℓ-indexes in the LCP interval left of i, right of i, or the next ℓ-index in the same LCP interval, respectively. After the removal of redundant down links, the three columns (**d**) can be stored as a single string cld of length n

Algorithm 8 GOROOT(v)

```
Input   v : ESA iterator
1: n ← |H| − 1                                ▷ n is the text length
2: v.l ← 0
3: v.r ← n
4: v.parentR ← ⊥
```

Algorithm 9 ISNEXTL(i)

```
Input   i : ℓ-index
1: j ← cld[i]
2: if i < j and H[i] = H[j] then
3:     return true
4: return false
```

Algorithm 10 GODOWN(v)

```
Input   v : ESA iterator
Output  boolean indicating success
1: if ISTERMINAL(v) then
2:     return false
3: if v.r ≠ v.parentR then
4:     v.parentR ← v.r
5:     v.r ← cld[v.r − 1]                     ▷ get up value of right boundary
6: else
7:     v.r ← cld[v.l]                         ▷ get down value of left boundary
8: return true
```

Algorithm 11 GORIGHT(v)

```
Input   v : ESA iterator
Output  boolean indicating success
1: if v.parentR ∈ {⊥, v.r} then
2:     return false
3: v.l ← v.r
4: if ISNEXTL(v.r) then
5:     v.r ← cld[v.r]                         ▷ v.r has a succeeding ℓ-index
6: else
7:     v.r ← v.parentR                        ▷ v.r is the last ℓ-index
8: return true
```

of r and the value of r with its next ℓ-value, if it exists, or the right boundary of the parent interval otherwise.

The functions GOROOT, GODOWN, GORIGHT (Algorithms 8, 10, and 11) move the iterator if possible and return false otherwise. The iterator starts in the root node with $[l, r) = [0, n)$ and stores the right boundary of the parent node in *parentR*. For all but the rightmost child, i.e., $r \neq parentR$, the first ℓ-index in ℓ-indexes(l, r) is the up value of r stored at cld$[r - 1]$; otherwise, it is the down value of l

stored only in this case at $\mathsf{cld}[l]$ (see GODOWN). It is straightforward to extend GODOWN to descend an edge beginning with a certain character. Other functions, e.g., ISTERMINAL and OCCURRENCES, are identical to the corresponding suffix array index functions.

2.4.2 *Lazy Suffix Tree*

A *lazy suffix tree* (LST) [21] is a suffix tree whose nodes are created on demand, i.e., when they are visited for the first time in a top-down traversal. Depending on the usage scenario, using a lazy suffix tree can significantly improve on the overall running time and memory consumption compared to an enhanced suffix array. Giegerich et al. [21] introduced the first lazy suffix tree data structure that utilizes the *Write-Only, Top-Down* (WOTD) algorithm [20] for the on-demand node expansion. In the following, we first describe this algorithm, then we generalize the original data structure to multiple sequences [56] and finally describe its top-down traversal.

2.4.2.1 The WOTD Algorithm

The basic idea of the WOTD algorithm is to determine the children of a branching suffix tree node by partitioning the set of corresponding suffixes by the character following the longest common prefix. Beginning with only the root node, the algorithm recursively expands a directed tree step-by-step up to the entire suffix tree. We consider a given non-empty text S of length n and a rooted, directed tree T in the following referred to as a *partial suffix tree* that in every state of the algorithm, is a subgraph of the suffix tree including its root. Let R be a function that maps any string $\alpha \in \Sigma^*$ to the set of suffixes of $S\$$ that begin with α:

$$R(\alpha) := \{\alpha\beta \mid \alpha\beta \text{ is a suffix of } S\$\} \setminus \{\$\}. \tag{2.15}$$

In the following, we denote by $\widehat{\alpha}$, if it exists, the tree node whose concatenation string is α. Given a branching suffix tree node $\widehat{\alpha}$, $R(\alpha)$ contains the concatenation strings of the leaves below $\widehat{\alpha}$. The children of $\widehat{\alpha}$ can be determined as follows: Divide $R(\alpha)$ into non-empty groups $R(\alpha c_1), R(\alpha c_2), \ldots, R(\alpha c_m)$ of suffixes, where each character $c_i \in \Sigma \cup \{\$\}$ for $i = 1, \ldots, m$ follows the common α-prefix. Let $\alpha c_i \beta_i$ be the longest common prefix of $R(\alpha c_i)$, which for singleton groups equals the only contained suffix. For non-singleton groups, $\widehat{\alpha c_i \beta_i}$ is a branching node in the suffix tree, as there are two suffixes of $S\$$ that differ in the character following their

Algorithm 12 WOTDEAGER($T, \widehat{\alpha}$)

Input T : partially constructed suffix tree
Input $\widehat{\alpha}$: suffix tree node

1: divide $R(\alpha)$ into subsets $R(\alpha c)$ of suffixes where character c follows the α-prefix
2: **for all** $c \in \Sigma \cup \{\$\}$ and $R(\alpha c) \neq \emptyset$ **do**
3: $\alpha c \beta \leftarrow \text{LCP}\, R(\alpha c)$
4: add $\widehat{\alpha c \beta}$ as a child of $\widehat{\alpha}$ in T
5: **if** $|R(\alpha c)| > 1$ **then**
6: WOTDEAGER($T, \widehat{\alpha c \beta}$)

Algorithm 13 CREATESUFFIXTREE(S)

Input S : string over the alphabet Σ
Output suffix tree of S

1: create tree T with only the root node $\widehat{\varepsilon}$
2: WOTDEAGER($T, \widehat{\varepsilon}$)
3: **return** T

common prefix $\alpha c_i \beta_i$. Singleton groups contain suffixes of $S\$$ which correspond to leaves $\widehat{\alpha c_i \beta_i}$ in the suffix tree. As every suffix with prefix αc_i also has a prefix $\alpha c_i \beta_i$, there is no branching node between $\widehat{\alpha}$ and $\widehat{\alpha c_i \beta_i}$. Hence, every $\widehat{\alpha c_i \beta_i}$ can be inserted as a child of $\widehat{\alpha}$ in T, which remains a partial suffix tree. This procedure, called *node expansion*, is recursively repeated for every newly inserted branching node; Algorithm 12 shows the corresponding pseudo-code. The WOTD algorithm begins with T consisting of only the root node and expands it and all its descendants, see Algorithm 13.

2.4.2.2 The Data Structure

A key property of the WOTD algorithm is that it constructs the suffix tree top-down, so that nodes from disjunctive subtrees can be expanded independently and in arbitrary order. That makes it possible to expand single nodes step-by-step instead of entire subtrees and allows turning the suffix tree construction into a lazy, on-demand construction. Such a lazy suffix tree requires a method to expand any suffix tree node and a data structure to represent a partial suffix tree whose nodes are either in an *expanded* or *unexpanded* state. Further, it requires $R(\alpha)$ for the expansion of nodes $\widehat{\alpha}$ and needs to provide the corresponding set of suffix start positions for all (even expanded) nodes $\widehat{\alpha}$:

$$l(\alpha) := \left\{ i \in [0, n) \;\middle|\; \exists_{\beta \in \Sigma^*}\, S_i\$ = \alpha\beta \right\} \tag{2.16}$$

to determine the text occurrences of a pattern. In the following, we propose an LST realization that differs from the original one by Giegerich [21] in two aspects: It maintains the lexicographical ordered of the children of its inner nodes and it generalizes to multiple sequences.

The data structure to represent a LST consists of the two strings, T and A. T is a string of integers that stores the tree structure. A is a permutation of $[0, n)$ that stores for each node $\widehat{\alpha}$ the positions $l(\alpha)$ where α occurs in the text as a contiguous block $A\,[i,j)$. There are three types of nodes: unexpanded inner nodes, expanded inner nodes, and leaves. Inner nodes are represented by two adjacent entries, e_1, e_2 and leaves by a single entry e_1 in T. To distinguish between inner nodes and leaves, a *Leaf bit* (L) is split off the first entry. The children of each expanded node are stored contiguously and in the same order as in the tree, where the last child of a node is marked by a *Last-Child bit* (LC) in the first entry. The leftmost child of a node $\widehat{\alpha}$ stores $|\mathrm{LCP}(R(\alpha))|$ in e_1, whereas all siblings store i the left boundary of their occurrence block $SA\,[i,j)$. Note that, analogously to the ESA, it is possible to retrieve i and j for each child, as adjacent children share a common boundary and the outer boundaries of the first and last child are equal to the parent node's boundaries. For expanded nodes $\widehat{\alpha}$, it stores *firstchild*($\widehat{\alpha}$), which refers to the first entry of the first child of $\widehat{\alpha}$ in T. Figure 2.7 shows T and A for different expansion stages of an LST. Algorithms 14–17 exemplarily show how to traverse the LST, and the following section describes the details of the function EXPANDNODE.

2.4.2.3 Node Expansion

Consider an unexpanded node $\widehat{\alpha}$ whose text occurrences are stored in $A\,[i,j)$. To expand $\widehat{\alpha}$, the elements $k \in A\,[i,j)$ are sorted by $S\,[k + |\alpha|]$ using counting sort. The result is a partition $[l_0, l_1), [l_1, l_2), \ldots, [l_{m-1}, l_m)$ with $l_0 = i$ and $l_m = j$, such that $A\,[l_0, l_1), A\,[l_1, l_2), \ldots, A\,[l_{m-1}, l_m)$ store the occurrences of the substrings $\alpha x_1 < \alpha x_2 < \cdots < \alpha x_m$ in the text $S\$$, with $x_i \in \Sigma \cup \{\$\}$. The intervals correspond to the children of the suffix tree node $\widehat{\alpha}$, and corresponding entries are contiguously and in the same order inserted at the end of T as described above, where singleton intervals become leaves and non-singleton intervals become unexpanded inner nodes. Finally, the former unexpanded node $\widehat{\alpha}$ will be converted into an expanded node by modifying its two entries. The whole LST is expanded in $\mathcal{O}\big(n^2 + |\Sigma|n\big)$ time in the worst and $\mathcal{O}\big(n \log_{|\Sigma|} n + |\Sigma|n\big)$ time in the average case [21].

2.4.2.4 Generalization to Multiple Sequences

The data structure described above is easily generalized to multiple sequences. Given a collection $\mathbb{S} = \{S^0, S^1, \ldots, S^{\kappa-1}\}$ of strings of lengths $n^0, n^1, \ldots, n^{\kappa-1}$, a suffix is represented by a pair of integers (i,j), with $i \in [0, \kappa)$ and $j \in [0, n^i)$. Hence,

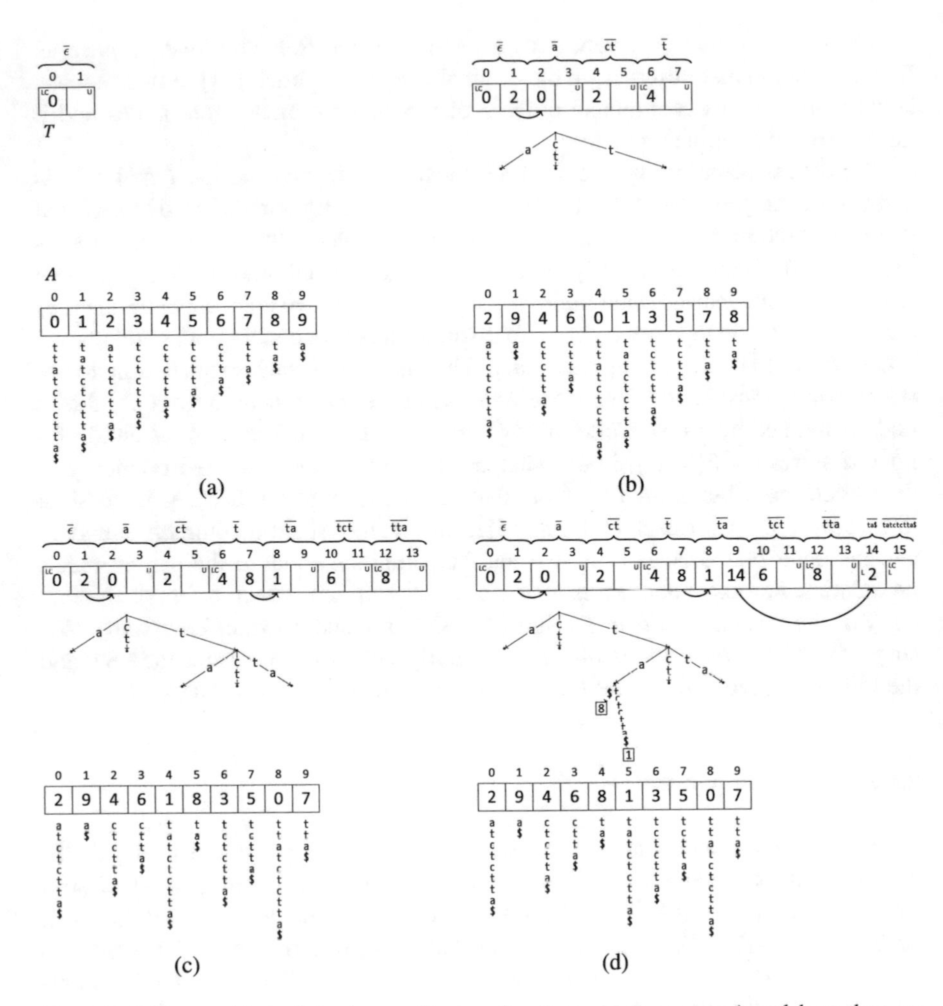

Fig. 2.7 Different states of the lazy suffix tree for S = ttatctctta$ and how they are represented by the data structure. Below each unexpanded node $\widehat{\alpha}$, the remaining suffixes $R(\alpha)$ are shown without their common α-prefix. In the beginning (**a**), the lazy suffix tree consists of only the unexpanded root node. (**b**) shows the result of the root node expansion, (**c**) and (**d**) the expansions of nodes $\widehat{\texttt{t}}$ and $\widehat{\texttt{ta}}$

we define A to be a string of pairs of length $n = \Sigma_{i \in [0,\kappa)} n^i$ and initialize it with:

$$A := (0,0)(0,1)\ldots(0,n^0-1)(1,0)\ldots(1,n^1-1)(2,0)\ldots(\kappa-1,n^{\kappa-1}-1). \quad (2.17)$$

T needs not to be changed, as all of its entries solely store prefix lengths, interval boundaries, or child pointers. In order to construct an LST, whose nodes have the same order as in the ESA-based suffix tree, the sentinel relation $\$^0 < \cdots < \$^{\kappa-1} < \min_{\text{lex}}(\Sigma)$ must be retained. This is achieved without introducing extra alphabet

Algorithm 14 GOROOT(v)

Input v : LST iterator

1: $v.l \leftarrow 0$
2: $v.r \leftarrow n$
3: $v.parentR \leftarrow n$
4: $v.node \leftarrow 0$ ▷ position in T
5: $\mathsf{LC} \leftarrow \mathsf{U} \leftarrow w - 1$ ▷ w is the number of
6: $\mathsf{L} \leftarrow w - 2$ ▷ bits of an entry in T

Algorithm 15 UPDATER(v)

Input v : LST iterator

1: **if** $T[v.node]$ & $2^{\mathsf{LC}} \neq 0$ **then**
2: $v.r \leftarrow v.parentR$
3: **else if** $T[v.node]$ & $2^{\mathsf{L}} \neq 0$ **then**
4: $v.r \leftarrow v.l + 1$
5: **else**
6: $v.r \leftarrow T[v.node + 2]$ & $(2^{\mathsf{L}} - 1)$

Algorithm 16 GODOWN(v)

Input v : LST iterator
Output boolean indicating success

1: **if** $T[v.node]$ & $2^{L} \neq 0$ **then**
2: **return false** ▷ *Leaf bit* is set
3: **if** $T[v.node + 1]$ & $2^{\mathsf{U}} \neq 0$ **then**
4: EXPANDNODE($v.node$)
5: $v.node \leftarrow T[v.node + 1]$ & $(2^{\mathsf{L}} - 1)$
6: $tmp \leftarrow v.r$
7: UPDATER(v)
8: $v.parentR \leftarrow tmp$
9: **return true**

Algorithm 17 GORIGHT(v)

Input v : LST iterator
Output boolean indicating success

1: **if** $v.r = v.parentR$ **then**
2: **return false**
3: $v.l \leftarrow v.r$
4: **if** $T[v.node]$ & $2^{\mathsf{L}} \neq 0$ **then**
5: $v.node \leftarrow v.node + 1$
6: **else**
7: $v.node \leftarrow v.node + 2$
8: UPDATER(v)
9: **return true**

characters in the implementation by using a stable counting sort with one extra bucket in front of all other buckets that represents all sentinels. When sorting the suffixes $R(\alpha)$ by their character at position $|\alpha|$, this sentinel bucket contains all pairs (i,j) with $j + |\alpha| = n^i$. As counting sort is stable, these pairs will have the same relation as in the initialization, i.e., in A a suffix $\alpha\$^i$ will be stored left of a suffix $\alpha\$^j$ with $i < j$. At last, these suffixes are appended to T as leaves below $\widehat{\alpha}$. They are appended in the same order as they occur in A, left of the remaining buckets.

2.5 *q*-Gram Index Realizations

The simplest but most commonly used index in HTS is the *q-gram index* (also known as the k-mer or inverted index) [7]. It is useful when the suffix tree structure is not required and the substrings of interest are short and of uniform length q. A q-gram is simply a sequence of q characters over a given alphabet Σ, e.g., $\Sigma = \{\mathtt{A}, \mathtt{C}, \mathtt{G}, \mathtt{T}\}$ in case of DNA. Given an input string (collection), a q-gram index allows the quick retrieval of all the occurrences of any q-gram from the input. An optimal q-gram index counts the number of occurrences of a q-gram in constant time and locates its occurrences in time proportional to their number.

2.5.1 *Direct Addressing*

The simplest optimal q-gram index consists of two tables, a table bkt to store the occurrences grouped by q-gram, and a *bucket* table pos to map each q-gram to its first entry in the occurrence table. If we, for example, consider a single sequence of length n, the occurrence table has $n - q + 1$ different entries between 0 and $n - q$, one for each position a q-gram can start at. The bucket table typically occupies an entry for each of the $|\Sigma|^q$ possible q-grams.

To look up the occurrences of a q-gram, it is first converted into a *hash code*, which is a number h between 0 and $|\Sigma|^q - 1$, e.g., its lexicographical rank in the set of all $|\Sigma|^q$ possible q-grams. Then, the bucket table entry at position h is used to determine the first entry of the group of occurrences in the occurrence table. Assuming that occurrences are grouped in the same order as q-grams are mapped to numbers, the end of the group is stored at entry $h + 1$ in the bucket table. Figure 2.8 gives a detailed example of such a q-gram index.

2.5.2 *Open Addressing*

As described above, the bucket table pos of the direct addressing q-gram index is a table of size $|\Sigma|^q + 1$, and the whole index consumes $\Theta(|\Sigma|^q + n)$ space. Thus, for

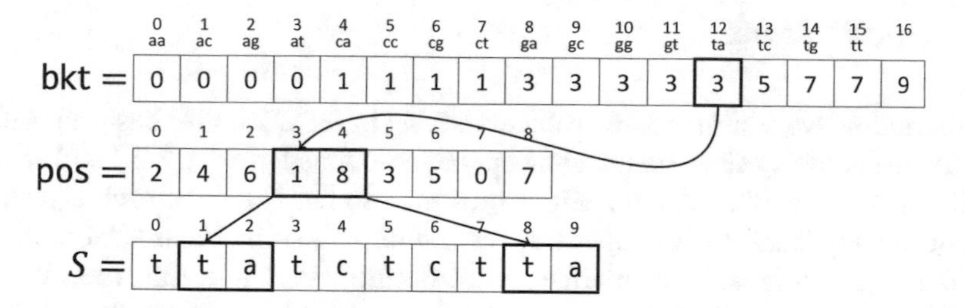

Fig. 2.8 Two-gram index of `ttatctctta`. To look up all occurrences of the 2-gram `ta`, first determine its numerical value code(`ta`) = 12. At positions 12 and 13 the bucket table bkt stores the boundaries 3 and 5 (excluding) of the group of occurrences in the occurrence table pos. Hence, 1 and 8 are the only occurrences of `ta` in the text

growing values of q, space consumption of the bucket table becomes prohibitive. In fact, table bkt is only needed to determine whether a q-gram bucket is non-empty and eventually to retrieve its interval pos. However, the number of non-empty buckets is not only bound by the number of possible different q-grams over Σ but also by the number of overlapping q-grams in the input string, i.e., it is at most $m = \min(|\Sigma|^q, n - q + 1)$.

In the following, we describe the *open addressing q-gram index* with a memory consumption of $\mathcal{O}(\alpha^{-1} n \log n)$ bits, for a fixed load factor α with $0 < \alpha \leq 1$, as proposed in [56]. Instead of addressing entries in bkt directly by q-gram code values, it uses an open addressing scheme [10] to map the q-gram codes (keys) of non-empty buckets to entries in bkt (values). The load factor determines the maximal ratio between used and available entries of the bucket table and provides a trade-off between the number of collisions and memory consumption. In order to not exceed a load factor of α, the open addressing index uses a bucket table of size $\lfloor \alpha^{-1} n \rfloor + 1$.

In addition to the bucket table bkt, the open addressing index uses a table C of size $\lfloor \alpha^{-1} n \rfloor$ with entries between -1 and $|\Sigma|^q - 1$, the so-called *code table*. A pseudo-random function hash : $[0, |\Sigma|^q) \rightarrow [0, \lfloor \alpha^{-1} n \rfloor)$ maps a q-gram code h to an entry in bkt. The index implementation can use arbitrary hash functions, e.g., the SSE4.2 CPU instruction _mm_crc32_u64 [28] and a modulo operation.

As hash may map different values of h to the same position i (collision), $C[i]$ stores the code that currently occupies the entry bkt[i] or equals -1 if empty. Whenever a collision with a different code occurs, the entries $C\left[(i + j) \bmod \lfloor \alpha^{-1} n \rfloor\right]$ for $j = 1^2, 2^2, 3^2, \ldots$ are probed for being empty or containing the correct code. Compared to linear probing, this quadratic probe sequence prevents primary clustering of buckets [10]. The buckets in pos need to be arranged in the same order as their code values appear in C. In this way, bkt[i] stores the begin and bkt[$i + 1$] end positions of the bucket in pos.

The direct and open addressing indexes are constructed efficiently using a modified counting sort algorithm [10]. They are easily extended to non-contiguous q-grams [6] and generalized to texts consisting of multiple sequences. For more details, we refer the reader to [56].

2.6 Applications

In this section, we consider core problems arising in HTS. We give basic algorithms that use the generic suffix trie traversal operations defined in Sect. 2.2.1 and are thus applicable to all suffix trie realizations presented so far. Note that such algorithms are equally applicable to any suffix tree realization, as a suffix trie is easily achieved by inserting non-branching internal nodes during suffix tree traversal. We first consider a simple depth-first traversal algorithm and subsequently, algorithms for exact and approximate string matching. Moreover, we show the experimental results of these algorithms.

As data structures, we consider the *suffix array* (SA), the FM-index with a two-levels DNA rank dictionary (FM), the fully constructed *lazy suffix tree* (LST), the *enhanced suffix array* (ESA), the *q-gram index with Direct Addressing* (QDA), and the *q-gram index with Open Addressing* (QOA) for fixed values of q. As text, we take the *C. elegans* reference genome (WormBase WS195), i.e., a collection of six DNA strings of about 100 Mbp total length. As patterns, we use sequences extrapolated from an Illumina sequencing run (SRA/ENA id: SRR065390). In the plots, we always show average runtimes per pattern. We run all experiments on a workstation running Linux 3.10.11, equipped with one Intel® Core i7-4770K CPU @ 3.50 GHz, 32 GB RAM, and a 2 TB HDD @ 7200 RPM.

2.6.1 Depth-First Traversal

Before turning to string matching algorithms, we introduce a simple algorithm that performs the top-down traversal of a suffix trie in depth-first order, the so-called *Depth-First Search* (DFS). The traversal is bounded, i.e., after reaching the nodes at depth d, it stops going down and goes right instead. A similar top-down traversal is used in Sect. 2.6.3 to solve approximate string matching.

The experimental evaluation of Algorithm 18 reported in Fig. 2.9 shows the practical performance of various suffix trie realizations. FM is the fastest index up to depth 10, while LST is the fastest for deeper traversals. Depth 14 marks the turning point, as the indexes become sparse: The tree indexes (ESA and LST) become significantly faster than FM.

Algorithm 18 DFS(x, d)

Input $x : \mathcal{T}_{\text{strie}}$ iterator
d : integer bounding the traversal depth

```
1: if d > 0 then
2:     if GODOWN(x) then
3:         do
4:             DFS(x, d − 1)
5:         while GORIGHT(x)
```

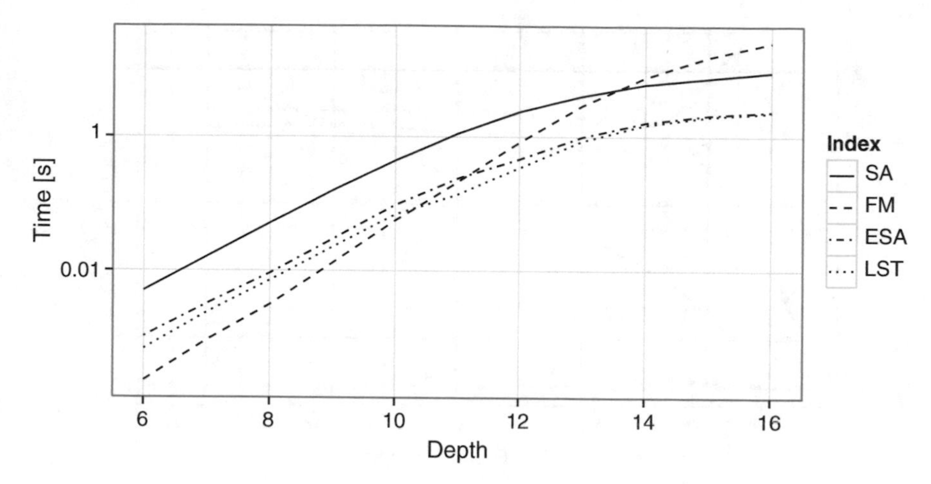

Fig. 2.9 Runtime of the depth-first traversal of various suffix trie implementations

Algorithm 19 EXACTSEARCH(t, p)

Input $t : \mathcal{T}_{\text{strie}}(T)$ iterator
p : pattern iterator
Output list of all occurrences of the pattern in the text

```
1: if ATEND(p) then
2:     report OCCURRENCES(t)
3: else if GODOWN(t, VALUE(p)) then
4:     GONEXT(p)
5:     EXACTSEARCH(t, p)
```

2.6.2 *Exact String matching*

Finding all text occurrences of a pattern is called the *exact string matching* problem:

Definition 17 (Exact String Matching) Given a string P of length m, called *pattern*, and a string T of length n, called *text*, the exact string matching problem is to find all occurrences of P in T [25].

Exact string matching is one of the most fundamental problems in stringology. The online variant of this problem has been extensively studied from the theoretical standpoint and is well solved in practice [16]. Here, we concentrate on the *offline* variant, which allows us to preprocess the text. The preprocessing consists in building a full-text index of the text beforehand to speed up subsequent searches.

In the following, we assume the text T to be indexed by its suffix trie $\mathcal{T}_{\text{strie}}(T)$. Algorithm 19 searches the pattern P by starting on the root node of $\mathcal{T}_{\text{strie}}(T)$ and following the path spelling the pattern. If the search ends up in a node v, then each leaf i in OCCURRENCES(v) points to a distinct suffix T_i such that $T[i, i+m)$ equals P.

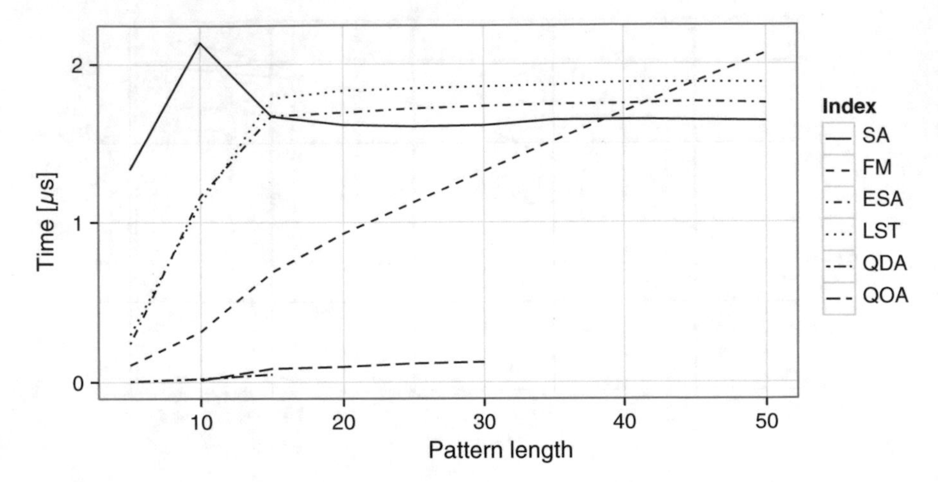

Fig. 2.10 Runtime of exact string matching on different full-text indexes

If GODOWN by character is implemented in constant time and OCCURRENCES in linear time, all occurrences of P into T are found in $\mathcal{O}(m + o)$ time, which is optimal, where m is the length of P and o is the number of its occurrences in T. Among the realizations seen in this chapter, Algorithm 19 runs in $\mathcal{O}(m\sigma + o)$ time on ESA and LST, in $\mathcal{O}(m \log n + o)$ time on SA, and in $\mathcal{O}(m + o \cdot \log^{1+\varepsilon} n)$ time on FM.

Figure 2.10 shows the results of the experimental evaluation of Algorithm 19. Note that we consider only the time to traverse the indexes, while we exclude the time of OCCURRENCES that retrieves the o occurrences of a pattern; this fact plays in favor of FM, achieving such a function in $\mathcal{O}(o \cdot \log^{1+\varepsilon} n)$ time with a sparse SA. On LST, ESA, and SA, the search time becomes practically constant for patterns of length above 14, i.e., when the tree becomes sparse. Conversely, on FM, the practical search time stays linear in the pattern length. ESA is always a better alternative to the fully constructed LST. SA shows a runtime peak around patterns of length 10 due to the fact that binary search Algorithm 1 converges more slowly for shorter patterns. Summing up, FM is the fastest suffix trie realization to match patterns up to length 35, while SA is the fastest for longer patterns.

It is interesting to see that FM is the sole index to exhibit in practice linear search time in the pattern length. This is due to the fact that it is based on backward search, as opposed to all other indexes which are based on forward search. Forward search algorithms refine an SA interval while they go down the suffix trie. For sparse trees, the interval quickly converges to the final interval, so the search benefits from cache locality. Conversely, backward search while going down computes SA intervals which are not necessarily nested. Hence, backward search suffers from poor cache locality.

q-Gram indexes deserve a separate discussion as they do not rely on Algorithm 19. Exact matching is limited to patterns of fixed lengths, i.e., QDA is limited to $q \leq 15$, while QOA is limited to $q \leq 30$. Within these ranges, q-gram indexes soundly outperform all suffix trie achievements in terms of speed: QDA is 16 times faster than FM for length 15, while QOA is 10 times faster than FM for length 30.

2.6.3 Approximate String Matching

The definition of distance functions between strings leads to a more challenging problem: *approximate string matching*.

Definition 18 (Hamming Distance) The *Hamming distance* $d_H : \Sigma^n \times \Sigma^n \rightarrow \mathbb{N}_0$ between two strings $x, y \in \Sigma^n$ counts the number of substitutions necessary to transform x into y [26].

Definition 19 (Approximate String Matching) Given a text T, a pattern P, and a *distance threshold* $k \in \mathbb{N}$, the approximate string matching problem is to find all the occurrences of P in T within distance k [19].
The approximate string matching problem under the Hamming distance is commonly called *the string matching with k-mismatches problem*. Frequently, the problem's input respects the condition $k \ll m \ll n$.

All the full-text indexes presented in this chapter solve *string matching with k-mismatches* via an explicit or implicit enumeration of the Hamming distance neighborhood of the pattern. Here, we focus on implicit enumeration via backtracking a suffix trie [54], which has been first proposed by Ukkonen [54]. Recently, various popular bioinformatics tools, e.g., Bowtie [32] and BWA [34], adopted variations of this method in conjunction with FM. Yet, the idea dates back to more than 20 years.

Algorithm 20 performs a top-down traversal on the suffix trie $\mathcal{T}_{\text{strie}}(T)$, spelling incrementally all distinct substrings of T. While traversing each branch of the trie, this algorithm incrementally computes the distance between the query and the spelled string. If the computed distance exceeds k, the traversal backtracks and proceeds on the next branch. Conversely, if the pattern P is completely spelled and the traversal ends up in a node v, each leaf l_i below v points to a distinct suffix T_i such that $d_H(T\,[i, i+m)\,, p) \leq k$.

Figure 2.11 shows the results of the experimental evaluation of Algorithm 20 for $k = 1$. Again, we consider only the time to traverse the indexes, while we exclude the time of OCCURRENCES. FM is always the fastest data structure; for instance, on patterns of length 30, SA is three times slower. ESA and LST are always slower than FM, even on longer patterns.

q-Gram indexes deserve again a separate discussion, as they support approximate string matching by explicitly enumerating the Hamming distance neighborhood of the pattern, rather than using Algorithm 20. Despite the explicit enumeration, q-gram indexes still outperform all suffix trie realizations in terms of speed: QDA is

Algorithm 20 KMISMATCHES(t, p, k)

Input $t : \mathscr{T}_{\text{strie}}(T)$ iterator
p : pattern iterator
k : number of mismatches

Output list of all occurrences of the pattern in the text

```
 1: if k = 0 then
 2:     EXACTSEARCH(t, p)
 3: else
 4:     if ATEND(p) then
 5:         report OCCURRENCES(t)
 6:     else if GODOWN(t) then
 7:         do
 8:             d ← δ(LABEL(t), VALUE(p))
 9:             GONEXT(p)
10:             KMISMATCHES(t, p, k − d)
11:             GOPREVIOUS(p)
12:         while GORIGHT(t)
```

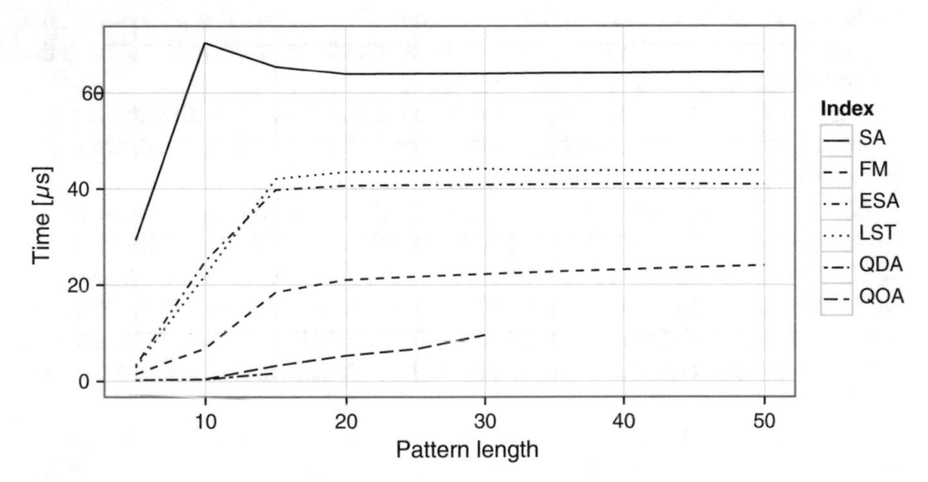

Fig. 2.11 Runtime of 1-mismatch search on different full-text indexes

11 times faster than FM for length 15, while QOA is 2.3 times faster than FM for length 30.

It is interesting to compare the runtimes of exact and approximate string matching algorithms. With FM, exact string matching for patterns of length 30 is 16 times faster than 1-mismatch; on average, the former problem takes 1.3 microseconds (ms), while the latter takes 21.0 ms. With ESA, exact string matching is 24 times faster than 1-mismatch; it spends 1.7 ms, compared to 40.8 ms. Finally, with QOA, exact string matching is 77 times faster than 1-mismatch.

2.7 Conclusion

In this chapter, we presented a comprehensive survey of state-of-the-art full-text indexes solving core problems in *High-Throughput Sequencing* (HTS) with different trade-offs. We considered three core problems: *Depth-First Search* (DFS), exact string matching, and approximate string matching via backtracking. These are fundamental primitives of HTS applications, including read error correction, assembly, and mapping. Our experimental evaluation provides, to the best of our knowledge, the first comprehensive picture of the practical performance of these full-text indexes on HTS data. Our evaluation points out that the best choice of a full-text index is application-specific and should be chosen wisely.

Almost all tools for HTS are based on one specific re-implementation of a certain full-text index along with its traversal algorithm. These tools are therefore fixed to a specific index, which is not necessarily the most efficient one. To enable developers to choose the best-suited index and easily integrate it, we propose frameworks that allow to abstract the algorithm from the underlying index and its implementation. Such a framework is *SeqAn* [12], the open-source software library for sequence analysis. It provides a generic interface for full-text indexes and thus allows to interchange them at no cost and includes all the full-text indexes described in this chapter.

References

1. Abouelhoda, M., Kurtz, S., Ohlebusch, E.: Replacing suffix trees with enhanced suffix arrays. J. Discrete Algorithms **2**, 53–86 (2004)
2. Adjeroh, D., Bell, T., Mukherjee, A.: The Burrows-Wheeler Transform: Data Compression, Suffix Arrays, and Pattern Matching. Springer Science & Business Media, Berlin (2008)
3. Arlazarov, V., Dinic, E., Kronrod, M., Faradzev, I.: On economical construction of the transitive closure of a directed graph. Dokl. Akad. Nauk **11**, 194 (1970)
4. Bauer, M.J., Cox, A.J., Rosone, G., Sciortino, M.: Lightweight LCP construction for next-generation sequencing datasets. In: Algorithms in Bioinformatics, pp. 326–337. Springer, Berlin (2012)
5. Bauer, M.J., Cox, A.J., Rosone, G.: Lightweight algorithms for constructing and inverting the bwt of string collections. Theor. Comput. Sci. **483**, 134–148 (2013)
6. Burkhardt, S., Kärkkäinen, J.: Better filtering with gapped q-grams. Fund. Inform. **56**(1,2), 51–70 (2003)
7. Burkhardt, S., Crauser, A., Ferragina, P., Lenhof, H.P., Rivals, E., Vingron, M.: q-gram based database searching using suffix arrays. In: Proceedings of the 3rd Annual International Conference on Computational Molecular Biology (RECOMB-99), pp. 77–83 (1999)
8. Burrows, M., Wheeler, D.J.: A block-sorting lossless data compression algorithm. Technical Report 124, Digital SRC Research Report (1994)
9. Cazaux, B., Lecroq, T., Rivals, E.: From indexing data structures to de Bruijn graphs. In: Combinatorial Pattern Matching, pp. 89–99. Springer, Berlin (2014)
10. Cormen, T.H., Leiserson, C.E., Rivest, R.L., Stein, C.: Introduction to Algorithms. MIT, Cambridge, MA (2001)

11. Crochemore, M., Grossi, R., Kärkkäinen, J., Landau, G.M.: A constant-space comparison-based algorithm for computing the Burrows–Wheeler transform. In: Combinatorial Pattern Matching, pp. 74–82. Springer, Berlin (2013)
12. Döring, A., Weese, D., Rausch, T., Reinert, K.: SeqAn an efficient, generic C++ library for sequence analysis. BMC Bioinf. **9**, 11 (2008)
13. Emde, A.K., Grunert, M., Weese, D., Reinert, K., Sperling, S.R.: MicroRazerS: rapid alignment of small RNA reads. Bioinformatics **26**(1), 123–124 (2010)
14. Emde, A.K., Schulz, M.H., Weese, D., Sun, R., Vingron, M., Kalscheuer, V.M., Haas, S.A., Reinert, K.: Detecting genomic indel variants with exact breakpoints in single- and paired-end sequencing data using splazers. Bioinformatics **28**(5), 619–627 (2012)
15. Farach-Colton, M., Ferragina, P., Muthukrishnan, S.: On the sorting-complexity of suffix tree construction. J. ACM **47**(6), 987–1011 (2000)
16. Faro, S., Lecroq, T.: The exact online string matching problem: a review of the most recent results. ACM Comput. Surv. **45**(2), 13 (2013)
17. Ferragina, P., Manzini, G.: Indexing compressed text. J. ACM **52**(4), 552–581 (2005)
18. Ferragina, P., Gagie, T., Manzini, G.: Lightweight data indexing and compression in external memory. Algorithmica **63**(3), 707–730 (2012)
19. Galil, Z., Giancarlo, R.: Data structures and algorithms for approximate string matching. J. Complexity **4**(1), 33–72 (1988)
20. Giegerich, R., Kurtz, S.: A comparison of imperative and purely functional suffix tree constructions. Sci. Comput. Program. **25**, 187–218 (1995)
21. Giegerich, R., Kurtz, S., Stoye, J.: Efficient implementation of lazy suffix trees. Softw. Pract. Exp. **33**(11), 1035–1049 (2003)
22. Gog, S., Beller, T., Moffat, A., Petri, M.: From theory to practice: plug and play with succinct data structures. In: Experimental Algorithms, pp. 326–337. Springer, Berlin (2014)
23. Grossi, R., Vitter, J.S.: Compressed suffix arrays and suffix trees with applications to text indexing and string matching. SIAM J. Comput. **35**(2), 378–407 (2005)
24. Grossi, R., Gupta, A., Vitter, J.S.: High-order entropy-compressed text indexes. In: Proceedings of the 14th Annual ACM-SIAM Symposium on Discrete Algorithms, SODA '03, pp. 841–850. Society for Industrial and Applied Mathematics, Philadelphia, PA (2003)
25. Gusfield, D.: Algorithms on Strings, Trees, and Sequences: Computer Science and Computational Biology. Cambridge University Press, New York (1997)
26. Hamming, R.W.: Error detecting and error correcting codes. Syst. Tech. J. **29**, 147–160 (1950)
27. Hon, W.K., Lam, T.W., Sadakane, K., Sung, W.K., Yiu, S.M.: A space and time efficient algorithm for constructing compressed suffix arrays. Algorithmica **48**(1), 23–36 (2007)
28. Intel: Intel® 64 and IA-32 Architectures Optimization Reference Manual. Intel Corporation, Santa Clara, CA (2011)
29. Jacobson, G.: Space-efficient static trees and graphs. In: 30th Annual Symposium on Foundations of Computer Science, 1989, pp. 549–554. IEEE, New York (1989)
30. Kasai, T., Lee, G., Arimura, H., Arikawa, S., Park, K.: Linear-time longest-common-prefix computation in suffix arrays and its applications. In: Proceedings of the 12th Annual Symposium on Combinatorial Pattern Matching, CPM '01, pp. 181–192. Springer, Berlin (2001)
31. Kehr, B., Weese, D., Reinert, K.: Stellar: fast and exact local alignments. BMC Bioinf. **12**(Suppl. 9), S15 (2011)
32. Langmead, B., Trapnell, C., Pop, M., Salzberg, S.: Ultrafast and memory-efficient alignment of short DNA sequences to the human genome. Genome Biol. **10**(3), R25 (2009)
33. Li, H., Durbin, R.: Fast and accurate short read alignment with burrows-wheeler transform. Bioinformatics **25**(14), 1754–1760 (2009)
34. Li, H., Handsaker, B., Wysoker, A., Fennell, T., Ruan, J., Homer, N., Marth, G., Abecasis, G., Durbin, R., 1000 Genome Project Data Processing Subgroup: The sequence alignment/map format and SAMtools. Bioinformatics **25**(16), 2078–2079 (2009)
35. Louza, F.A., Telles, G.P., Ciferri, C.D.D.A.: External memory generalized suffix and LCP arrays construction. In: Combinatorial Pattern Matching, pp. 201–210. Springer, Berlin (2013)

36. Manber, U., Myers, E.: Suffix arrays: a new method for on-line string searches. In: SODA '90, pp. 319–327. SIAM, Philadelphia (1990)
37. Manber, U., Myers, E.: Suffix arrays: a new method for on-line string searches. SIAM J. Comput. **22**(5), 935–948 (1993)
38. Mantaci, S., Restivo, A., Rosone, G., Sciortino, M.: An extension of the Burrows–Wheeler transform. Theor. Comput. Sci. **387**(3), 298–312 (2007)
39. Manzini, G.: An analysis of the Burrows–Wheeler transform. J. ACM **48**(3), 407–430 (2001)
40. McCreight, E.M.: A space-economical suffix tree construction algorithm. J. ACM **23**(2), 262–272 (1976)
41. Morrison, D.R.: Patricia – practical algorithm to retrieve information coded in alphanumeric. J. ACM **15**(4), 514–534 (1968)
42. Navarro, G., Mäkinen, V.: Compressed full-text indexes. ACM Comput. Surv. **39**(1), 2:1–2:61 (2007)
43. Ohlebusch, E.: Bioinformatics Algorithms: Sequence Analysis, Genome Rearrangements, and Phylogenetic Reconstruction. Oldenbusch, Bremen (2013)
44. Puglisi, S., Smyth, W., Turpin, A.: A taxonomy of suffix array construction algorithms. In: Holub, J. (ed.) Proceedings of Prague Stringology Conference '05, Prague, pp. 1–30 (2005)
45. Rausch, T., Emde, A.K., Weese, D., Döring, A., Notredame, C., Reinert, K.: Segment-based multiple sequence alignment. Bioinformatics **24**(16), i187–192 (2008)
46. Rausch, T., Koren, S., Denisov, G., Weese, D., Emde, A.K., Döring, A., Reinert, K.: A consistency-based consensus algorithm for de novo and reference-guided sequence assembly of short reads. Bioinformatics **25**(9), 1118–1124 (2009)
47. Schulz, M.H., Weese, D., Rausch, T., Döring, A., Reinert, K., Vingron, M.: Fast and adaptive variable order Markov chain construction. In: Crandall, K., Lagergren, J. (eds.) Algorithms in Bioinformatics. Lecture Notes in Computer Science, vol. 5251, pp. 306–317. Springer, Berlin (2008)
48. Schulz, M.H., Weese, D., Holtgrewe, M., Dimitrova, V., Niu, S., Reinert, K., Richard, H.: Fiona: a parallel and automatic strategy for read error correction. Bioinformatics **30**(17), i356–i363 (2014)
49. Shi, F.: Suffix arrays for multiple strings: a method for on-line multiple string searches. In: Jaffar, J., Yap, R. (eds.) Concurrency and Parallelism, Programming, Networking, and Security. Lecture Notes in Computer Science, vol. 1179, pp. 11–22. Springer, Berlin (1996). doi:10.1007/BFb0027775. http://dx.doi.org/10.1007/BFb0027775
50. Simpson, J.T., Durbin, R.: Efficient de novo assembly of large genomes using compressed data structures. Genome Res. **22**(3), 549–556 (2012)
51. Siragusa, E.: Approximate string matching for high-throughput sequencing. Ph.D. thesis, Freie Universität Berlin (2015)
52. Siragusa, E., Weese, D., Reinert, K.: Fast and accurate read mapping with approximate seeds and multiple backtracking. Nucleic Acids Res. **41**(7), e78 (2013)
53. Siragusa, E., Weese, D., Reinert, K.: Scalable string similarity search/join with approximate seeds and multiple backtracking. In: Proceedings of the Joint EDBT/ICDT 2013 Workshops, pp. 370–374. ACM, New York (2013)
54. Ukkonen, E.: Approximate string-matching over suffix trees. In: Combinatorial Pattern Matching, pp. 228–242. Springer, Berlin (1993)
55. Ukkonen, E.: On-line construction of suffix trees. Algorithmica **14**(3), 249–260 (1995)
56. Weese, D.: Indices and applications in high-throughput sequencing. Ph.D. thesis, Freie Universität Berlin (2013)
57. Weese, D., Schulz, M.H.: Efficient string mining under constraints via the deferred frequency index. In: Proceedings of the 8th Industrial Conference on Data Mining (ICDM '08). LNAI, vol. 5077, pp. 374–388. Springer, Berlin (2008)
58. Weiner, P.: Linear pattern matching algorithms. In: Proceedings of the 14th Symposium on Switching and Automata Theory, SWAT '73, pp. 1–11. IEEE Computer Society, Washington (1973)

Chapter 3
Searching and Indexing Circular Patterns

Costas S. Iliopoulos, Solon P. Pissis, and M. Sohel Rahman

3.1 Introduction

Circular DNA sequences can be found in viruses, as plasmids in archaea and bacteria, and in the mitochondria and plastids of eukaryotic cells. Hence, circular sequence comparison finds applications in several biological contexts [4, 5, 32]. This motivates the design of efficient algorithms [6] and data structures [21] that are devoted to the specific comparison of circular sequences, as they can be relevant in the analysis of organisms with such structure [17, 18]. In this chapter, we describe algorithms and data structures for *searching* and *indexing* patterns with a circular structure.

In order to provide an overview of the described algorithms and data structures, we begin with a few definitions generally following [10]. A *string* x of length n is an array $x[0 \ldots n-1]$, where $x[i]$, $0 \leq i < n$, is a letter drawn from some fixed *alphabet* Σ of size $\sigma = |\Sigma| = \mathcal{O}(1)$. The string of length 0 is called *empty string* and is denoted by ε. A string x is a *factor* of a string y, or x *occurs* in y, if there exist two strings u and v, such that $y = uxv$. If $u = \varepsilon$ then x is a *prefix* of y. If $v = \varepsilon$ then x is a *suffix* of y. Every occurrence of a string x in a string y can be characterized by a position in y. Thus, we say that x of length n occurs at the *starting position* i in

This chapter was written when Rahman was on a sabbatical leave from BUET.

C.S. Iliopoulos • S.P. Pissis (✉)
Department of Informatics, King's College London, The Strand, London WC2R 2LS, UK
e-mail: c.iliopoulos@kcl.ac.uk; solon.pissis@kcl.ac.uk

M.S. Rahman
AℓEDA Group, Department of Computer Science and Engineering, Bangladesh University of Engineering and Technology (BUET), Dhaka 1205, Bangladesh
e-mail: msrahman@cse.buet.ac.bd

M. Elloumi (ed.), *Algorithms for Next-Generation Sequencing Data*,
DOI 10.1007/978-3-319-59826-0_3

y when $y[i \ldots i+n-1] = x$. We define the *i-th prefix* of x as the prefix ending at position i, $0 \leq i < n$, i.e., the prefix $x[0 \ldots i]$. On the other hand, the *i-th suffix* is the suffix starting at position i, $0 \leq i < n$, i.e., the suffix $x[i \ldots n-1]$.

A circular string can be viewed as a linear string where the left-most character is concatenated to the right-most one. Hence, a circular string of length m can be seen as m different linear strings, all of which are considered as being equivalent. Given a string x of length m, we denote by $x^i = x[i \ldots m-1]x[0 \ldots i-1]$, $0 < i < m$, the *i-th rotation* of x and $x^0 = x$.

We first consider the problem of finding occurrences of a string *pattern* x of length m with a circular structure in a string *text* t of length n with a linear structure. This is the problem of *circular string matching*. A naïve solution with *quadratic* in m time complexity consists in applying a classical algorithm for searching a finite set of strings after having built the dictionary of the m rotations of x [10]. This problem has also been considered in [25], where an $\mathcal{O}(n)$-time algorithm was presented. The approach presented in [25] consists in preprocessing x by constructing the suffix automaton of the string $x' = xx$, noting that every rotation of x is a factor of x'. Then, by feeding t into the automaton, the lengths of the longest factors of x' occurring in t can be found by the links followed in the automaton in time $\mathcal{O}(n)$. In [15], the authors presented an *average-case* time-optimal algorithm for circular string matching, by showing that the average-case lower time bound for *single* string matching of $\mathcal{O}(n \log_\sigma m/m)$ also holds for circular string matching. Recently, in [9], the authors presented two fast average-case algorithms based on word-level parallelism. The first algorithm requires average-case time $\mathcal{O}(n \log_\sigma m/w)$, where w is the number of bits in the computer word. The second one is based on a mixture of word-level parallelism and q-grams. The authors showed that with the addition of q-grams and by setting $q = \mathcal{O}(\log_\sigma m)$, an average-case optimal time of $\mathcal{O}(n \log_\sigma m/m)$ is achieved.

Given a set $\mathcal{D}$ of d pattern strings, the problem of dictionary matching is to index $\mathcal{D}$ such that for any online query of a text t, one can quickly find the occurrences of any pattern of $\mathcal{D}$ in t. This problem has been well-studied in the literature [1, 8], and an index taking optimal space and simultaneously supporting time-optimal queries is achieved [7, 19]. In some applications in computational molecular biology [5] and also in pattern recognition [28], we are interested in searching for not only the original patterns in $\mathcal{D}$ but also *all* of their rotations. This is the problem of *circular dictionary matching* [2, 20, 21].

The aim of constructing an index to provide efficient procedures for answering *online* text queries related to the content of a fixed dictionary of circular patterns was studied in [20]. The authors proposed a variant of the suffix tree (for an introduction to suffix trees, see [33]), called *circular suffix tree* and showed that it can be compressed into succinct space. With a tree structure augmented to a circular pattern matching index called *circular suffix array*, the circular suffix tree can be used to solve the circular dictionary matching problem efficiently. Assuming that the lengths of any two patterns in $\mathcal{D}$ is bounded by a constant, an online text query can be answered in time $\mathcal{O}((n + Occ) \log^{1+\varepsilon} n)$, for any fixed $\varepsilon > 0$, where Occ is the number of occurrences in the output. In [21], the authors proposed the first algorithm for space-efficient construction of the circular suffix tree, which requires time $\mathcal{O}(M \log M)$ and $\mathcal{O}(M \log \sigma + d \log M)$ bits of working space, where M is the total length of the dictionary patterns. In [2], the authors proposed an algorithm to solve the circular dictionary matching problem by constructing an indexing data structure in time and space $\mathcal{O}(M)$ and answering a subsequent online text query in average-case time $\mathcal{O}(n)$, assuming that the shortest pattern is sufficiently long.

Analogously, the aim of constructing an index to provide efficient procedures for answering *online* circular pattern queries related to the content of a fixed text was studied in [24]. The authors proposed two indexing data structures. The first one can be constructed in time and space $\mathcal{O}(n \log^{1+\varepsilon} n)$, and an online pattern query can be answered in time $\mathcal{O}(m \log \log n + Occ)$. However, it involves the construction of two suffix trees as well as a complex range-search data structure. Hence, it is suspected that, despite a good theoretical time bound, the practical performance of this construction would not be very good both in terms of time and space. The second indexing data structure, on the other hand, uses the suffix array (for an introduction to suffix arrays, see [26]), which is more space-efficient than the suffix tree and does not require the range-search data structure. It is conceptually much simpler and can be built in time and space $\mathcal{O}(n)$; an online pattern query can then be answered in time $\mathcal{O}(m \log n + Occ)$.

In this chapter, we revisit the aforementioned problems for searching and indexing circular patterns; we formally define them as follows:

CIRCULARDICTIONARYMATCHING (CDM)
Input: a fixed set $\mathcal{D} = \{x_0, x_1 \dots, x_{d-1}\}$ of patterns of total length M and, subsequently, an online query of a text t of length n, such that $n > |x_j|$, $0 \leq j < d$
Output: all factors u of t such that $u = x_j^i$, $0 \leq j < d$, $0 \leq i < |x_j|$

CIRCULARPATTERNINDEXING (CPI)
Input: a fixed text t of length n and, subsequently, an online query of a pattern x of length m, such that $n > m$
Output: all factors u of t such that $u = x^i$, $0 \leq i < m$

Notice that in the CDM and CPI problem settings, once the relevant indexing data structures are constructed, an unlimited number of online queries can be handled. We present details of the following solutions to these problems:

- An algorithm to solve the CDM problem by constructing an indexing data structure in time and space $\mathcal{O}(M)$ and answering a subsequent online text query in average-case time $\mathcal{O}(n)$, assuming that the shortest pattern in $\mathcal{D}$ is sufficiently long [2].
- An algorithm to solve the CPI problem by constructing an indexing data structure in time and space $\mathcal{O}(n)$ and answering a subsequent online pattern query in time $\mathcal{O}(m \log n + Occ)$ [24].

The rest of the chapter is organized as follows: In Sect. 3.2, we present a solution to the CDM problem; in Sect. 3.3, we present a solution to the CPI problem; finally, in the last section, we summarize the presented results and present some future proposals.

3.2 Circular Dictionary Matching

In this section, we focus on the implementation of an index to provide efficient procedures for answering online text queries related to the content of a fixed dictionary of circular patterns. This is the circular dictionary matching (CDM) problem defined in Sect. 3.1. In this setting, we will be given a fixed dictionary of circular patterns beforehand for preprocessing, and, subsequently, one or more online text queries may follow, one at a time. All text queries would be against the dictionary supplied to us for preprocessing.

Here, we describe an algorithm to solve the CDM problem by constructing an indexing data structure in time and space $\mathcal{O}(M)$ and answering a subsequent online text query in average-case time $\mathcal{O}(n)$, assuming that the shortest pattern in $\mathcal{D}$ is sufficiently long. We start by giving a brief outline of the *partitioning* technique in general and then show some properties of the version of the technique we use for our algorithms. We then describe average-case suboptimal algorithms for circular string matching and show how they can be easily generalized to solve the CDM problem efficiently.

3.2.1 Properties of the Partitioning Technique

The partitioning technique, introduced in [34], and in some sense earlier in [29], is an algorithm based on *filtering out* candidate positions that could never give a solution to speed up string matching algorithms. An important point to note about this technique is that it reduces the search space but does not, by design, verify potential occurrences. To create a string matching algorithm, filtering must be combined with some verification technique. The technique was initially proposed for approximate string matching, but here we show that this can also be used for circular string matching.

The idea behind the partitioning technique is to partition the given pattern in such a way that at least one of the fragments must occur exactly in any valid approximate occurrence of the pattern. It is then possible to search for these fragments exactly to give a set of *candidate* occurrences of the pattern. It is then left to the verification portion of the algorithm to check if these are valid approximate occurrences of the pattern. It has been experimentally shown that this approach yields very good practical performance on large-scale datasets [16], even if it is not theoretically optimal.

For circular string matching, for an efficient solution, we cannot simply apply well-known string matching algorithms, as we must also take into account the rotations of the pattern. We can, however, make use of the partitioning technique and, by choosing an appropriate number of fragments, ensure that at least one fragment must occur in any valid occurrence of a rotation. Lemma 1, together with the following fact, provides this number.

Fact 1 ([2]) *Let x be a string of length m. Any rotation of x is a factor of $x' = x[0 \dots m-1]x[0 \dots m-2]$, and any factor of length m of x' is a rotation of x.*

Lemma 1 ([2]) *Let x be a string of length m. If we partition $x' = x[0 \dots m-1]x[0 \dots m-2]$ into four fragments of length $\lfloor (2m-1)/4 \rfloor$ and $\lceil (2m-1)/4 \rceil$, at least one of the four fragments is a factor of any factor of length m of x'.*

3.2.2 *Circular String Matching via Filtering*

In this section, we consider the circular string matching problem.

CircularStringMatching (CSM)
Input: a pattern x of length m and a text t of length $n > m$
Output: all factors u of t such that $u = x^i$, $0 \leq i < m$

We present average-case suboptimal algorithms for circular string matching via filtering [2, 3]. They are based on the partitioning technique and a series of practical and well-established data structures.

3.2.2.1 Longest Common Extension

First, we describe how to compute the longest common extension, denoted by lce, of two suffixes of a string in constant time. In general, the *longest common extension* problem considers a string x and computes, for each pair (i, j), the length of the longest factor of x that starts at both i and j [23]. lce queries are an important part of the algorithms presented later on.

Let SA denote the array of positions of the lexicographically sorted suffixes of string x of length n, i.e., for all $1 \leq r < n$, we have $x[\mathsf{SA}[r-1] \dots n-1] < x[\mathsf{SA}[r] \dots n-1]$. The inverse iSA of the array SA is defined by $\mathsf{iSA}[\mathsf{SA}[r]] = r$, for all $0 \leq r < n$. Let $\mathsf{lcp}(r, s)$ denote the length of the longest common prefix of the strings $x[\mathsf{SA}[r] \dots n-1]$ and $x[\mathsf{SA}[s] \dots n-1]$ for all $0 \leq r, s < n$ and 0 otherwise. Let LCP denote the array defined by $\mathsf{LCP}[r] = \mathsf{lcp}(r-1, r)$, for all $1 < r < n$, and $\mathsf{LCP}[0] = 0$. Given an array A of n objects taken from a well-ordered set, the range minimum query $\mathsf{RMQ}_{\mathsf{A}}(l, r) = \text{argmin}\, \mathsf{A}[k]$, $0 \leq l \leq k \leq r < n$ returns the position of the minimal element in the specified subarray $\mathsf{A}[l \dots r]$. We perform the following linear-time and linear-space preprocessing:

- Compute arrays SA and iSA of x [27];
- Compute array LCP of x [13];
- Preprocess array LCP for range minimum queries; we denote this by $\mathsf{RMQ}_{\mathsf{LCP}}$ [14].

With the preprocessing complete, the lce of two suffixes of x starting at positions p and q can be computed in constant time in the following way [23]:

$$\mathsf{LCE}(x, p, q) = \mathsf{LCP}[\mathsf{RMQ}_{\mathsf{LCP}}(\mathsf{iSA}[p] + 1, \mathsf{iSA}[q])]. \quad (3.1)$$

Example 1 Let the string x = abbababba. The following table illustrates the arrays SA, iSA, and LCP for x.

i	0	1	2	3	4	5	6	7	8
x[i]	a	b	b	a	b	a	b	b	a
SA[i]	8	3	5	0	7	2	4	6	1
iSA[i]	3	8	5	1	6	2	7	4	0
LCP[i]	0	1	2	4	0	2	3	1	3

We have $\mathsf{LCE}(x, 2, 1) = \mathsf{LCP}[\mathsf{RMQ}_{\mathsf{LCP}}(\mathsf{iSA}[2] + 1, \mathsf{iSA}[1])] = \mathsf{LCP}[\mathsf{RMQ}_{\mathsf{LCP}}(6, 8)] = 1$, implying that the lce of bbababba and bababba is 1.

3.2.2.2 Algorithm CSMF

Given a pattern x of length m and a text t of length $n > m$, an outline of algorithm CSMF for solving the CSM problem is as follows:

1. Construct the string $x' = x[0 \dots m-1]x[0 \dots m-2]$ of length $2m-1$. By Fact 1, any rotation of x is a factor of x'.
2. The pattern x' is partitioned in four fragments of length $\lfloor (2m-1)/4 \rfloor$ and $\lceil (2m-1)/4 \rceil$. By Lemma 1, at least one of the four fragments is a factor of any rotation of x.
3. Match the four fragments against the text t using an Aho-Corasick automaton [11]. Let $\mathscr{L}$ be a list of size Occ of tuples, where $< p_{x'}, \ell, p_t > \in \mathscr{L}$ is a 3-tuple such that $0 \leq p_{x'} < 2m - 1$ is the position where the fragment occurs in x', ℓ is the length of the corresponding fragment, and $0 \leq p_t < n$ is the position where the fragment occurs in t.
4. Compute SA, iSA, LCP, and $\mathsf{RMQ}_{\mathsf{LCP}}$ of $T = x't$. Compute SA, iSA, LCP, and $\mathsf{RMQ}_{\mathsf{LCP}}$ of $T_r = \mathsf{rev}(tx')$, that is the reverse string of tx'.
5. For each tuple $< p_{x'}, \ell, p_t > \in \mathscr{L}$, we try to extend to the right via computing

$$\mathscr{E}_r \leftarrow \mathsf{LCE}(t, p_{x'} + \ell, 2m - 1 + p_t + \ell);$$

in other words, we compute the length $\mathscr{E}_r$ of the longest common prefix of $x'[p_{x'} + \ell \dots 2m - 1]$ and $t[p_t + \ell \dots n - 1]$, both being suffixes of T. Similarly, we try to extend to the left via computing $\mathscr{E}_l$ using lce queries on the suffixes of T_r.

6. For each $\mathscr{E}_l$ and $\mathscr{E}_r$ computed for tuple $< p_{x'}, \ell, p_t > \in \mathscr{L}$, we report all the valid starting positions in t by first checking if the total length $\mathscr{E}_l + \ell + \mathscr{E}_r \geq m$; that is the length of the full extension of the fragment is greater than or equal to m, matching at least one rotation of x. If that is the case, then we report positions

$$\max\{p_t - \mathscr{E}_\ell, p_t + \ell - m\}, \ldots, \min\{p_t + \ell - m + \mathscr{E}_r, p_t\}.$$

Example 2 Let us consider the pattern $x =$ GGGTCTA of length $m = 7$ and the text

$$t = \texttt{GATACGATACCTAGGGTGATAGAATAG}.$$

Then, $x' =$ GGGTCTAGGGTCT (Step 1). x' is partitioned in GGGT, CTA, GGG, and TCT (Step 2). Consider $< 4, 3, 10 > \in \mathscr{L}$, that is, fragment $x'[4 \ldots 6] =$ CTA, of length $\ell = 3$, occurs at starting position $p_t = 10$ in t (Step 3). Then,

$$T = \texttt{GGGTCTAGGGTCTGATACGATACCTAGGGTGATAGAATAG}$$

and

$$T_r = \texttt{TCTGGGATCTGGGGATAAGATAGTGGGATCCATAGCATAG}$$

(Step 4). Extending to the left gives $\mathscr{E}_l = 0$, since $T_r[9] \neq T_r[30]$, and extending to the right gives $\mathscr{E}_r = 4$, since $T[7 \ldots 10] = T[26 \ldots 29]$ and $T[11] \neq T[30]$ (Step 5). We check that $\mathscr{E}_l + \ell + \mathscr{E}_r = 7 = m$, and therefore, we report position 10 (Step 6):

$$p_t - \mathscr{E}_\ell = 10 - 0 = 10, \ldots, p_t + \ell - m + \mathscr{E}_r = 10 + 3 - 7 + 4 = 10;$$

that is, $x^4 =$ CTAGGGT occurs at starting position 10 in t.

Theorem 1 ([3]) *Given a pattern x of length m and a text t of length $n > m$, algorithm* CSMF *requires average-case time $\mathscr{O}(n)$ and space $\mathscr{O}(n)$ to solve the CSM problem.*

3.2.2.3 Algorithm CSMF-Simple

In this section, we present algorithm CSMF-Simple [2], a more space-efficient version of algorithm CSMF. Algorithm CSMF-Simple is very similar to algorithm CSMF. The only differences are:

- Algorithm CSMF-Simple does not perform Step 4 of algorithm CSMF.
- For each tuple $< p_{x'}, \ell, p_t > \in \mathscr{L}$, Step 5 of algorithm CSMF is performed without the use of the pre-computed indexes. In other words, we compute $\mathscr{E}_r$ and

$\mathscr{E}_\ell$ by simply performing letter-to-letter comparisons and counting the number of mismatches that occurred. The extension stops right before the first mismatch.

Fact 2 ([2]) *The expected number of letter comparisons required for each extension in algorithm* CSMF-Simple *is less than* 3.

Theorem 2 ([2]) *Given a pattern x of length m and a text t of length $n > m$, algorithm* CSMF-Simple *requires average-case time $\mathcal{O}(n)$ and space $\mathcal{O}(m)$ to solve the CSM problem.*

3.2.3 Circular Dictionary Matching via Filtering

In this section, we give a generalization of our algorithms for circular string matching to solve the problem of circular dictionary matching. We denote this algorithm by CDMF. Algorithm CDMF follows the same approach as before but with a few key differences. In circular dictionary matching, we are given a set $\mathscr{D} = \{x_0, x_1, \ldots, x_{d-1}\}$ of patterns of total length M and we must find all occurrences of the patterns in $\mathscr{D}$ or any of their rotations. To modify algorithm CSMF to solve this problem, we perform Steps 1 and 2 for every pattern in $\mathscr{D}$, constructing the strings $x'_0, x'_1, \ldots, x'_{d-1}$ and breaking them each into four fragments in the same way specified in Lemma 1. From this point, the algorithm remains largely the same (Steps 3–4); we build the automaton for the fragments from every pattern and then proceed in the same way as algorithm CSMF. The only extra consideration is that we must be able to identify, for every fragment, the pattern from which it was extracted. To do this, we alter the definition of $\mathscr{L}$ such that it now consists of tuples of the form $< p_{x'_j}, \ell, j, p_t >$, where j identifies the pattern from where the fragment was extracted, $p_{x'_j}$ and ℓ are defined identically with respect to the pattern x_j, and p_t remains the same. This then allows us to identify the pattern for which we must perform verification (Steps 5–6) if a fragment is matched. The verification steps are then the same as in algorithm CSMF with the respective pattern.

In a similar way as in algorithm CSMF-Simple, we can apply Fact 2 to obtain algorithm CDMF-Simple and achieve the following result:

Theorem 3 ([2]) *Given a set $\mathscr{D} = \{x_0, x_1, \ldots, x_{d-1}\}$ of patterns of total length M drawn from alphabet Σ, $\sigma = |\Sigma|$, and a text t of length $n > |x_j|$, where $0 \leq j < d$, drawn from Σ, algorithm* CDMF-Simple *requires average-case time $\mathcal{O}((1 + \frac{d|x_{\max}|}{\sigma^{\frac{2|x_{\min}|-1}{4}}})n + M)$ and space $\mathcal{O}(M)$ to solve the CDM problem, where $x_{\min}$ and $x_{\max}$ are the minimum- and maximum-length patterns in $\mathscr{D}$, respectively.*

Algorithm CDMF achieves average-case time $\mathcal{O}(n + M)$ *if and only if*

$$\frac{4d|x_{\max}|}{\sigma^{\frac{2|x_{\min}|-1}{4}}}n \leq cn$$

for some fixed constant c. So we have

$$\frac{4d|x_{\max}|}{\sigma^{\frac{2|x_{\min}|-1}{4}}} \leq c$$

$$\log_\sigma\left(\frac{4d|x_{\max}|}{c}\right) \leq \frac{2|x_{\min}|-1}{4}$$

$$4(\log_\sigma 4 + \log_\sigma d + \log_\sigma |x_{\max}| - \log_\sigma c) \leq 2|x_{\min}| - 1$$

Rearranging and setting c such that $\log_\sigma c \geq 1/4 + \log_\sigma 4$ gives a sufficient condition for our algorithm to achieve average-case time $\mathcal{O}(n + M)$:

$$|x_{\min}| \geq 2(\log_\sigma d + \log_\sigma |x_{\max}|).$$

Corollary 1 ([2]) *Given a set* $\mathcal{D} = \{x_0, x_1, \ldots, x_{d-1}\}$ *of patterns of total length* M *drawn from alphabet* Σ, $\sigma = |\Sigma|$ *and a text* t *of length* $n > |x_j|$, *where* $0 \leq j < d$, *drawn from* Σ, *algorithm* CDMF-Simple *solves the CDM problem in average-case time* $\mathcal{O}(n + M)$ *if and only if* $|x_{\min}| \geq 2(\log_\sigma d + \log_\sigma |x_{\max}|)$, *where* $x_{\min}$ *and* $x_{\max}$ *are the minimum- and maximum-length patterns in* $\mathcal{D}$, *respectively.*

3.2.4 Key Results

We described algorithm CDMF-Simple to solve the CDM problem by constructing an indexing data structure—the Aho-Corasick automaton of the patterns' fragments—in time and space $\mathcal{O}(M)$ and answering a subsequent online text query in average-case time $\mathcal{O}(n)$, assuming that the shortest pattern in $\mathcal{D}$ is sufficiently long.

3.3 Circular Pattern Indexing

In this section, we focus on the implementation of an index to provide efficient procedures for answering online circular patterns queries related to the content of a fixed text. This is the *Circular Pattern Indexing* (CPI) problem defined in Sect. 3.1. In this setting, we will be given a fixed text beforehand for preprocessing, and, subsequently, one or more online pattern queries may follow, one at a time. All pattern queries would be against the text supplied to us for preprocessing.

Here, we describe an algorithm to solve the CPI problem by constructing an indexing data structure in time and space $\mathcal{O}(n)$ and answering a subsequent online pattern query in time $\mathcal{O}(m \log n + Occ)$ [24]. For the rest of this section, we make use of the following notation. For rotation x^i of string x, we denote $x^i = x^{i,f}x^{i,\ell}$ where $x^{i,f} = x[i \ldots m-1]$ and $x^{i,\ell} = x[0 \ldots i-1]$.

3.3.1 The **CPI-II** *Data Structure*

In this section, we discuss a data structure called CPI-II. CPI-II and another data structure, called CPI-I, were presented in [24] to solve the CPI problem. CPI-II consists of the suffix array SA and the inverse suffix array iSA of text t. Both SA and iSA of t of length n can be constructed in time and space $\mathcal{O}(n)$ [27]. To handle the subsequent queries, CPI-II uses the following well-known results from [18] and [22]. The result of a query for a pattern x on the suffix array SA of t is given as a pair (s, e) representing an interval $[s \dots e]$, such that the output set is $\{\mathsf{SA}[s], \mathsf{SA}[s+1], \dots, \mathsf{SA}[e]\}$. In this case, the interval $[s \dots e]$ is denoted by Int^t_x.

Lemma 2 ([18]) *Given a text t of length n, the suffix array of t, and the interval* $[s \dots e] = Int^t_x$ *for a pattern x, for any letter* α*, the interval* $[s' \dots e'] = Int^t_{x\alpha}$ *can be computed in time* $\mathcal{O}(\log n)$.

Lemma 3 ([22]) *Given a text t of length n, the suffix array of t, the inverse suffix array of t, interval* $[s' \dots e'] = Int^t_{x'}$ *for a pattern* x'*, and interval* $[s'' \dots e''] = Int^t_{x''}$ *for a pattern* x''*, the interval* $[s \dots e] = Int^t_{x'x''}$ *can be computed in time* $\mathcal{O}(\log n)$.

Remark 1 At this point, a brief discussion on how $\mathrm{Int}^t_{x'}$ is combined with $\mathrm{Int}^t_{x''}$ to get $\mathrm{Int}^t_{x'x''}$ is in order. We need to find the interval $[s \dots e]$. The idea here is to find the smallest s and the largest e such that both the suffixes $t[\mathsf{SA}[s] \dots n-1]$ and $t[\mathsf{SA}[e] \dots n-1]$ have $x'x''$ as their prefixes. Therefore, we must have that $[s \dots e]$ is a subinterval of $[s' \dots e']$. Now, suppose that $|x'| = m'$. By definition, the lexicographic order of the suffixes $t[\mathsf{SA}[s'] \dots n-1], t[\mathsf{SA}[s'+1] \dots n-1], \dots, t[\mathsf{SA}[e'] \dots n-1]$ is increasing. Then, since they share the same prefix, x', the lexicographic order of the suffixes $t[\mathsf{SA}[s'] + m' \dots n-1], t[\mathsf{SA}[s'+1] + m' \dots n-1], \dots, t[\mathsf{SA}[e'] + m' \dots n-1]$ is increasing as well. Therefore, we must have that $\mathsf{iSA}[\mathsf{SA}[s'] + m'] < \mathsf{iSA}[\mathsf{SA}[s'+1] + m'] < \cdots < \mathsf{iSA}[\mathsf{SA}[e'] + m']$. Finally, we just need to find the smallest s such that $s' \leq \mathsf{iSA}[\mathsf{SA}[s] + m'] \leq e'$ and the largest e such that $s'' \leq \mathsf{iSA}[\mathsf{SA}[e] + m'] \leq e''$.

Lemma 4 ([22]) *Given a text t of length n, the suffix array of t, the inverse suffix array of t, interval* $[s \dots e] = Int^t_x$ *for a pattern x, and an array* Count *such that* Count$[\alpha]$ *stores the total number of occurrences of all* $\alpha' \leq \alpha$*, where '*$\leq$*' implies lexicographically smaller or equal, for any letter* α*, the interval* $[s' \dots e'] = Int^t_{\alpha x}$ *can be computed in time* $\mathcal{O}(\log n)$.

CPI-II handles a query as follows. It first computes the intervals for all the prefixes and suffixes of x and stores them in the arrays $\mathsf{Pref}[0 \dots m-1]$ and $\mathsf{Suf}[0 \dots m-1]$, respectively. To compute the intervals for prefixes, Lemma 2 is used as follows. CPI-II first computes the interval for the letter $x[0]$ and stores the interval in $\mathsf{Pref}[0]$. This can be done in time $\mathcal{O}(\log n)$. Next, it computes the interval for $x[0 \dots 1]$ using $\mathsf{Pref}[0]$ in time $\mathcal{O}(\log n)$ by Lemma 2. Then, it computes the interval for $x[0 \dots 2]$ using $\mathsf{Pref}[1]$ and so on. So, in this way, the array $\mathsf{Pref}[0 \dots m-1]$ can be computed in time $\mathcal{O}(m \log n)$. Similarly, CPI-II computes the intervals for suffixes in $\mathsf{Suf}[0 \dots m-1]$. Using Lemma 4, this can be done in time $\mathcal{O}(m \log n)$

as well. Then CPI-II uses Lemma 3 to combine the intervals of $x^{j,f}$ and $x^{j,\ell}$ for $0 \leq j \leq m-1$ from the corresponding indexes of the Pref and Suf arrays. Thus, all the intervals for all the rotations of x are computed. Finally, CPI-II carefully reports all the occurrences in this set of intervals as follows. To avoid reporting a particular occurrence more than once, it sorts the m intervals according to the start of the interval requiring time $\mathcal{O}(m \log m)$, and then it does a linear traversal on the two ends of the sorted intervals to get an equivalent set of *disjoint* intervals. Clearly, CPI-II data structure can be constructed in time and space $\mathcal{O}(n)$ and can answer the relevant queries in time $\mathcal{O}(m \log n + Occ)$ per query.

Further improvement on the above running time seems possible. In fact, we can improve both from the time and space point of view, albeit not asymptotically, if we use the FM-index [12]. Using the *backward-search technique*, we can compute array Suf in time $\mathcal{O}(m)$. However, any analogous result for Lemma 2 still eludes us. We are also unable to improve the time requirement for Lemma 3. Notably, some results on bidirectional search [30] may turn out to be useful in this regard which seems to be a good candidate for future investigations.

3.3.2 A Folklore Indexing Data Structure

Before concluding this section, we briefly shed some light on another sort of folklore indexing data structure to solve the CPI problem. The data structure is based on the suffix tree and Weiner's algorithm for its construction [33]. Suppose we have computed the suffix tree ST_t of $t\$$, where $\$ \notin \Sigma$ using Weiner's algorithm. Then an online pattern query can be answered as follows:

1. Let $x' = xx\#$, where $\# \notin \Sigma$ and $\# \neq \$$. Set $\ell = |x'|$.
2. Let $\alpha = x'[\ell - 1]$. Extend ST_t such that it becomes the suffix tree of $t' = \alpha t\$$. While extending, keep track of whether the new suffix diverges from a node v with $depth(v) \geq m$ of the original tree ST_t. Let us call these nodes *output nodes*.
3. Set $\ell = \ell - 1$. If $\ell > 0$, then go to Step 2.
4. Output the starting positions of the suffixes corresponding to t from the output nodes.
5. Undo the steps so that we again get ST_t and become ready for any further query.

The correctness of the above approach heavily depends on how Weiner's algorithm constructs the suffix tree. We are not going to provide all the details here, but very briefly, Weiner's algorithm first handles the suffix $t[n-1 \ldots n-1]$, followed by $t[n-2 \ldots n-1]$, and so on. This is why feeding x' to ST_t (cf. the loop comprising Steps 2 and 3 above) works perfectly. At first glance, it seems that the above data structure would give us a time-optimal query of $\mathcal{O}(m + Occ)$ with a data structure construction time of $\mathcal{O}(n)$. However, there is a catch. The linear-time construction algorithm of Weiner (and in fact all other linear-time constructions, e.g., [31]) depends on an amortized analysis. Hence, while we can certainly say that the construction of the suffix tree for the string $xx\#t\$$ can be done in time $\mathcal{O}(m+n)$, given the suffix tree for $t\$$, we cannot always claim that extending it for $xx\#t\$$ can be done in time $\mathcal{O}(m)$.

3.3.3 Key Results

In Sect. 3.3.1, we described an algorithm to solve the CPI problem by constructing an indexing data structure—the CPI-II data structure—in time and space $\mathcal{O}(n)$ and answering a subsequent online pattern query in time $\mathcal{O}(m \log n + Occ)$.

3.4 Final Remarks and Outlook

In this chapter, we described algorithms and data structures for searching and indexing patterns with a circular structure. We considered two different settings for this task: in the first one, we are given a fixed set of patterns, and we must construct a data structure to answer subsequent online text queries; in the second, we are given a fixed text, and we must construct a data structure to answer subsequent online pattern queries. This type of task finds applications in computational molecular biology and in pattern recognition.

For the first setting, we described an algorithm to solve the problem by constructing an indexing data structure in time and space $\mathcal{O}(M)$ and answering a subsequent online text query in average-case time $\mathcal{O}(n)$, assuming that the shortest pattern is sufficiently long. For the second setting, we described an algorithm to solve the problem by constructing an indexing data structure in time and space $\mathcal{O}(n)$ and answering a subsequent online pattern query in time $\mathcal{O}(m \log n + Occ)$.

Despite both theoretical and practical motivations for searching and indexing circular patterns, we noticed a lack of significant research effort on this particular topic, especially from the perspective of computational molecular biology. Sequence comparison algorithms require an *implicit assumption* on the input data: the left- and right-most positions for each sequence are relevant. However, this is *not* the case for circular structures. To this end, we conclude this chapter posing an exciting question that could alter the perspective of current knowledge and state of the art for sequence comparison:

Would dropping this implicit assumption yield more significant alignments and thereby new biological knowledge, for instance in phylogenetic analyses, for organisms with such structure?

We believe that the circular string matching paradigm could bring forth a complete new era in this context.

Acknowledgements Part of this research has been supported by an INSPIRE Strategic Partnership Award, administered by the British Council, Bangladesh, for the project titled "Advances in Algorithms for Next Generation Biological Sequences."

References

1. Aho, A.V., Corasick, M.J.: Efficient string matching: an aid to bibliographic search. Commun. ACM **18**(6), 333–340 (1975)
2. Athar, T., Barton, C., Bland, W., Gao, J., Iliopoulos, C.S., Liu, C., Pissis, S.P.: Fast circular dictionary-matching algorithm. Math. Struct. Comput. Sci. **FirstView**, 1–14 (2015)
3. Barton, C., Iliopoulos, C.S., Pissis, S.P.: Circular string matching revisited. In: Proceedings of the Fourteenth Italian Conference on Theoretical Computer Science (ICTCS 2013), pp. 200–205 (2013)
4. Barton, C., Iliopoulos, C.S., Pissis, S.P.: Fast algorithms for approximate circular string matching. Algorithms Mol. Biol. **9**(9) (2014)
5. Barton, C., Iliopoulos, C.S., Kundu, R., Pissis, S.P., Retha, A., Vayani, F.: Accurate and efficient methods to improve multiple circular sequence alignment. In: Bampis, E. (ed.) Experimental Algorithms. Lecture Notes in Computer Science, vol. 9125, pp. 247–258. Springer International Publishing, Cham (2015)
6. Barton, C., Iliopoulos, C.S., Pissis, S.P.: Average-case optimal approximate circular string matching. In: Dediu, A.H., Formenti, E., Martín-Vide, C., Truthe, B. (eds.) Language and Automata Theory and Applications. Lecture Notes in Computer Science, vol. 8977, pp. 85–96. Springer, Berlin/Heidelberg (2015)
7. Belazzougui, D.: Succinct dictionary matching with no slowdown. In: Proceedings of the 21st Annual Conference on Combinatorial Pattern Matching, CPM'10, pp. 88–100. Springer, Berlin/Heidelberg (2010)
8. Chan, H.L., Hon, W.K., Lam, T.W., Sadakane, K.: Compressed indexes for dynamic text collections. ACM Trans. Algorithms **3**(2) (2007)
9. Chen, K.-H., Huang, G.-S., Lee, R.C.-T.: Bit-parallel algorithms for exact circular string matching. Comput. J **57**(5), 731–743 (2014)
10. Crochemore, M., Hancart, C., Lecroq, T.: Algorithms on Strings. Cambridge University Press, New York (2007)
11. Dori, S., Landau, G.M.: Construction of Aho Corasick automaton in linear time for integer alphabets. Inf. Process. Lett. **98**(2), 66–72 (2006)
12. Ferragina, P., Manzini, G.: Opportunistic data structures with applications. In: Proceedings of the FOCS, pp. 390–398. IEEE Computer Society, Los Alamitos, CA (2000)
13. Fischer, J.: Inducing the LCP-array. In: Dehne, F., Iacono, J., Sack, J.R. (eds.) Algorithms and Data Structures. Lecture Notes in Computer Science, vol. 6844, pp. 374–385. Springer, Berlin/Heidelberg (2011)
14. Fischer, J., Heun, V.: Space-efficient preprocessing schemes for range minimum queries on static arrays. SIAM J. Comput. **40**(2), 465–492 (2011)
15. Fredriksson, K., Grabowski, S.: Average-optimal string matching. J. Discrete Algorithms **7**(4), 579–594 (2009)
16. Frousios, K., Iliopoulos, C.S., Mouchard, L., Pissis, S.P., Tischler, G.: REAL: an efficient REad ALigner for next generation sequencing reads. In: Proceedings of the First ACM International Conference on Bioinformatics and Computational Biology, BCB '10, pp. 154–159. ACM, New York (2010)
17. Grossi, R., Iliopoulos, C.S., Mercas, R., Pisanti, N., Pissis, S.P., Retha, A., Vayani, F.: Circular sequence comparison with q-grams. In: Pop, M., Touzet, H. (eds.) Proceedings of Algorithms in Bioinformatics - 15th International Workshop, WABI 2015, Atlanta, GA, Sept 10–12, 2015. Lecture Notes in Computer Science, vol. 9289, pp. 203–216. Springer, Berlin (2015)

18. Gusfield, D.: Algorithms on Strings, Trees, and Sequences - Computer Science and Computational Biology. Cambridge University Press, Cambridge (1997)
19. Hon, W.K., Ku, T.H., Shah, R., Thankachan, S.V., Vitter, J.S.: Faster compressed dictionary matching. In: Proceedings of the 17th International Conference on String Processing and Information Retrieval, SPIRE'10, pp. 191–200. Springer, Berlin/Heidelberg (2010)
20. Hon, W.K., Lu, C.H., Shah, R., Thankachan, S.V.: Succinct indexes for circular patterns. In: Proceedings of the 22nd International Conference on Algorithms and Computation, ISAAC'11, pp. 673–682. Springer, Berlin/Heidelberg (2011)
21. Hon, W.K., Ku, T.H., Shah, R., Thankachan, S.: Space-efficient construction algorithm for the circular suffix tree. In: Fischer, J., Sanders, P. (eds.) Combinatorial Pattern Matching. Lecture Notes in Computer Science, vol. 7922, pp. 142–152. Springer, Berlin/Heidelberg (2013)
22. Huynh, T.N.D., Hon, W.K., Lam, T.W., Sung, W.K.: Approximate string matching using compressed suffix arrays. Theor. Comput. Sci. **352**(1–3), 240–249 (2006)
23. Ilie, L., Navarro, G., Tinta, L.: The longest common extension problem revisited and applications to approximate string searching. J. Discrete Algorithms **8**(4), 418–428 (2010)
24. Iliopoulos, C.S., Rahman, M.S.: Indexing circular patterns. In: Proceedings of the 2nd International Conference on Algorithms and Computation, WALCOM'08, pp. 46–57. Springer, Berlin/Heidelberg (2008)
25. Lothaire, M. (ed.): Applied Combinatorics on Words. Cambridge University Press, New York (2005)
26. Manber, U., Myers, E.W.: Suffix arrays: a new method for on-line string searches. SIAM J. Comput. **22**(5), 935–948 (1993)
27. Nong, G., Zhang, S., Chan, W.H.: Linear suffix array construction by almost pure induced-sorting. In: Proceedings of the 2009 Data Compression Conference, DCC '09, pp. 193–202. IEEE Computer Society, Washington, DC (2009)
28. Palazón-González, V., Marzal, A.: Speeding up the cyclic edit distance using LAESA with early abandon. Pattern Recogn. Lett. (2015). http://dx.doi.org/10.1016/j.patrec.2015.04.013
29. Rivest, R.L.: Partial-match retrieval algorithms. SIAM J. Comput. **5**(1), 19–50 (1976). doi:10.1137/0205003
30. Schnattinger, T., Ohlebusch, E., Gog, S.: Bidirectional search in a string with wavelet trees and bidirectional matching statistics. Inf. Comput. **213**, 13–22 (2012)
31. Ukkonen, E.: On-line construction of suffix trees. Algorithmica **14**(3), 249–260 (1995)
32. Uliel, S., Fliess, A., Unger, R.: Naturally occurring circular permutations in proteins. Protein Eng. **14**(8), 533–542 (2001)
33. Weiner, P.: Linear pattern matching algorithms. In: Proceedings of the 14th Annual Symposium on Switching and Automata Theory (SWAT 1973), pp. 1–11. IEEE Computer Society, Washington, DC (1973)
34. Wu, S., Manber, U.: Fast text searching: allowing errors. Commun. ACM **35**(10), 83–91 (1992)

Chapter 4
De Novo NGS Data Compression

Gaetan Benoit, Claire Lemaitre, Guillaume Rizk, Erwan Drezen, and Dominique Lavenier

4.1 Introduction

During the last decade, the fast evolution of sequencing technologies has led to an explosion of DNA data. Every field of life science has been impacted. All of them offer new insights to approach many biological questions. The low sequencing cost is also an important factor that contributes to their successes. Hence, today, getting molecular information from living organisms is no longer a bottleneck. Sequencing machines can generate billions of nucleotides. On the other hand, managing this mass of data to extract relevant knowledge is becoming a real challenge.

The first step is simply to deal with the information files output by the sequencers. The size of these files can be huge and often ranges from tens to hundreds of gigabytes. Files have to be stored on the disk storage computing environment (1) to perform the genomic analysis and (2) to be archived. In both cases, the required space storage is high and asks for important hardware resources. Compressing *Next Generation Sequencing* (NGS) data is thus a natural way of significantly reducing the cost of the storage/archive infrastructure.

Another important point is the exchange of data. The first level is Internet communication. Practically, transferring files of terabytes through the Internet takes time and is fastidious (the larger the files, the longer the transfer time and the higher the probability to be interrupted). A 100 Mbit/s connection will require several hours to transfer a file of 100 GB. A second level is the internal computer network. Data are generally moved on specific storage bays. When required for processing, they are transferred to the computing nodes. If many jobs involving many different genomic datasets are run in parallel, the I/O computer network can be saturated, leading to

G. Benoit • C. Lemaitre • G. Rizk • E. Drezen • D. Lavenier (✉)
INRIA/IRISA, Rennes, France
e-mail: gaetan.benoit@inria.fr; claire.lemaitre@inria.fr; guillaume.rizk@inria.fr; erwan.drezen@inria.fr; dominique.lavenier@irisa.fr

M. Elloumi (ed.), *Algorithms for Next-Generation Sequencing Data*,
DOI 10.1007/978-3-319-59826-0_4

a drastic degradation of the overall computer performance. One way to limit the potential I/O traffic jam is to directly work with compressed data. It can be faster to spend time in decompressing data than waiting for uncompressed ones.

Compressing NGS data can also be an indirect way to speed up data processing. Actually, the sequencing coverage provides a significant redundancy that is exploited by DNA data compressors. In other words, similar sequences are more or less grouped together to optimize their description. In many treatments such as sequence comparison, genome assembly or *Single Nucleotide Polymorphism* (SNP) calling (through mapping operation), the aim is to find similarity between sequences. Thus, even if the final objective is different, the way data are manipulated is conceptually very similar. In that sense, a NGS data compressor can be viewed as a preprocessing step before further analysis. But to be efficient, the compressor should be able to directly transmit this preprocessing information to other NGS tools.

Hence, based on the above argumentation, the challenge of the NGS data compression is manifold:

1. Provide a compact format to limit computer storage infrastructure.
2. Provide a fast decoding scheme to lighten I/O computer network traffic.
3. Provide a format to directly populate a data structure suitable for further NGS data processing. Of course, these different criteria are more or less antagonistic, and compression methods have to make compromises or deliberately privilege one aspect.

Generic compression methods, such as GZIP [8], do not fit the above requirements. Compression is performed locally (i.e., without vision of the complete dataset) and, thus, cannot exploit the coverage redundancy brought by NGS data. However, due to the text format (ASCII) of the NGS files and the reduced alphabet of the DNA sequences, a compression rate ranging from 3 to 4 is generally obtained. This is far from negligible when such volume of data is involved. Furthermore, the great advantage of GZIP is that it is well recognized, stable, and that many NGS tools already take it as input.

Also, independently of not exploiting the global redundancy, GZIP-like methods cannot perform lossy compression. Indeed, NGS data include different types of information that can be more or less useful according to downstream processing. Each DNA fragment comes with additional information related to the technological process and the quality control. In some cases, it can be acceptable to suppress or degrade this information to improve the compression ratio. Hence, some compressors propose lossy options that essentially operate on these metadata [2, 4].

Another way to increase the compression ratio is to rely on external knowledge. If a reference genome is known, NGS data from the same or from a close organism can advantageously benefit from that information. In that case, the compression consists in finding the best mapping for all DNA fragments. Fragments that map are replaced by the coordinates where the best match has been found on the reference together with the differences. This chapter will not discuss any more of this type of compression as it is dedicated to de novo compression, i.e., compression without reference. But this approach must not be forgotten. Actually, it could be of great interest in the future as the number of sequenced genomes is growing every day.

The rest of the chapter is structured as follows: Sect. 4.2 gives an overview of standard string compression methods. Actually, as it will be seen in Sect. 4.3, NGS files include metadata that can be processed with generic compression methods. It is thus interesting to have a brief overview of these techniques. Section 4.3 is devoted to methods that have been specifically developed to compressed NGS data. Lossless and lossy compression are presented. Section 4.4 provides an evaluation of the different NGS compressors and proposes a few benchmarks. Section 4.5 concludes this chapter.

4.2 Generic Text Compression

All the methods compressing NGS data files somehow use techniques from generic text compression. These techniques can be adapted or tuned for compressing each component of the NGS files. Therefore, it is essential to know the bases of generic text compression techniques and how the popular methods work.

Text compression consists in transforming a source text in a compressed text whose bit sequence is smaller. This smaller bit sequence is obtained by providing fewer bits to the most frequent data, and vice versa. The compression can be lossless or lossy. Lossless techniques rebuild the original text from the compressed text, whereas lossy techniques only retrieve an approximation of the source. In some cases, the loss of information can be desirable, but this section focuses only on lossless methods.

As shown in Fig. 4.1, a text compressor is composed of an optional transformation algorithm, a model, and a coder. Transformation algorithms usually produce a shorter representation of the source or a reordering of the source. They can help the compressor in different ways: increasing the compression rate, improving its speed, or reducing its memory usage. The model chooses which data is more frequent in the source and estimates a probability distribution of the symbols of this source. The coder outputs a code for each symbol based on the probabilities of the model: the highest probable symbols are encoded by the fewest bit codes.

There are three main kinds of lossless text compression techniques: statistical-based, dictionary-based, and those based on the *Burrows-Wheeler Transform* (BWT). The first uses a complex modeling phase to accurately predict the symbols that often appear in the source. The more accurate are the predictions, the better will be the compression rate. The second and the third are transformation algorithms. Dictionary-based methods build a dictionary containing sequences of the source; each occurrence of these sequences is then replaced by its respective address in the dictionary. Finally, BWT-based methods rearrange the symbols of the source to improve the compression.

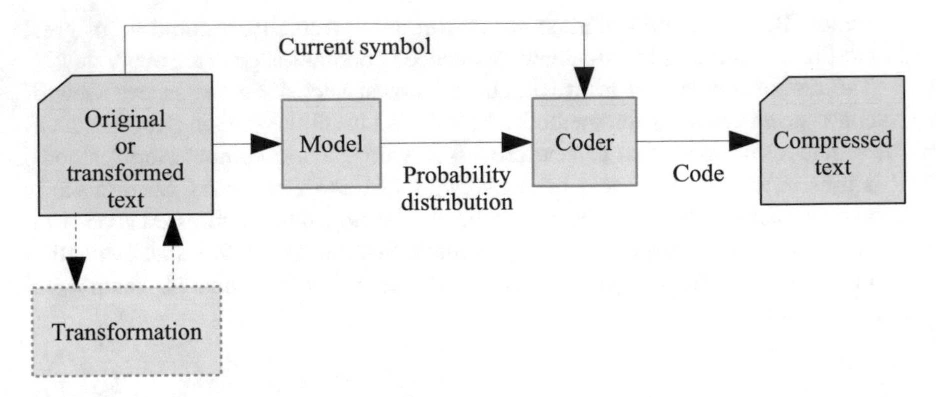

Fig. 4.1 Standard workflow of a text compressor. Firstly, the original text can be transformed to help further compression. The model then makes a probability distribution of the text, and the coder encodes the symbols based on their probability of occurrences

4.2.1 Coding

Coders output a code for each symbol depending on the probability distribution computed by the model. The codes must be chosen to provide an unambiguous decompression. The Shannon information theory [24] tells us that the optimal code of a symbol s of probability p has a bit length of $-\log_2(p)$.

The two most used coders are Huffman coding [11] and arithmetic coding [28]. Huffman coding starts by sorting the symbols based on their probabilities. A binary tree is then constructed, assigning a symbol to each of its leaves. The code of the symbols is obtained by traversing the tree from the root to the leaves. The tree is unbalanced because the codes produced by the encoder must be unambiguously decoded by the decompressor. Frequent symbols will be stored closed to the root, ensuring that frequent data will have the shortest codes, and vice versa.

In the rare case where the probability of a given symbol is a power of 1/2, Huffman coding produces optimal code length. In the other cases, this method does not strictly respect the Shannon entropy because the code length are integers.

Arithmetic coding does not bear this limitation and produces codes near the optimal, even when the entropy is a fractional number. Instead of splitting the source in symbols and assigning them a code, this method can assign a code to a sequence of symbols. The idea is to assign to each symbol an interval. The algorithm starts with a current interval [0, 1). The current interval is split in N subintervals where N is the size of the alphabet of the source. The size of the subintervals is proportional to the probability of each symbol. The subinterval of the current symbol to encode is then selected. The boundary of the current interval is narrowed to the range of the selected subinterval.

Splitting and selecting intervals continue until the last symbol of the source is processed. The final subinterval specifies the source sequence. The larger is this final interval, the fewer bits will be required to specify it, and vice versa.

Even if Huffman coding can be faster than arithmetic coding, arithmetic coders are most often preferred. The gain in compression rate is really worth despite the speed penalty.

4.2.2 *Modeling*

The model gathers statistics on the source and estimates a probability distribution of the symbols. A model can be static or dynamic. In the static case, a first pass over the source determines the probability distribution, which is used to compress the text in a second pass. The distribution must be stored in the compressed file to allow the decompressor to retrieve the original text. In the dynamic case, the distribution is updated each time a symbol is encoded. The compression is then executed in a single pass over the data. The distribution does not need to be transmitted to the decompressor because the information can be retrieved each time a symbol is decoded. This is a great advantage compared to static models because the distribution can sometimes be heavy. Arithmetic coders perform well on dynamic models, whereas Huffman coders are better suited to static models.

4.2.2.1 Basic Modeling

The most used model is a variation of Markov chains, the finite context models. In an order-N context model, the probability of the current symbol is given using the preceding N symbols. This kind of model relies on the fact that the same symbol will often appear after a specific context. A simple example can demonstrate the effectiveness of this model. In an English text, an $order-0$ model (no context) may assign a probability of 2.7% to the letter "u," whereas an $order-1$ model (context of size 1) can assign a probability of 97% to "u" within the context "q", meaning that "q" is almost always followed by "u." A symbol with such a high probability of occurrence can be encoded on one bit or less by the coders which are presented in the following section.

4.2.2.2 Statistical-Based Approach

These methods perform a complex modeling phase in order to maximize the efficiency of the coder, usually an arithmetic coder. We have seen in the previous section that order-N context models predict better the probability of the symbols. Increasing N increases the specificity of a context and, thus, the probabilities of the symbols following it. However, up to a point, the compression rate will decrease because each context has been seen only once in the source and no prediction can be made.

The *Prediction by Partial Matching* (PPM) [18] aims at solving this problem by using a variable order model. The probability of the current symbol is determined by finding the longest context such that it has already been seen. If the context of size N does not exist in the distribution, the model is updated and the $N-1$ context is tried out. The compressor reports the change of context by creating an escape symbol. These operations are repeated until the context of size 0 is reached. In this case, the same probability is assigned to the 256 possible symbols. The context mixing approach [17] goes further by combining the predictions of several statistical models. A lot of mixing strategies exist; the simplest one is to average the probabilities of each model. The resulting prediction is often better than the prediction of a single model.

The PAQ series [16] is a set of compressors which have implemented a lot of complex models. These tools achieve the best compression rates, but they are not so popular, mainly because of their lower speed.

4.2.3 Transforming

Transformation algorithms produce a shorter representation of the source or a reordering of the symbols to help further compression. After the transformation, a modeling and a coding phase are always required, but simpler models can be used.

4.2.3.1 Dictionary-Based Approaches

Dictionary-based methods choose sequences of symbols of the source and replace them by short code words. The idea is to store the sequences once in a data structure, called the dictionary, their code being simply their index in the dictionary.

Lempel and Ziv implemented LZ77 [31], the first algorithm based on dictionary coding. There are numerous variations of this algorithm, the key point being how they choose the sequences and how they store them in the dictionary. LZW [27] uses a dictionary of 4096 entries, each sequence is thus encoded on 12 bits. During compression, the longest sequence matching with the dictionary is searched. When such a match is found, the algorithm outputs the index of the sequence and a new entry is inserted in the dictionary, which is the concatenation of the sequence and the character following this sequence. The dictionary is initially filled with the 256 possible values of a byte in the 256 first entries. It allows the decompressor to rebuild the dictionary from the compressed data without having to store anything about it in the compressed file.

LZ variants are implemented in popular compression tools, such as GZIP [8] which uses a combination of LZ77 and Huffman coding. Their compression speed tends to be slow because of the construction of the dictionary, but their decompression speed, on the contrary, is very fast. This makes them very useful algorithms for text archiving.

4.2.3.2 BWT-Based Approaches

The BWT [3] is not directly a compression technique but a reordering of the symbols of the source. The transformation tends to group the identical symbols together when the source contains redundant sequences. A simple compression technique, such as *Run-Length Encoding* (RLE), can then be used to reduce the sequences of repeated symbols. RLE replaces these sequences by a couple representing the number of occurrences and the repeated symbol, for example, the string BBBBBBBCDDD will become 7B1C3D.

The BWT is obtained by sorting all the rotations of a sequence by alphabetical order. The last symbol of each rotation is concatenated to form the BWT. One of the main advantages of this technique is the reversibility property of the BWT. A compressor can then forget the original text and work only on the BWT.

The popular compressor BZIP2 [23] is based on the BWT. Its compression rate is better than the dictionary-based methods, but the execution time is slower for both compression and decompression.

4.3 NGS Data Compression

This section focuses on the compression of raw data provided by high throughput sequencing machines. After a short description of the NGS data format that mainly contains quality and DNA sequence information, lossless and lossy compression techniques are presented.

4.3.1 Introduction

The sequencing machines are highly parallel devices that process billions of small DNA biomolecules. A sequencing run is a complex biochemical protocol that simultaneously "reads" all these biomolecules. This reading is performed by very sensitive captors that amplify and transform the biochemical reactions into electronic signals. These signals are analyzed to finally generate text files that contain string sequences over the A, C, G, T alphabet, each of them representing one nucleotide. In addition, depending of the quality of the signal, a quality score is associated to each nucleotide.

The standard file format output by the sequencing machines is the FASTQ format. This is a list of short DNA strings enhanced with quality information. Each sequence has also a specific identifier where different information related to the sequencing process is given. An example of an FASTQ file with only two DNA sequences is:

```
@FCC39DWACXX:4:1101:12666:1999#CTAAGTCG/1
NGCCGAAACTTAGCGACCCCGGAAGCTATTCAATTACATGTGCTATCGGTAACATAGTTA
+
BPacceeeggggfhifhiiihihfhiifhhigiiiiiiiffgZegh`gh`geggbeeeee
@FCC39DWACXX:4:1101:16915:1996#CTAAGTCG/1
NTGACTTCGGTTAAAATGTTAAGTTATGGACGAGTTTGAGTTTGTGATTTTAATCTTTCA
+
BPacceeeggggiiiiihhhhihhiiihiiihicggihichhihiiihihiiiiiiih
```

The line starting with the character '@' is the sequence header. It gives the metadata associated to the DNA sequence and to the sequencing process. The following line is the DNA sequence itself and is represented as a list of nucleotides. The 'N' character can also appear for nucleotides that have not been successfully sequenced. The line '+' acts as a separator. The last line represents the quality. The number of symbols is equal to the length of the DNA sequence. A quality value ranging from 0 to 40 is associated to each symbol, each value being encoded by a single ASCII symbol.

Hence, we may consider that NGS files contain three different types of information. Each of them has its own properties that can be exploited to globally optimize the compression. Headers are very similar to each other, with a common structure and fixed fields. They can therefore be efficiently compressed with generic text compression methods described in the previous section. The DNA sequence stream contains redundant information provided by the sequencing coverage, but this information is spread over the whole file. Therefore, sequences must first be grouped together to benefit from this redundancy. Finally, quality compression is probably the most challenging part since there are no immediate features to exploit.

Actually, the quality information may highly limit the compression rate. So, it is legitimate to wonder whether this information is essential to preserve or, at least, if some loss in quality would be acceptable. A first argument is that quality information is often skipped by a majority of NGS tools. A second one is that the computation of the quality scores with such precision (from 0 to 40) does not reflect the real precision of sequencers. The range could be reduced to lower the number of possible values with very minor impact to the incoming processing stages but with a great added value on compression. Recent studies have even shown that reducing the quality scoring can provide better results on SNP calling [30]. Removing the quality, re-encoding the quality score on a smaller scale, or modifying locally the score to smooth the quality information are also interesting ways to increase the compression rate.

The two next subsections describe in detail quality and NGS sequence compression methods. The header part compression is no longer considered since generic text compression techniques, as described Sect. 4.2, perform well.

4.3.2 *Quality Compression*

For each nucleotide, the FASTQ format includes a quality score. It gives the "base-calling" confidence of the sequencer, i.e., the probability that a given nucleotide has been correctly sequenced. In its current format, quality scores Q are expressed in an integer scale ranging from 0 to 40, derived from the probability p of error with $Q = -10\log_{10} p$. These scores are usually named *Phred scores*, after the name of the first software that employed it. In the uncompressed FASTQ file, these scores are encoded with ASCII characters. Each DNA sequence is accompanied by a quality sequence of the same length; therefore, quality strings take the same space as DNA in the original file. Its compression is more challenging than DNA compression in several aspects. Firstly, its alphabet is much larger ($\approx$40 symbols instead of four); secondly, it does not feature significant redundancy across reads that could be exploited; and lastly, it often looks like a very noisy signal.

We distinguish between two different compression schemes: *lossless* and *lossy*. *Lossless* mode ensures that the decompressed data exactly match the original data. On the contrary, *lossy* compressors store only an approximate representation of the input data but achieves much higher compression rates. *Lossy* compression is traditionally used when the input data is a continuous signal already incorporating some kind of approximation (such as sound data). At first, NGS compression software were mostly using *lossless* compression, then several authors explored various *lossy* compression schemes. *Lossy* compression schemes can themselves be roughly divided in two categories: schemes that work only from the information contained in the quality string and schemes that also exploit the information contained in the DNA sequence to make assumptions about which quality score can safely be modified.

4.3.2.1 Lossless Compression

Context Model and Arithmetic Coding

It has been observed that in most sequencing technologies, there is a correlation between quality values and their position, and also a strong correlation with the nearby preceding qualities. The use of an appropriate context model can therefore help predict a given quality value. QUIP uses a third-order context model followed by arithmetic coding to exploit the correlation of neighboring quality scores, DSRC2 uses an arithmetic coder with context length up to 6, and FQZCOMP also uses a mix of different context models to capture the different kind of correlations [2, 13, 20].

Some contexts used by FASTQZ to predict a quality Q_i are, for example, Q_{i-1} (immediate preceding quality score), $\max(Q_{i-2}, Q_{i-3})$, whether Q_{i-2} equals Q_{i-3} or not, and $\min(7, i/8)$, i.e., reflecting that the score is dependent on position. These contexts are heuristics and may perform differently on datasets generated by varying sequencing technologies. The context model is then generally followed by an arithmetic coder, coding highly probable qualities with fewer bits.

Transformations

Wan et al. explored different lossless transformations that can help further compression: *min shifting* and *gap translating* [26]. With *min shifting*, quality values are converted to $q - q_{\min}$, with $q_{\min}$ the minimum quality score within a block. With *gap translating*, qualities are converted to $q - q_{\text{prec}}$. The objective of such transformations is to convert quality to lower values, possibly coded with fewer bits. This can act as a preprocessing step for other methods.

4.3.2.2 Lossy Compression

One major difficulty of quality compression comes from the wide range of different qualities. The first simple idea to achieve a better compression ratio is to reduce this range to fewer different values.

Wan et al. explored the effects of three lossy transformations: *unibinning*, *log binning*, and *truncating* [26]. In the first transformation, the error probabilities are equally distributed to N bins. For example with $N = 5$, the first bin spans quality scores with probability of error from 0% to 20%. All quality values in that bin will be represented by the same score. In the second transformation, the logarithm of error probabilities is equally divided into N bins. This respects the spirit of the original *Phred* score; if original *Phred* scores are represented with values from 0 to 63, a log binning with $N = 32$ amounts to convert each uneven score to its lower nearest even value. Several software use similar discretization schemes, such as FASTQZ and FQZCOMP in their lossy mode [2].

Other approaches exploit information in the read sequences to modify quality values. Janin et al. assume that if a given nucleotide can be completely predicted by the context of the read, then its corresponding quality value can be aggressively compressed or even completely discarded. They first transform the set of reads to their BWT to make such predictions from the *Longest Common Prefix* (LCP) array [12].

Yu et al. also exploit the same idea but without computing the time-consuming BWT of reads [30]. Instead, they compute a dictionary of commonly occurring k-mers throughout a read set. Then, they identify in each read k-mers within small Hamming distances of commonly occurring k-mers. Position corresponding to differences from common k-mers are assumed to be SNP or sequencing errors, and their quality values are preserved. Other quality scores are discarded, i.e., replaced by a high-quality score. However, their method scales to a very large dataset only if the dictionary of common k-mers is computed on a sample of the whole dataset. LEON [1] also exploits a similar idea. It counts the frequency of all k-mers in the read set and automatically computes the probable solidity threshold above which k-mers are considered to be error-free. Then, if any given position in a read is covered by a sufficient number of solid k-mer, its quality value is discarded. If the initial quality score is low, then a higher number of solid k-mers is required in order to replace the quality score with a high value. The underlying idea is that replacing a low quality score with a high score may be dangerous for downstream NGS analysis and therefore should be conducted carefully. These approaches are obviously *lossy* since they change the original qualities. However, modifying the quality values based on the information extracted from the reads means that some quality scores are actually *corrected*. This can be viewed as an improvement instead of a loss and explains the improvements of downstream NGS analysis discussed in [30].

4.3.3 DNA Sequence Compression

The simplest compression method for DNA sequences consists in packing multiple nucleotides in a single byte. The DNA alphabet is a four letter alphabet A, C, G and T, and each base can thus be 2 bit encoded. The N symbol (that actually rarely appears) does not necessarily need to be encoded because its quality score is always 0. During decompression, the symbol is simply inserted when a null quality score is seen. Packing four nucleotides per byte represents, however, a good compression rate, decreasing the size of the DNA sequences by a factor of 4.

To get a better compression ratio, two main families of compression techniques have been developed: reference-based methods and reference-free methods, also called de novo methods. The idea of the reference-based methods is to store only the differences between already known sequences. On the contrary, The de novo methods do not use external knowledge. They extract the compression rate by exploiting the redundancy provided by the sequencing coverage. The compression rate of the reference-based methods can be significantly better, but they can be only used if similar sequences are already stored in databases. This is generally not the case when sequencing new species.

This chapter only focuses on de novo compression methods that, from an algorithmic point of view, are more challenging. They have first to quickly extract redundancy from the dataset and have also to deal with sequencing errors that actually disrupt this redundancy. De novo DNA sequence compression methods fall into three categories:

1. Approaches using a context model to predict bases according to their context
2. Methods that reorder the reads to boost generic compression approaches
3. Approaches inspired from the reference genome methods.

Historically, the first DNA compressors were only improvements of generic text compression algorithms and belonged to the context model and reordering categories. It is only recently that new methods appeared that fully exploit algorithms and data structures tailored for the analysis of NGS data, such as the *de Bruijn Graph*, mapping schemes, or the BWT.

4.3.3.1 Statistical-Based Methods

These methods start with a modeling phase in order to learn the full genome structure, i.e., its word composition. With high-order context models and a sufficient sequencing coverage, the models are able to accurately predict the next base to encode.

G-SQZ [25] is the first tool to focus on the de novo compression of NGS data. It achieves a low compression rate by simply using an order 0 model followed by a Huffman coder. QUIP [13] and FQZCOMP [2] obtained, by far, a better compression rate by using a higher-order context model of length up to 16 and an arithmetic coder. FASTQZ [2] grants more time to the modeling phase by using a mix of multiple context models. DSRC2 [7] can be seen as an improvement of GZIP adapted for NGS data. Its best compression rate is obtained by using a context model of order 9 followed by a Huffman coder. But the proposed fully multi-threaded implementation makes it the fastest of all NGS compression tools.

The efficiency of these methods is strongly correlated with the complexity and the size of the target genome. On large genomes, such as the human one, an order 16 context model will not be sparsely populated and will see more than one of the four possible nucleotides after a lot of context of length 16, resulting in a poor compression rate. A solution will be to use higher-order models, but a context model is limited by its memory usage of 4^N for the simplest models. These methods also cannot deal with sequencing errors. These errors cause a lot of unique contexts which will just degrade the compression rate.

4.3.3.2 Read Reordering Methods

The purpose of reordering reads is to group together similar reads in order to boost the compression rate obtained by a generic compression tool, such as GZIP. In fact, dictionary- and BWT-based compressors only work on small blocks of data before clearing their indexing structure. This is a solution to prevent a too high memory usage. In the case of DNA sequence compression, such tools miss the whole redundancy of the source because similar reads can be anywhere in the dataset. If the reordering is accurate, some dedicated transformation algorithms can also be used to encode a read using one of its previous reads as reference.

RECOIL [29] constructs a similarity graph where the nodes store the sequences. A weighted edge between two nodes represents their number of shared k-mers. Subsets of similar reads are retrieved by finding maximal paths in the graph. Finally, the largest matching region between similar reads is determined by extending their shared k-mers. FQC [21] also identifies similar reads by their shared k-mers. But this time, the reads are clustered together. And for each cluster, a consensus sequence is built by aligning all of its reads. The compression algorithm uses the consensus as reference to encode the reads. Both methods are time or memory consuming. They have been applied only on small datasets and are hardly scalable to today's volumes of data.

BEETL [6] proposes an optimization of the BWT which is able to scale on billions of sequences. Classical BWT algorithms cannot handle as many sequences because of their memory limitations. Here, much of the memory is saved during the process by making use of sequential reading and writing files on disk. Cox et al. [6] also showed that the compression ratio is influenced by the sequence order. In fact, each sequence in the collection is terminated by a distinct end-marker. Depending on the assignment of these end-markers to each sequence, hence depending on the order of the sequences in the collection, the BWT can result in longer runs of the same character. The most compressible BWT is obtained by reversing the sequences and sorting them by lexicographic order. To avoid a time- and memory-consuming preprocessing step of sorting all sequences, the algorithm makes use of a bit-array (called SAP) indicating whether a given suffix is the same as the previous one (except from the end-marker). This enables to sort only small subsets of sequences after the construction of an initial BWT, in order to locally update the BWT.

SCALCE [10] and ORCOM [9] first partition all reads in several bins stored on disk. Each bin is identified by a core string with the idea that reads sharing a large overlap must have the same core string. In SCALCE, the core substrings are derived from a combinatorial pattern matching technique that aims to identify the *building blocks* of a string, namely the *Locally Consistent Parsing* (LCP) method [22]. In ORCOM, the chosen core string is called a minimizer, which is the lexicographically smallest substring of size m of a read. When the disk bins are built, they are sorted by lexicographical order, with respect to the core string position, to move the overlapping reads close to each other. SCALCE uses generic compression tools, such as GZIP, to compress each bin independently, whereas ORCOM uses a dedicated compression technique: each read is encoded using one of its previous reads as reference, the best candidate being the one which has the fewest differences with the current read and is determined by alignment anchored on the minimizer. The differences are encoded using an order 4 PPM model followed by an arithmetic coder.

Reordering methods are well suited to exploit the redundancy of high throughput sequencing data. Some of them can avoid memory problems on large datasets by reordering the reads on disk. For a long time, the main drawback of these methods has been their compression and decompression speed, but this problem has been solved recently by ORCOM which is faster and obtains a better compression rate than the other methods. However, it is important to notice that reordering methods is a form of lossy compression. In fact, most of these tools do not restore the original read order, and this can be an issue for datasets containing paired reads (which are the most common datasets). All NGS tools relying on paired reads information (read mapping, de novo assembly, structural variation analysis. . .) can therefore not work on data compressed/decompressed with these tools.

4.3.3.3 Assembly-Based Methods

Contrary to the two above categories, methods falling in this third category do not try to find similarities between the reads themselves but between each read and a go-between, called a reference. These approaches are largely inspired from the ones relying on a reference genome.

Reference-based methods require external data to compress and decompress the read file, that is, a known reference sequence assumed to be very similar from the one from which reads have been produced. The idea is to map each read to the reference genome and the compressed file only stores for each read its mapping position and the potential differences in the alignment instead of the whole read sequence. However, as already mentioned, relying on external knowledge is extremely constraining, and this knowledge is not available for numerous datasets (e.g., de novo sequencing or metagenomics). To circumvent these limitations, the scheme of some de novo methods is to infer the reference from the read data by itself and then apply the reference-based approaches.

Conceptually, there are no differences between assembly-based methods and reference-based methods. In the first case, however, an assembly step is needed to build a reference sequence. Once this reference is obtained, identical mapping strategies can then be used. An important thing to note is that the reference sequence may not necessarily reflect a biological reality. It is hidden from the user point of view and is only exploited for its compression features.

Two methods follow this strategy: QUIP [13] and LEON [1]. The main difference lies in the data structure holding the reference: in QUIP the reference is a set of sequences, whereas in LEON it is a graph of sequences, namely a *de Bruijn Graph*.

In QUIP, the reference set of sequences is obtained by de novo assembly of the reads. Since de novo assembly is time and memory consuming, only a subset of the reads (the first 2.5 million by default) are used for this first step. The reads are then mapped on the assembled sequences, called contigs, using a seed-and-extend approach. A perfect match of size 12 between the contigs and the current read is first searched. The remaining of the alignment is then performed; the chosen contig is the one which minimizes the Hamming distance. The read is then represented as a position in the chosen contig and the eventual differences.

In LEON, a *de Bruijn Graph* is built from the whole set of reads. The *de Bruijn Graph* is a common data structure used for de novo assembly. It stores all words of size k (k-mers) contained in a given set of sequences as nodes and puts a directed edge between two nodes when there is a $k - 1$ overlap between the k-mers. When k-mers generated by sequencing errors are filtered out and the value of k is well chosen (usually around 30), the *de Bruijn Graph* is a representation of the reference genome, where each chromosome sequence can be obtained by following a path in the graph. In LEON each read is encoded by a path in such a graph, that is, an anchoring k-mer and a list of bifurcations to follow when the path encounters nodes with out-degree greater than 1.

The main difference between these two methods is that QUIP executes a complete assembly of the reads, whereas LEON stops at an intermediate representation of the assembly which is the *de Bruijn Graph*. In order to decompress the data, the reference must be stored in the compressed file. A priori, storing contigs may seem lighter than storing the whole graph. In fact, a naive representation storing each base on 2 bits costs 60 bits per 30-mer which is not acceptable for a compression purpose.

The LEON's strategy is possible thanks to a light representation of the *de Bruijn Graph* [5, 19] which stores each k-mer on only a tenth of bits. The mapping results should be better for LEON than QUIP since a part of the reads cannot be aligned on contigs: those being at contigs extremities and those which are not assembled.

4.3.4 Summary

Table 4.1 shows a summary of the compressors presented in this section. Tools are classified in three main categories: statistical-based methods, reordering methods, and assembly-based methods. A brief overview of the compression techniques used

Table 4.1 Summary of compression methods and software, classified according to their method for the DNA stream

Software	Year	Methods	Quality	Random access	Multi-threaded	Remarks
Statistical-based methods						
G-SQZ [25]	2010	Huf		Yes	No	
DSRC [7, 20]	2014	Markov, Huf	Lossless /lossy	Yes	Yes	
FASTQZ [2]	2013	CM, AC	Lossless/lossy	No	No	
FQZCOMP [2]	2013	Markov, CM, AC	Lossless/lossy	No	No	
Reordering methods						
RECOIL [29]	2011	Generic		No	No	FASTA only
FQC [21]	2014	PPM, AC	Lossless	–	–	No impl
BEETL [6]	2012	BWT, PPM		Yes	No	FASTA only
SCALCE [10]	2012	Generic	Lossless/lossy	No	Yes	
ORCOM [9]	2014	Markov, PPM, AC		No	Yes	Seq only
Assembly-based methods						
QUIP [13]	2012	Markov, AC	Lossless	No	No	
LEON [1]	2014	Markov, AC	Lossless/lossy	No	Yes	
Quality only methods						
RQS [30]	2014	k-mer dictionary	Lossy	No	No	
BEETL [12]	2013	BWT, LCP	Lossy	No	No	
LIBCSAM [4]	2014	Block smoothing	Lossy	No	No	

Random access is the ability for the compressor to retrieve a sequence without the need to decompress the whole file.
Huf: Huffman coding, Markov: context model, CM: context mixing, AC: arithmetic coding, Generic: generic compression tools such as GZIP and BZIP2, PPM: prediction by partial matching, Seq: sequence, LCP: longest common prefix array, FASTA: FASTQ format without the quality stream, No impl: no implementation available

is also given. In the Random access column, one can identify the tools that propose the additional feature of quickly retrieving a given read without decompressing the whole file, such as G-SQZ and DSRC.

BEETL is more than a compressor because its BWT representation can be used for popular indexation structures such as the FM-index. Consequently, BEETL can quickly deliver a list of sequences containing a given *k*-mer.

Some compressors such as G-SQZ and RECOIL are already outdated. They are not multi-threaded and do not scale on the current datasets. FQZCOMP and FASTQZ can use a maximum of three cores to compress the three streams in parallel. This solution was not qualified as multi-threaded in Table 4.1 since it is limited and it provides unbalanced core activity, the header stream being actually much faster to compress than the two other streams. To take full advantage of multi-core architectures, a solution is to process the data by blocks and compress them independently, as this is the case in DSRC or LEON. Depending on the method, implementing such a solution may not be straightforward if one wants to preserve good compression factors.

Compression tools are not generally focusing on one specific kind of NGS data. One exception is PATHENC which seems to be specialized for RNA-seq data, although it can still be used on any data [13]. A comparative study of compression performance on different kinds of NGS data (whole genome, exome, RNA-seq, metagenomic, ChIP-Seq) is conducted in the LEON and QUIP papers [1, 13]. It shows without surprise that the best compression is obtained on data with the highest redundancy.

Some tools cannot compress the FASTQ format. RECOIL and BEETL only accept the FASTA format which is a FASTQ format without quality information. ORCOM only processes DNA sequences and discards header and quality streams. FQC cannot be tested because there is no implementation of the method available.

4.4 Evaluation of NGS Compressors

This section aims to illustrate the compression methods presented in the previous section by evaluating current NGS compressors. The capacity to compress or decompress files is only tested, even if some tools have advanced functionalities such as the possibility to randomly access sequences from the compress data structure. For each tool, the evaluation process followed this protocol:

1. Clear the I/O, cache system'
2. Compress the original file *A*
3. clear the I/O cache system
4. decompress the compressed file(s) into file *B*
5. generate metrics

NGS compressors deal with large files, making them I/O data intensive. The way the operating systems handle the read and write file operations may have a significant impact on the compression and decompression time measurements. More precisely, if many compressors are sequentially tested on the same NGS file, the first one will have to read data from the disk, whereas the next ones will benefit of the I/O cache system. Thus, to fairly compare each software and to have a similar evaluation environment, we systematically clear the I/O cache before running compression and decompression. All tools were run on a machine equipped with a 2.50 GHz Intel E5-2640 CPU with 12 cores and 192 GB of memory. All tools were set to use up to eight threads.

4.4.1 *Metrics*

Remember that the FASTQ format is made of:

- (d) a genomic data stream
- (q) a quality data stream as phred scores associated to the genomic data
- (h) an header data stream as a textual string associated to each read

We now define several metrics allowing compressor software to be fairly evaluated.

Compression factor (CF): This is the principal metric. We define a compression factor (CF) as:

$$\mathrm{CF}(s) = \frac{\text{size of stream } s \text{ in the original file}}{\text{size of stream } s \text{ in the compressed file}}$$

More generally, we define a global compression factor for the whole file:

$$\mathrm{CF} = \frac{\text{size of the original file}}{\text{size of all streams in the compressed file(s)}}$$

Stream correctness (SC): Depending of the nature of the stream, the correctness is computed differently. For the DNA data stream, we check that the DNA sequences after the compression/decompression steps are identical. The idea is to perform a global checksum on the DNA sequences only. The DNA data stream correctness is calculated as follows:

$$\mathrm{SC}(d) = \begin{cases} \text{ok if } \mathrm{Checksum}(A, d) = \mathrm{Checksum}(B, d) \\ \text{ko if } \mathrm{Checksum}(A, d) \neq \mathrm{Checksum}(B, d) \end{cases}$$

Note that A is the original file and B is the decompressed file.

For the header and the quality streams, the correctness depends on the lossy or lossless options. It also depends on the way software handle header compression. We therefore define the stream correctness metric as the following percentage (0 means correct stream):

$$\mathrm{SC}(s) = \frac{\sum_i nb \text{ of mismatches of } (ra_i, rb_i) \text{ in stream } s}{\text{size of stream } s} \times 100$$

where

$$\begin{cases} s &= \text{one of the streams } (q),\ or\ (h) \\ ra_i &= i\text{th read of } A \\ rb_i &= i\text{th read of } B \end{cases}$$

Execution time: This is the execution time for compression or decompression expressed in second.

Memory peak: This is the maximum memory used (in MBytes) to compress or decompress a file.

4.4.2 Benchmarks

4.4.2.1 High Coverage Benchmarks

NGS compressors are tested with the three following high coverage datasets extracted from the SRA database:

- Whole genome sequencing of the bacteria *E. coli* (genome size $\sim$ 5 Mbp): 1.3 GB, 116× coverage (SRR959239)
- Whole genome sequencing of the worm *C. elegans* (genome size $\sim$ 100 Mbp): 17.4 GB, 70× coverage (SRR065390)
- Whole genome sequencing of a human individual (genome size $\sim$ 3 Gbp): 732 GB, 102× coverage (SRR345593/4)

They are real datasets from high throughput sequencing machines (here, Illumina, with around 100 bp reads) and are representative of NGS files that are generated daily. Table 4.2 summarizes the evaluation. Columns have the following meaning:

- **Prog**: name of the NGS compressor. Tools allowing lossy compression on the quality stream are labeled with a * suffix.
- **Factor**: total compression factor, followed by the compression factor for each stream (header, DNA sequence and quality). Note that some tools like GZIP do not have specific stream factors; in such a case, we display only the total factor. The tool ORCOM is specific because it compresses only the DNA sequence stream.
- **Correctness**: results on header, genomic data, and quality streams are reported (0 means correct stream).

Table 4.2 High coverage benchmarks for *E. coli*, durability, and human FASTQ files

Prog	Factor				Correctness				Compress		Decompress	
	Main	hdr	seq	qlt	Sum	hdr	seq	qlt	Time (s)	mem. (MB)	Time (s)	mem. (MB)
WGS E. coli—1392 MB—116×												
GZIP	3.9	–	–	–	ok	0	0	0	188	1	25	1
BZIP2	4.9	–	–	–	ok	0	0	0	164	9	83	6
DSRC	6.0	–	–	–	ok	0	0	0	21	1995	35	2046
DSRC*	7.6	–	–	–	ok	0	0	81	12	1942	30	1996
FQZCOMP	9.9	35.2	12.0	5.7	ok	0	0	0	64	4424	66	4414
FQZCOMP*	17.9	35.2	12.0	19.6	ok	0	0	95	64	4165	69	4159
FASTQZ	10.3	39.9	13.8	5.6	ko	0	75	0	290	1347	320	1347
FASTQZ*	13.4	39.9	13.8	8.6	ko	0	75	40	248	1347	286	1347
LEON	8.4	45.1	17.5	3.9	ok	0	0	0	106	404	45	277
LEON*	30.9	45.1	17.5	59.3	ok	0	0	96	49	390	44	239
QUIP	8.4	29.8	8.5	5.3	ok	0	0	0	206	990	205	807
SCALCE	8.9	21.0	12.9	5.0	ok	35	75	57	85	1993	47	1110
ORCOM	–	–	33.51	–	ok	100	75	100	10	2212	15	181
FASTQZ	10.9	40.8	23.5	5.7	ko	0	75	0	306	1316	301	1318
WGS C. elegans—67 GB—70×												
GZIP	3.8	–	–	–	ok	0	0	0	2218	1	301	1
BZIP2	4.6	–	–	–	ok	0	0	0	1808	9	1216	6
DSRC	5.8	–	–	–	ok	0	0	0	200	5355	342	5626
DSRC*	7.9	–	–	–	ok	0	0	86	109	5156	271	4993
FQZCOMP	8.1	54.2	7.6	5.2	ok	0	0	0	931	4424	927	4414
FQZCOMP*	12.8	54.2	7.6	15.0	ok	0	0	86	921	4169	996	4155
FASTQZ	7.9	61.9	7.3	5.1	ok	0	0	0	4044	1533	3934	1533

(continued)

Table 4.2 (continued)

Prog	Factor				Correctness				Compress		Decompress	
	Main	hdr	seq	qlt	Sum	hdr	seq	qlt	Time (s)	mem. (MB)	Time (s)	mem. (MB)
FASTQZ*	10.3	61.9	7.3	8.7	ok	0	0	76	3703	1533	3312	1533
LEON	7.3	48.6	12.0	3.7	ok	0	0	0	1168	1885	446	434
LEON*	21.3	48.6	12.0	32.9	ok	0	0	86	704	1886	442	417
QUIP	6.5	54.3	4.8	5.2	ok	0	0	0	823	782	764	773
SCALCE	7.7	16.5	10.0	4.7	ok	38	73	58	1316	5285	526	1112
ORCOM	–	–	24.2	–	ok	94	73	100	283	9488	324	1826
FASTQZ	10.4	63.4	19.2	5.2	ok	0	0	0	3831	1500	3441	1500
WGS Human–732 GB—102×												
GZIP	3.26	–	–	–	ok				104,457	1	9124	1
BZIP2	4.01	–	–	–	ok				69,757	7	34,712	4
DSRC	4.64	–	–	–	ok				1952	408	2406	522
DSRC*	6.83	–	–	–	ok				1787	415	2109	506
FQZCOMP	5.35	23.2	4.53	4.2	ok				18,695	79	29,532	67
FQZCOMP*	8.34	23.2	4.53	14.9	ok				20,348	80	24,867	67
LEON	5.65	27.54	9.17	3.03	ok				61,563	9607	23,814	
LEON*	15.63	27.54	9.17	27.1	ok				38,860	10,598	21,687	5870
QUIP	5.25	16.95	4.47	4.2	ok				52,855	798	46,595	791
ORCOM	–	–	19.2	–	ok				29,364	27,505	10,889	62,009

Programs with a * are run in lossy compression mode. The overall compression factor is given (main) followed by the compression factor of each stream. The correctness (expressed as a % of differences) is also given for each stream

- **Compression**: system metrics for the compression; execution time (in seconds) and the memory peak (in MBytes)
- **Decompression**: same metrics as above.

The first comment on these results is that, in general, the specific NGS compressors perform better than the generic ones (GZIP and BZIP2), in terms of both compression rate and execution time. The main reason is that these tools exploit the features of NGS data and especially the redundancy provided by the sequencing coverage. This redundancy is spread over the whole file and cannot be locally captured by generic text compressors.

We can also observe that the header streams are always well compressed. As already mentioned, no specific methods have been developed for this data stream, which is mainly composed of repetitive motifs. The generic methods provide near optimal compression, and this stream does not constitute the critical part of the NGS data compression challenge.

Concerning the DNA stream, the compression factor depends on the read dataset. It is expected to depend on the read coverage, but the table highlights that it depends also on the size and complexity of the sequenced genome from which the reads are generated; compression factors are the worst for the human dataset, even when compared to the *C. elegans* dataset which has a lower coverage (thus less redundancy). This can be explained by the higher complexity of this genome, i.e., mainly its amount of repeated sequences. Most compressors try to learn a model representative of the genome, that is, its composition in words of size k that would enable to predict a given word knowing its preceding one: the more the genome contains large repeated sequences (larger than k), the more difficult it is to predict its sequence.

Interestingly, FASTQZ used with a reference genome does not always have the best compression ratio. On the *C. elegans* dataset, ORCOM outperforms FASTQZ on the DNA sequence stream. However, strictly speaking, the comparison is not perfectly fair since ORCOM reorders the reads. The decompressed set of reads remains the same but is stored in a different order. For technologies that provide pair-end or mate-pair reads in two different files, this strategy cannot be used.

When the lossy option is activated, the compression ratio can be significantly improved. Techniques that do not consider independently compression for the DNA sequence and the quality streams but try to exhibit correlations between the two streams are particularly powerful as demonstrated by the LEON compressor.

4.4.2.2 Metagenomic Benchmark

This test aims to demonstrate that NGS compressors are not well suited for any kind of NGS datasets. In particular, they can perform poorly on NGS metagenomic data. The main reason is that these datasets generally contain low redundancy. Table 4.3 confirms that the compression rate of DNA sequences is low and is similar to generic compressors. The lossy mode, however, allows FASTQ files to be significantly compressed.

Table 4.3 Metagenomic benchmark: 15 GB (SRR1519083)

	Factor				Check				Compression		Decompression	
Prog	Main	hdr	seq	qlt	Sum	hdr	seq	qlt	Time (s)	mem. (MB)	Time (s)	mem. (MB)
GZIP	3.4	–	–	–	ok	0	0	0	2206	1	153	1
BZIP2	4.4	–	–	–	ok	0	0	0	1578	8	701	4
DSRC	5.0	–	–	–	ok	0	0	0	33	345	25	307
DSRC*	6.8	–	–	–	ok	0	0	78	29	330	39	285
FQZCOMP	5.8	32.6	4.5	4.2	ok	0	0	0	360	78	582	67
FQZCOMP*	8.9	32.6	4.5	14.8	ok	0	0	90	343	80	456	67
FASTQZ	6.1	30.0	4.9	4.3	ko				3318	1527		
FASTQZ*	7.3	30.0	4.9	6.6	ko				2776	1527		
LEON	4.8	36.8	4.3	3.1	ok	0	0	0	955	1870	438	1544
LEON*	9.0	36.8	4.3	19.8	ok	0	0	91	606	1876	460	1511
QUIP	5.7	25.6	4.7	4.1	ok	0	0	0	966	774	908	773
ORCOM	–	–	6.9	–	ok	94	74	89	230	9442	178	1698

Programs with a * are run in lossy compression mode

4.4.2.3 SNP Calling Evaluation

This last part evaluates the loss in quality—or the gain in quality—of downstream processing after a lossy compression. The chosen bioinformatics task is the SNP calling. The evaluation protocol is the following:

- Compress file *A* in lossy mode.
- Decompress file *A* into file *B*.
- Process file *B* for SNP calling.
- Compute Recall/Precision metrics.

In this experiment, SNPs are called with BWA aligner followed by *samtools mpileup* [14, 15]. Precision/Recall are computed from a validated set of SNPs from human chromosome 20, coming from the "1000 genomes project", on individual HG00096, read set SRR062634.

Five lossy compression tools, representative of all categories of lossy compression methods are compared. A lossless method was also included in the test, as well as the results obtained when discarding all quality scores; in this case, all quality values were replaced by a high value, "H," for the SNP calling procedure. Results are shown in Table 4.4.

First, it can be seen that qualities are indeed useful for SNP calling; the *no quality* test has a significant drop in precision, from 85.02 in lossless to 57.7%. Then, the tools FASTQZ, LIBCSAM, and FQZCOMP achieve a better compression factor than lossless, with only a slight degradation in precision/recall. Lastly, tools using information coming from the reads, such as LEON and RQS, achieve both a high compression factor and an improvement in precision and recall compared to the original dataset. This can be explained by the fact that during compression, some qualities are in fact *corrected* using read information.

Table 4.4 Recall and precision of SNP calling after quality values were modified by several lossy compressors

HG00096 chrom 20			
Prog	Precision	Recall	CF
No quality	57.73	68.66	–
Lossless	85.02	67.02	*2.95
FASTQZ	85.46	66.63	5.4
LIBCSAM	84.85	67.09	8.4
FQZCOMP	85.09	66.61	8.9
LEON	85.63	67.17	11.4
RQS	85.59	67.15	12.4

Programs with a * are run in lossy compression mode. CF is the compression factor. No quality means that qualities were discarded by compression, all replaced by "H." For the *lossless* line, the best compression factor obtained by lossless compression tools is given (obtained here with FQZCOMP)

4.5 Conclusion

Due to the recent emergence of NGS data, de novo NGS compression is still an open and active research field. A large number of methods have already been proposed and have shown that NGS data redundancy can bring a high compression rate. In addition, if lossy compression of the quality information is permitted, much better compression can be reached with, in some cases, an improvement in the postprocessing quality.

Generally, performances of the NGS compressors are first evaluated by their compression rate. In lossless mode, the compression rate ranges from 5 to 10. The differences come both from the compression methods and the data themselves. For NGS files with a high sequencing coverage, the maximum compression rate on DNA fragments is achieved, but the global compression rate is limited by the quality information that doesn't present the same redundancy property. On the other hand, if lossy mode is allowed, the compression rate can be drastically improved as demonstrated, for example, by the LEON compressor.

Compression and decompression time are another parameter to take into consideration. NGS compressors behave very differently on that point. DSRC, for example, is very fast to compress NGS files but has a moderate compression rate compared to other concurrent software. Its biggest advantage is that it is ten times faster than generic compressors with a better compression rate. As usual, there is a trade-off between speed and quality: the longer the compression time, the better the compression rate.

The nature of the NGS data may also dictate the choice of the compressors. Metagenomic data, for instance, are not very redundant. As shown in the previous section, compressors perform poorly on these data. In that case, a fast compressor is probably a much better choice since methods optimized to extract redundancy will systematically fail.

However, even if specific compression methods devoted to NGS data have demonstrated their superiority over generic tools (such as GZIP for example), no tool has emerged as a recognized and routinely used software. The main reasons are:

- **Durability**: tools are often research laboratory prototypes with a first objective of demonstrating a new compression strategy. The maintainability of these tools over the time is not guaranteed nor their ascendant compatibility. In that case, the decompression of a NGS file can be impossible with higher versions of the compression/decompression software.
- **Robustness**: our tests have shown that, in some cases, NGS compression tools were not able to process large files or that files, after decompression, were differing from the original ones. This is probably due to the youth of these tools that are not yet fully debugged.
- **Flexibility**: the literature shows there exists no universal NGS compression tool. Each tool has its own special features that do not cover all needs. For example, some compressors only compress DNA sequences, others take only as input reads of fixed length, and others do not propose the lossy/lossless quality option, etc.

Finally, the huge size of the NGS files leads to important compression/decompression time. The compression of a 700 GB file (see previous section) takes several hours. Decompression is often a little bit faster but remains a time-consuming task. One way to avoid these compression/decompression steps is to have bioinformatics tools that directly exploit the compressed format. This is the case for the GZIP format: many tools are able to internally perform this decompression step thanks to the availability of a GZIP library that can be easily included in the source codes. However, even if this task becomes transparent for users, it is still performed. The next step would be to directly exploit the data structure of the compressed files, i.e., without explicitly reconstructing the original list of DNA fragments. As a matter of fact, the majority of the treatments performed on NGS files are detection of small variants (such as SNP calling), mapping on a reference genome, assembly, etc. All these treatments would greatly benefit from the data processing done during the compression step, especially the management of redundancy. The next generation of compressors should act on this direction and propose formats to favor such usage.

References

1. Benoit, G., Lemaitre, C., Lavenier, D., Drezen, E., Dayris, T., Uricaru, R., Rizk, G.: Reference-free compression of high throughput sequencing data with a probabilistic de Bruijn graph de Bruijn graph. BMC Bioinf. **16**, 288 (2015)
2. Bonfield, J.K., Mahoney, M.V.: Compression of fastq and sam format sequencing data. PLoS One **8**(3), e59190 (2013)
3. Burrows, M., Wheeler, D.: A block sorting lossless data compression algorithm. Technical Report 124, Digital Equipment Corporation (1994)
4. Cánovas, R., Moffat, A., Turpin, A.: Lossy compression of quality scores in genomic data. Bioinformatics **30**(15), 2130–2136 (2014)

5. Chikhi, R., Rizk, G.: Space-efficient and exact de bruijn graph representation based on a bloom filter. Algorithms Mol. Biol. **8**(1), 22 (2013)
6. Cox, A.J., Bauer, M.J., Jakobi, T., Rosone, G.: Large-scale compression of genomic sequence databases with the burrows-wheeler transform. Bioinformatics **28**(11), 1415–1419 (2012)
7. Deorowicz, S., Grabowski, S.: Compression of DNA sequence reads in fastq format. Bioinformatics **27**(6), 860–862 (2011)
8. Deutsch, P., Gailly, J.: Zlib compressed data format specification version 3.3. RFC 1950 (1996)
9. Grabowski, S., Deorowicz, S., Roguski, Ł.: Disk-based compression of data from genome sequencing. Bioinformatics **31**(9), 1389–1395 (2014)
10. Hach, F., Numanagic, I., Alkan, C., Sahinalp, S.C.: Scalce: boosting sequence compression algorithms using locally consistent encoding. Bioinformatics **28**(23), 3051–3057 (2012)
11. Huffman, D.: A method for the construction of minimum-redundancy codes. In: Proceedings of the Institute of Radio Engineers (1952)
12. Janin, L., Rosone, G., Cox, A.J.: Adaptive reference-free compression of sequence quality scores. Bioinformatics **30**(1), 24–30 (2014)
13. Jones, D.C., Ruzzo, W.L., Peng, X., Katze, M.G.: Compression of next-generation sequencing reads aided by highly efficient de novo assembly. Nucleic Acids Res. **40**(22), e171 (2012)
14. Li, H., Durbin, R.: Fast and accurate short read alignment with burrows–wheeler transform. Bioinformatics **25**(14), 1754–1760 (2009)
15. Li, H., Handsaker, B., Wysoker, A., Fennell, T., Ruan, J., Homer, N., Marth, G., Abecasis, G., Durbin, R., Subgroup, G.P.D.P.: The sequence alignment/map format and samtools. Bioinformatics **25**(16), 2078–2079 (2009). doi:10.1093/bioinformatics/btp352
16. Mahoney, M.: (2000) http://mattmahoney.net/dc/
17. Mahoney, M.: Adaptive weighing of context models for lossless data compression. Florida Tech. Technical Report (2005)
18. Moffat, A.: Implementing the PPM data compression scheme. IEEE Trans. Commun. **38**, 1917–1921 (1990)
19. Rizk, G., Lavenier, D., Chikhi, R.: DSK: k-mer counting with very low memory usage. Bioinformatics **29**(5), 652–653 (2013)
20. Roguski, L., Deorowicz, S.: DSRC 2-industry-oriented compression of FASTQ files. Bioinformatics **30**(15), 2213–2215 (2014)
21. Saha, S., Rajasekaran, S.: Efficient algorithms for the compression of fastq files. In: 2014 IEEE International Conference on Bioinformatics and Biomedicine (2014)
22. Sahinalp, S.C., Vishkin, U.: Efficient approximate and dynamic matching of patterns using a labeling paradigm. In: Proceedings of the 37th Annual Symposium on Foundations of Computer Science, FOCS '96, Washington, DC, pp. 320–328. IEEE Computer Society, Los Alamitos (1996). http://dl.acm.org/citation.cfm?id=874062.875524
23. Seward, J.: (1996) bzip2 : http://www.bzip.org/1.0.3/html/reading.html
24. Shannon, C., Weaver, W.: The Mathematical Theory of Communication. University of Illinois Press, Urbana (1949)
25. Tembe, W., Lowey, J., Suh, E.: G-SQZ: compact encoding of genomic sequence and quality data. Bioinformatics **26**(17), 2192–2194 (2010)
26. Wan, R., Anh, V.N., Asai, K.: Transformations for the compression of fastq quality scores of next-generation sequencing data. Bioinformatics **28**(5), 628–635 (2012)
27. Welch, T.: A technique for high-performance data compression. Computer **6**, 8–19 (1984)
28. Witten, I., Neal, R., Cleary, J.: Arithmetic coding for data compression. Commun. ACM **30**, 520–540 (1987)
29. Yanovsky, V.: Recoil - an algorithm for compression of extremely large datasets of dna data. Algorithms Mol. Biol. **6**, 23 (2011)
30. Yu, Y.W., Yorukoglu, D., Berger, B.: Traversing the k-mer landscape of ngs read datasets for quality score sparsification. In: Research in Computational Molecular Biology, pp. 385–399. Springer, Berlin (2014)
31. Ziv, J., Lempel, A.: A universal algorithm for sequential data compression. IEEE Trans. Inf. Theory **23**(3), 337–343 (1977)

Chapter 5
Cloud Storage-Management Techniques for NGS Data

Evangelos Theodoridis

5.1 Introduction

Current scientific advancements in both computer and biological sciences are bringing new opportunities to intra-disciplinary research topics. On one hand, computers and big-data analytics cloud software tools are being developed, rapidly increasing the capability of processing from terabyte data sets to petabytes and beyond. On the other hand, the advancement in molecular biological experiments is producing huge amounts of data related to genome and RNA sequences, protein and metabolite abundance, protein–protein interactions, gene expression, and so on. In most cases, biological data are forming big, versatile, complex networks. Sequencing has seen major scientific progress in recent years and has leveraged the development of novel bioinformatic applications. Consequently, bioinformatics and life sciences applications in general are facing a rapidly increasing demand for data-handling capacity. In many cases, from low-level applications (such as systems biology) to high-level integrated applications (such as systems medicine), the amounts of data to store, transfer, and process meet congestion in many current technologies.

Current advances in high-throughput sequencing are leading to sequence data set generation that is growing at an exponential rate. Many large-scale projects are producing consistantly vast data sets. Projects such as the Human Genome Project [25], 1000 Genomes Project [1], and Genome 10K Project [20] are gathering sequencing data from the community. Technologies such as Ilumina [26], Roche [39], and others are capable of producing giga base pairs per sequence device per day. According to [19], the total second-generation sequencing capacity surpasses 13 peta base pairs per year (as recorded in 2011) and is increasing five times

E. Theodoridis (✉)
Computer Technology Institute, Patras, Greece
e-mail: theodori@cti.gr

M. Elloumi (ed.), *Algorithms for Next-Generation Sequencing Data*,
DOI 10.1007/978-3-319-59826-0_5

per year. Moreover, other large-scale data are emerging from high-throughput technologies, such as gene expression data sets, protein 3D structures, protein–protein interactions, and others, which are also generating large volumes of data.

The massive data volumes being generated by these new technologies require new data management and storage techniques. The bioinformatics research community needs new and improved approaches to facilitate *Next Generation Sequencing* (NGS) data management and analysis. According to the NGS research community [10, 17, 38], this is a complex topic requiring bioinformaticians, computer scientists, and biomedical scientists to join efforts to bring NGS data management and analysis to new levels of efficiency and integration. In [19, 27, 32], the reader can find interesting surveys on sequence alignment and biomedical analytics applications developed on cloud infrastructures.

In this chapter, we survey cutting-edge big-data management cloud technologies for storing and processing large biological and NGS data sets. We focus on platforms based on batch-processing big-data technologies, such as Apache Hadoop, Google BigQuery, Apache Spark, and Hive. We highlight the special characteristics of each technology and platform. Moreover, we highlight the features of these platforms that make them suitable for storing, indexing, and processing large amounts of NGS data.

The structure of this chapter is as follows. In Sect. 5.2, we survey the existing big-data management cloud technologies for storing and processing large biological and NGS data sets. We focus on platforms based on batch-processing big-data technologies, such as *Apache Hadoop* [3], *Google BigQuery* [22], *Apache Spark* [8], and *Hive* [4], also providing an overall comparison of them. In Sect. 5.3, we discuss the open challenges in processing and indexing NGS data and how they can be overtaken with current technologies. Finally, in Sect. 5.4, we conclude the chapter by identifying a number of deficiencies still existing in this domain and open topics for future developments.

5.2 Current Technologies

In this section, we review the basic characteristics of existing big-data management cloud technologies for storing based on batch processing.

5.2.1 Apache Hadoop

Apache Hadoop [3] tries to deal with a fundamental problem, inherited natively from how computer systems are built and interconnected. The problem is that storage capacities of a single machine have increased massively but the access speed of data is still considerably small. As a consequence, storing large volumes of data in a single machine is somehow prohibited. So, the primitive idea behind

Apache Hadoop is the distribution of data (replicated or not) to several storage machines to make parallel data access possible. Even so, bottlenecks might appear at the communication level when someone tries to process large volumes of data originating from several storage machines. So, the second principle idea of *Apache Hadoop* is to keep the computations as close as possible to the storage machines, distributing the computation and avoiding the gathering of retrieved data sets to a single machine.

Building a system with these primitives has to deal in practice with hardware failures. A common way of avoiding data loss is through replication: redundant copies of the data are kept by the system so that in the event of failure another copy is available. A parametrized file distribution and replication mechanism is supported by the *Hadoop Distributed File System* (HDFS). The second major issue is joining and processing data originated from multiple data sources. *Apache Hadoop*, *MapReduce* has a programming model that abstracts the problem from the disk accessed, transforming it into a computation over sets of keys and values and splitting the computation in two parts: the map, the reduce, and the interface between the two steps where the combination of data occurs. In cases where it is necessary to perform a query against the entire data set (or at least a major part of it), *MapReduce* is the appropriate programming model. *MapReduce* is a batch query processor and is quite powerful when the computational problem can split on partitions of data sets and finally combine the subsolutions of each partition to the final one (Fig. 5.1).

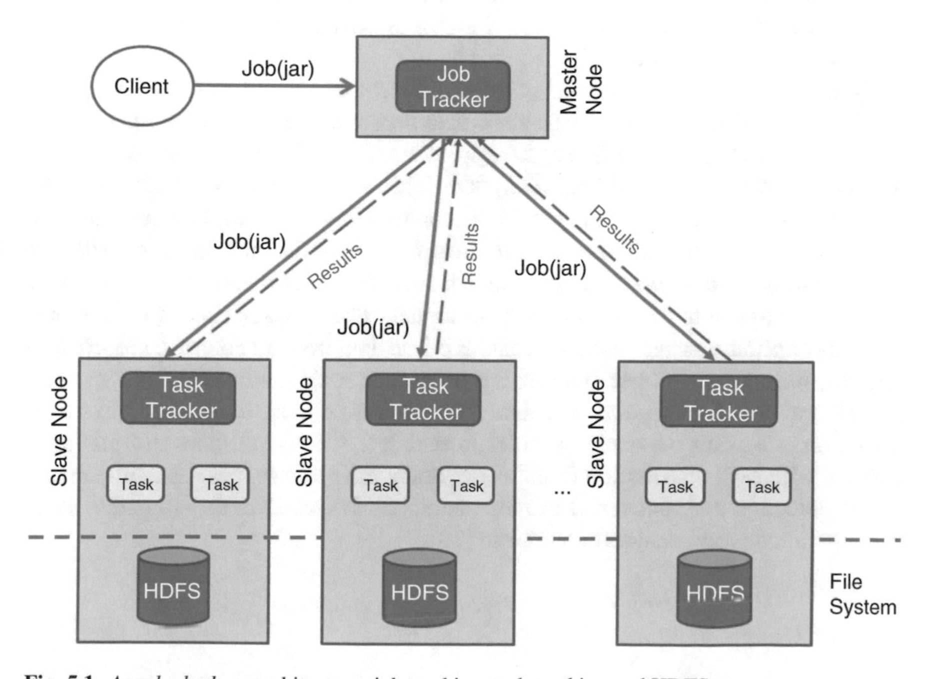

Fig. 5.1 *Apache hadoop* architecture: job tracking, task tracking and HDFS

In comparison to traditional *Relational Database Management Systems* (RDBMS) with a set of disks, *Apache Hadoop* is more appropriate for batch processing [48]. If the data access pattern is dominated by seeks, it usually takes more time to read or write large portions of the data set than streaming all the data out of the disks. On the contrary, in the case of reading or updating a small part of the data set, traditional RDBMS, by using indexing structures such as B-Trees or hashing tables, outperforms the others. For accessing or updating most of a data set, a B-Tree/Hash Table is less efficient than *MapReduce*. *MapReduce* suits applications where the data are written once and read many times, whereas relational databases are good for data sets that are continually updated. *MapReduce* is suited when someone wants to analyze the whole data set in contract with RDBMS, which is good for point queries or updates, where the data set has been indexed to deliver low-latency retrieval and update times of a relatively small amount of data.

Moreover, *MapReduce* and RDBMS differentiate greatly in relation to how much structure the support has in the data sets on which they operate. RDBMS follow a quite strict schema definition while on the other hand semi-structured or unstructured data sets such as XML, plain text, or image data files are more well suited for *Apache Hadoop* mechanisms. *MapReduce* works well on unstructured or semi-structured data, since they are designed to interpret the data at processing time. Relational data sets are normalized to retain their integrity and remove redundancy while in the case of unstructured data and *MapReduce* developers ad hoc select attributes of data to be analyzed. Normalization poses problems for *MapReduce*, since it makes reading a record a nonlocal operation, and one of the central assumptions that *MapReduce* makes is that it is possible to perform (high-speed) streaming reads and writes [47].

Furthermore, *Apache Hadoop* is quite different compared to Grid Computing architectures, which perform large-scale data processing using such APIs as *Message Passing Interface* (MPI). These approaches distribute the work across a set of machines that accesses a shared file system. This fits better to extremely compute-intensive jobs, but network congestion becomes a big issue when machines need to access large data volumes. *MapReduce* tries to collocate the data with the computations, so data access is fast since it is local. A distributed MPI framework gives great control to the programmers to transfer data in their cluster, but requires that they explicitly program the mechanics of the data flow. In contrast, *MapReduce* operates only at the higher level as the programmer thinks in terms of functions of key and value pairs, and the data flow is implicit. In conclusion, from the programmer's point of view, the order in which the tasks run does not matter in contract with MPI distribution frameworks where the programmer has to implement failure detection and recovery. For more details on *MapReduce* design patterns and computation design, readers can refer to [29].

5.2.2 Apache Pig

Apache Pig [6] relies on *Apache Hadoop* mechanics and introduces an extra level of abstraction for processing large data sets. *Apache Pig* consists of two major components, a language called *Pig Latin* [36] designed for expressing data flows and an execution environment to run *Pig Latin* programs. The execution can be done either locally in a single JVM or distributed on a *Apache Hadoop* cluster. A *Pig Latin* program is made up of a series of operations, or transformations, that are applied to the input data to produce output. *Apache Pig* internally expresses the transformations into a series of *MapReduce* jobs transparently to the programmer, which allows programmers to focus on the data rather than the workflow of the execution. *Apache Pig* was designed to be extensible [34]. Processing statements in *Pig Latin* consist of operators, which take inputs and emit outputs. The inputs and outputs are structured data expressed in bags, maps, tuples, and scalar data. *Apache Pig* resembles a data-flow graph, where the directed vertices are the paths of data and the nodes are operators (operations such as FILTER, GROUP, and JOIN) that process the data. In *Pig Latin*, each statement executes as soon as all data reach them, in contrast to a traditional program that executes as soon as it encounters the statement. Most parts of the execution path are customizable: loading, storing, filtering, grouping, and joining can be programmable by user-defined functions. *Apache Pig*, like *Apache Hadoop*, is suitable for batch processing of data accessing a major part of the analyzed data set. For more details on processing design patterns on *Apache Pig*, rely on [34].

5.2.3 Apache Hive

Apache Hive [4] is a framework for data warehousing on top of *Apache Hadoop*. *Apache Hive* supports analysis of large data sets stored in *Apache Hadoop's* HDFS and compatible file systems. Hive provides an SQL dialect, called *Hive Query Language* (HQL), for querying data stored in an *Apache Hadoop* cluster [45]. *Apache Hive* translates most queries to *MapReduce* jobs, thereby exploiting the scalability of Hadoop, while presenting a familiar SQL abstraction. To accelerate queries, Hive provides indexes, including bitmap indexes. Hive is most suited for data warehouse applications, when the data are not changing rapidly, where fast response times are not required. Due to inherited characteristics from *Apache Hadoop*, queries in *Apache Hive* have higher latency because of the start-up overhead for *MapReduce* jobs. Therefore, Hive does not provide functionalities required for transaction processing, so it is best suited for data warehouse applications, where a large data set is maintained and mined for aggregates and reporting.

5.2.4 Apache Spark

Apache Spark [8] is also an open-source data analytics cluster computing framework. *Apache Spark* is built on top of the HDFS but does not follow the *MapReduce* programming paradigm. The framework provides primitives for in-memory cluster computing that allow user programs to load data into a cluster's memory and query it repeatedly, making it well suited to machine-learning algorithms. The framework is based on two key components, *Resilient Distributed Data sets* [51], a distributed memory abstraction that lets programmers perform in-memory computations on large clusters in a fault-tolerant manner, and a directed acyclic graph computation execution engine. This technology supports two kinds of operations: transformations and actions. Transformations create new data sets from the input (e.g., map or filter operations are transformations), whereas actions return a value after executing calculations on the data set (e.g., reduce or count operations are actions).

5.2.5 Google BigQuery

BigQuery [22] has evolved on top of *Dremel* [28], which is an interactive ad hoc query system for analysis of read-only nested data. It supports multi-level execution trees and columnar data layout and execution of aggregation queries This framework is proprietary and is coupled with the *Google Storage*. Storage and computing resources automatically scale to the size of the data and analysis requirements. The technical implementation is different from that of a *relational database management system* (RDBMS) and traditional data warehousing solutions. *BigQuery* supports data inserts, deletion (through dropping of tables), and the ability to add columns to table schema after the data have already been loaded. It does not support direct updates to the stored data. The updates can be handled by using temporary tables until the final ones having the updates. The query is processed by many servers in a multi-level execution tree structure, with the final results aggregated at the root. *BigQuery* stores the data in a columnar format so that only data from the columns being queried are read. This approach affects I/O performance and performs better compared to traditional relational database models because the entire record is not accessed. For more details, refer to [23].

5.2.6 Bioinformatics BigData Processing Technologies

In this section, we review current NGS platforms based on the aforementioned large data processing platforms.

- ***Seal:*** [37] is a tool for short read pair mapping and duplicate removal. It is currently structured in two sequential methods implemented with *MapReduce* on

Apache Hadoop. The first one is *PairReadsQseq*, which is a utility that converts the sequencing files produced by Illumina sequencing machines into a file format that places entire read pairs on a single line. The next is *Seqal*, which implements read alignment and performs duplicate read removal. *PairReadsQseq* groups mate pairs from sequencing data files into the same record, producing files where each line consists of five tab-separated fields: id, sequence and ASCII-encoded base qualities for read 1 and 2. *Seqal*, takes input pairs in the produced format and produces mapped reads in SAM format. The read alignment is implemented in the map function by integrating *Burrows-Wheeler Aligner* [12]. For each pair of reads, the aligner produces a pair of alignment records. The user can choose to filter these by whether or not the read is mapped and by mapping quality. Then, the reads may be directly output to SAM files or put through a reduce phase where duplicates are removed; the choice is made through a command line option. Seqal identifies duplicate reads by noting that they are likely to map to the same reference coordinates. When a set of duplicate pairs is found, only the one with the highest average base quality is kept.

- ***Hadoop-BAM*** [30] is a library for manipulating aligned NGS data in the *Apache Hadoop*. Its primary goal is to integrate analysis applications and files containing sequence alignment data (BAM, SAM [40]) that are stored and processed on HDFS. The library provides a proper API for implementing map and reduce functions that can directly operate on BAM records, and it based on the *Picard* SAM *Java* development library [35]. The library supports two modes of access to BAM files. The first one uses a precomputed index that maps byte offsets to BAM records to allow random access. The second one locates boundaries between compressed blocks via BGZF numbers and then on a lower level detects BAM block boundaries via redundancies in the BAM file format. Moreover, *Hadoop-BAM* provides a *Picard*-compatible *Java* API to programmers, facilitating coding without considering the issues of BGZF compression, block boundary detection, BAM record boundary detection, or parsing of raw binary data.
- ***BioPig*** [31] is a framework for genomic data analysis using *Apache Pig* ([6], Sect. 5.2.2) and *Apache Hadoop*. The software builds a .jar file that can then be used on *Apache Hadoop* clusters or through the *Apache Pig* query language. The framework also integrates a set of libraries, which adds an abstraction layer for processing sequence data. Basically, the framework contains three core functional modules: *BioPigIO* module, which reads and writes sequence files in the FASTA or FASTQ format, *BioPigAggregation* module, which is a wrapper for common bioinformatics programs, such as BLAST, CAP3, and *kmerMatch*, a set of tools for the compression and encoding of sequences.
- ***DistMap*** [33] is a platform for sorting/mapping reads in a *MapReduce* framework on a local *Apache Hadoop* cluster. *DistMap* supports a large number of different short read mapping methods, such as BWA [12], GSNAP [50], *TopHat* [46], and many others [33]. It accepts reads in FASTQ format as input and provides mapped reads in an SAM/BAM format. *DistMap* supports both paired- and single-end reads, thereby allowing the mapping of read data produced by dif-

ferent sequencing platforms. Moreover, *DistMap* supports two different queuing systems of *Apache Hadoop*, the *Capacity* and *Fair* scheduler [3].

- ***SeqPig*** [42] is a collection of tools that facilitates analysis of sequence data sets. *SeqPig* scripts use the *Apache Pig* [6] (Sect. 5.2.2), which automatically parallelizes and distributes data-processing tasks. The main goal of this platform is to provide a set of easy tools that can be used by bioinformaticians without deep knowledge of *Apache Hadoop* and *MapReduce* internals. It extends*Apache Pig* with a number of features and functionalities specialized for processing sequencing data. Specifically, it provides data input and output functionalities, functions to access fields and transform data, and methods for frequent tasks such as pile-up, read coverage, and various base statistics.
- ***SparkSeq*** [49] platform is built on top of *Apache Spark* [8] (Sect. 5.2.4), targeting analysis for next-generation sequencing data. It can be used to build genomic analysis pipelines in *Scala* and run them in an interactive way. *SparkSeq* employs *Picard SAM JDK* [35] and *Hadoop-BAM* [30] (Sect. 5.2.6) for analyzing NGS data sets.
- ***Crossbow*** [18] is a software pipeline, which combines *Bowtie* [11], a short read aligner that aligns short DNA sequences (reads) to the human genome, and *SoapsSNP* [44] for whole-genome re-sequencing analysis. The platform runs in the cloud on top of Hadoop in a local installation or on *Amazon* cloud services.
- ***CloudDOE*** [15] is a support tool that guides developers to setup technical details of *Apache Hadoop* processing tasks with a user-friendly graphical interface. Users are guided through the user interface to deploy a *Apache Hadoop* cloud in their infrastructure within in-house computing environments and to execute bioinformatics methods. Some of these NGS applications are *CloudBurst* [41], *CloudBrush* [13], and *CloudRS* [14]. Finally, the tool may facilitate developers to deploy *ApacheHadoop* tools on a public cloud.
- ***SeqWare*** [43] *SeqWare* project is a software infrastructure designed to analyze genomics data sets. It consists of an infrastructure focused on enabling the automated, end-to-end analysis of sequence data from raw base calling to analyzed variants ready for interpretation by users. *SeqWare* is a tool agnostic, it is a framework for building analysis workflows and does not provide specific implementations out of the box. *SeqWare* currently provides five main tools: *MetaDB* provides a common database to store metadata used by all components; *Portal*is a web application to manage samples, record computational events, and present results back to end users; *Pipeline* is a workflow engine capable of wrapping and combining other tools (BFAST, BWA, SAMtools, etc.); *Web Service* is a programmatic API that lets people build new tools on top of the project; the Query Engine is a NoSQL database designed to store and query variants and other events inferred from sequence data.

5.3 Discussion and Open Challenges

Although most of the aforementioned technologies have been developed in the last 10 years, they have long way to go before they can be used widely and easily off the shelf in many bioinformatics applications. Setting up and programming these frameworks require much effort and expertise to develop parallelized execution applications. As we showed in the previous section, there have been some efforts to simplify the development process (e.g., with SQL interfaces that generate *Apache Hadoop* jobs in the background). Nevertheless, *MapReduce* is designed to place each line as an individual record by default. In most cases, sequence file formats involve multiple lines per sequence, so programmers have to dig in internal storage to map multiple line formats to single line formats.

Moreover, most of the aforementioned technologies are designed for use as back-end data-processing techniques, storing output data in files in most cases. Currently, front-end visualization functionalities are lacking for either visualizing and browsing the results or setting up and deploying the jobs. In most cases, this is done through a shell command line. As discussed before, there have been some efforts to create developer-friendly GUIs such as *CloudDOE* [15] or *Cloudgene* [16], but these solutions are still in development with experimental status (beta) and provide limited visualization functionalities.

Furthermore, one of the most significant challenges is the exchange of NGS data over the web and internet. In many cases, researchers would like to process large volumes of genomic data and acquire metadata and extra information for various online data sources. Getting all these data into and out of the cloud might be a very time-consuming process. Software tools are needed that will orchestrate the data-gathering workflow until the final processing takes place. Moreover, modern transfer techniques, such as multi-part upload of *Amazon S3* [2], can speed up the data transfer process considerably. Another interesting direction is to adapt platforms to support various compression/decompression functionalities so that processing can be performed directly on compressed data sets.

According to [21], data integration is a persisting challenge driven by technology development producing increasing amounts and types of data. While the availability of genomics data is reasonably well provided for by publicly accessible and well-maintained data repositories, improved annotation standards and requirements are needed in data repositories to enable better integration and reuse of public data. The data exploitation aspect of data integration is quite significant as it involves using prior knowledge, and how it is stored and expressed, the development of statistical methods to analyze heterogeneous data sets, and the creation of data exploratory tools that incorporate both useful summary statistics and new visualization tools.

The massive data volumes being generated by NGS technologies require new data management and storage techniques. According to the NGS research community [10, 17, 38], this is a complex topic requiring bioinformaticians, computer scientists, and biomedical scientists to join efforts to bring NGS data management and analysis to new levels of efficiency and integration. In recent NGS network community workshops, important working topics were identified such as scalability, automation, timeliness, storage, privacy and security, standardization, and metadata definition. In more detail, some of their highlighted topics are:

- decrease the complexity and improve the efficiency of data accessibility
- many processing pipeline tasks must be automated
- standardization is needed to generate automated processes
- scaling up of data processing and storage needs by integrating new distributed processing technologies
- metadata and protocols have to be standardized and implemented along the whole processing pipeline
- standardization of data, processes, and software APIs
- intuitive user interfaces have to be developed so that they can be used by non-IT experts
- reproducibility and traceability of processing pipelines

Currently, there is a new phase of emerging big-data analytics platforms focusing on real-time processing, stream processing, and interactive calls as well. Such platforms include the *Apache Spark* [8] (Sect. 5.2.4), *Apache Tez* [9], *Apache Flink* [5], and *Apache Samza* [7]. Most of the technologies utilize *Apache Hadoop Yarn* [24], a successor of the *MapReduce* mechanism. The basic idea of *YARN*, in comparison to the classic *Apache Hadoop*, is to split up the two major responsibilities of the *JobTracker*, i.e., resource management and job scheduling/monitoring, into separate processes aiming to provide scalability, efficiency, and flexibility. Moreover, it provides the capability of developing applications that are not based on batch processing of the whole data set but on interactive queries on top of the data set. All these new technologies bring new opportunities to the NGS community along with the extra challenges of integrating them or adapting the existing tools.

5.4 Conclusions

In this chapter, we provided a basic introduction to and presentation of key batch-processing big-data technologies and platforms, such as *Apache Hadoop*, *Google BigQuery*, *Apache Spark*, and *Apache Hive*. We highlighted the special characteristics along with the various advantages and disadvantages that each platform demonstrates. Moreover, we reviewed platforms and tools built on top of these technologies for indexing and processing large amounts of NGS data. We have identified a number of deficiencies still existing in this domain and several topics for future developments. The large volumes of produced NGS data require sophisticated data management and storage techniques. Although the new big-data management technologies have been largely adopted by the bioinformatics community, much work remains to be done to the increase maturity of tools, e.g., the automatic setup of analysis environments, application design and interfacing with the analysis layer, and design and development of visualization tools.

References

1. 1000 genomes project (2013). http://www.1000genomes.org/
2. Amazon S3 multipart upload. http://aws.amazon.com/blogs/aws/amazon-s3-multipart-upload/
3. Apache Hadoop. http://hadoop.apache.org/
4. Apache Hive. https://hive.apache.org/
5. Apache Flink. http://flink.incubator.apache.org
6. Apache Pig. http://pig.apache.org/
7. Apache Samza. http://samza.incubator.apache.org/
8. Apache Spark. https://spark.apache.org/
9. Apache Tez. http://tez.apache.org/
10. Bongcam-Rudloff, E., et al.: The next NGS challenge conference: data processing and integration. EMBnet. J. **19**(A), p-3 (2013)
11. Bowtie. http://bowtie-bio.sourceforge.net/index.shtml
12. Burrows-Wheeler Aligner. http://bio-bwa.sourceforge.net/
13. Chang, Y.J., Chen, C.C., Chen, C.L., Ho, J.M.: A de novo next generation genomic sequence assembler based on string graph and MapReduce cloud computing framework. BMC Genomics **13**, 1–17 (2012)
14. Chen, C.C., Chang, Y.J., Chung, W.C., Lee, D.T., Ho, J.M.: CloudRS: an error correction algorithm of high-throughput sequencing data based on scalable framework. In: BigData Conference, pp. 717–722. IEEE (2013)
15. Chung, W.-C., et al.: CloudDOE: a user-friendly tool for deploying Hadoop clouds and analyzing high-throughput sequencing data with MapReduce. PLoS One **9**(6), e98146 (2014). doi:10.1371/journal.pone.0098146
16. CloudGENE A graphical MapReduce platform for cloud computing. http://cloudgene.uibk.ac.at/index.html
17. COST Action BM1006: next generation sequencing data analysis network. http://www.seqahead.eu/
18. Crossbow. http://bowtie-bio.sourceforge.net/crossbow/index.shtml
19. Daugelaite, J., O' Driscoll, A., Sleator, R.D.: An overview of multiple sequence alignments and cloud computing in bioinformatics. ISRN Biomath. **2013**, 14 pp. (2013). doi:10.1155/2013/615630. Article ID 615630
20. Genome 10K Community of Scientists: Genome 10K: a proposal to obtain whole-genome sequence for 10,000 vertebrate species. J. Hered. **100**, 659–674 (2009)
21. Gomez-Cabrero, D., Abugessaisa, I., Maier, D., Teschendorff, A., Merkenschlager, M., Gisel, A., Ballestar, E., Bongcam-Rudloff, E., Conesa A., Tegnér, J.: Data integration in the era of omics: current and future challenges. BMC Syst. Biol. **8**(Suppl. 2), I1 (2014)
22. Google BigQuery. https://developers.google.com/bigquery/
23. Google BigQuery. https://cloud.google.com/developers/articles/getting-started-with-google-bigquery
24. Hadoop Yarn. http://hadoop.apache.org/docs/current/hadoop-yarn/hadoop-yarn-site/YARN.html
25. Human genome project information (2013). http://web.ornl.gov/sci/techresources/HumanGenome/
26. Illumina. https://www.illumina.com/
27. Lin, Y.-C., Yu, C.-S., Lin, Y.-J.: Enabling large-scale biomedical analysis in the cloud. BioMed. Res. Int. **2013**, 6 pp. (2013). doi:10.1155/2013/185679. Article ID 185679
28. Melnik, S., Gubarev, A., Long, J.J., Romer, G., Shivakumar, S., Tolton, M., Vassilakis, T.: Dremel: interactive analysis of web-scale datasets. Proc. VLDB Endow. **3**, 330–339 (2010)
29. Miner, D., Shook, A.: Mapreduce Design Patterns: Building Effective Algorithms and Analytics for Hadoop and Other Systems, 1st edn. O'Reilly Media, Inc., Sebastopol (2012)
30. Niemenmaa, M., et al.: Hadoop-BAM: directly manipulating next generation sequencing data in the cloud. Bioinformatics **28**(6), 876–877 (2012)

31. Nordberg, H., Bhatia, K., Wang, K., Wang, Z.: BioPig: a Hadoop-based analytic toolkit for large-scale sequence data. Bioinformatics **29**(23), 3014–3019 (2013)
32. O'Driscoll, A., Daugelaite, J., Sleator, R.D.: Big data', Hadoop and cloud computing in genomics. J. Biomed. Inform. **46**(5), 774–781 (2013)
33. Pandey, R.V., Schlötterer, C.: DistMap: a toolkit for distributed short read mapping on a Hadoop cluster. PLoS One **8**(8), e72614 (2013)
34. Pasupuleti, P.: Pig Design Patterns. Packt Publishing, Birmingham (2014)
35. Picard Tools. http://picard.sourceforge.net/
36. Pig Latin. http://pig.apache.org/docs/r0.13.0/basic.html
37. Pireddu, L., Leo, S., Zanetti, G.: SEAL: a distributed short read mapping and duplicate removal tool. Bioinformatics **27**(15), 2159–2160 (2011). doi:10.1093/bioinformatics/btr325. http://biodoop-seal.sourceforge.net/
38. Regierer, B., et al.: ICT needs and challenges for big data in the life sciences. A workshop report-SeqAhead/ISBE Workshop in Pula, Sardinia, 6 June 2013. EMBnet. J. **19**(1), pp-31 (2013)
39. Roche/454 http://www.454.com/
40. SAMtools http://www.htslib.org/
41. Schatz, M.C.: CloudBurst: highly sensitive read mapping with MapReduce. Bioinformatics **25**, 1363–1369 (2009)
42. Schumacher, A., et al.: SeqPig: simple and scalable scripting for large sequencing data sets in Hadoop. Bioinformatics **30**(1), 119–120 (2014)
43. SeqWare https://seqware.github.io/
44. SoapsSNP http://bowtie-bio.sourceforge.net/index.shtml
45. Thusoo, A., Sarma, J.S., Jain, N., Shao, Z., Chakka, P., Anthony, S., Liu, H., Wyckoff, P., Murthy, R.: 2009. Hive: a warehousing solution over a map-reduce framework. Proc. VLDB Endow. **2**(2), 1626–1629 (2009)
46. Trapnell, C., Pachter, L., Salzberg, S.L.: TopHat: discovering splice junctions with RNA-Seq. Bioinformatics (2009). doi:10.1093/bioinformatics/btp120
47. Venner, J.: Pro Hadoop, 1st edn. Apress, Berkely, CA (2009)
48. White, T.: Hadoop: The Definitive Guide, 1st edn. O'Reilly Media, Inc., Sebastopol (2009)
49. Wiewiórka, M.S., et al.: SparkSeq: fast, scalable, cloud-ready tool for the interactive genomic data analysis with nucleotide precision. Bioinformatics (2014) doi:10.1093/bioinformatics/btu343. First published online: May 19 (2014)
50. Wu, T.D., Nacu, S.: Fast and SNP-tolerant detection of complex variants and splicing in short reads. Bioinformatics **26**, 873–881 (2010)
51. Zaharia, M., et al.: Resilient distributed datasets: a fault-tolerant abstraction for in-memory cluster computing. In: Proceedings of the 9th USENIX Conference on Networked Systems Design and Implementation (NSDI'12) (2012)

Part II
Error Correction in NGS Data

Chapter 6
Probabilistic Models for Error Correction of Nonuniform Sequencing Data

Marcel H. Schulz and Ziv Bar-Joseph

6.1 Introduction

Over the last few years, *next-generation sequencing* (NGS) technologies have revolutionized our ability to study genomic data. While these techniques have initially been used to study DNA sequence data [12], they are now widely used to study additional types of dynamic and condition-specific biological data, including that in genome-wide studies of transcriptomics and *chromatin immunoprecipitation* (ChIP). The sequencing of polyadenylated RNAs (RNA-Seq) is rapidly becoming standard practice [25] due to its ability to accurately measure RNA levels [13], detect alternative splicing [15] and RNA editing [17], and determine allele- and isoform-specific expression [19] and its usage in de novo transcriptome analysis [5, 21]. Similarly, the sequencing of metagenomes has gained a lot of attention for measuring the composition of bacterial species in different parts of the human body and correlating these to metabolic function [11] or disease states [28] or studies of microbial evolution [6] to name a few.

M.H. Schulz (✉)
Excellence Cluster for Multimodal Computing and Interaction, Saarland University, Saarbrücken, Germany

Computational Biology and Applied Algorithms, Max Planck Institute for Informatics, Saarbrücken, Germany
e-mail: mschulz@mmci.uni-saarland.de

Z. Bar-Joseph
Machine Learning Department and Lane Center for Computational Biology, Carnegie Mellon University, Pittsburgh, PA, USA
e-mail: zivbj@cs.cmu.edu

M. Elloumi (ed.), *Algorithms for Next-Generation Sequencing Data*,
DOI 10.1007/978-3-319-59826-0_6

6.1.1 De Novo Read Error Correction

NGS experiments generate tens to hundreds of millions of short sequence reads. Depending on the specific technology used, these reads are typically between 50 and 450 base pairs long, though the length is expected to increase. The standard pipeline for analyzing these experiments starts with aligning these reads to the known genomes to identify their origin and to correct errors. While several methods have been proposed to facilitate these analyses, in many cases, alignments to a genome are either not possible or can miss important events. For most model organisms, we do not currently have a complete sequenced genome. Analysis of nonsequenced organisms with varying evolutionary distances from current models is important for studying evolution and development [7]. However, assembly and annotation of complete genomes is time- and effort-consuming, and to date, less than 250 of the more than eight million estimated eukaryotic species have been fully sequenced at the chromosome level.[1]

These experiments in which all or several reads cannot be accurately mapped to the genome present several challenges for read error correction, which has a large impact on the downstream analysis, especially assembly. For example, methods for the de novo assembly of transcripts from RNA-Seq data or genomes from metagenomics or single-cell data are commonly based on the *de Bruijn graph* [1, 5, 21, 24]. The de Bruijn graph is known to be susceptible to sequencing errors as it relies on exact k-mer overlaps. It was shown that correcting for sequencing errors significantly improves the performance of de novo assembly [8, 10, 16] as we will also illustrate at the end of this chapter using the *Oases* de novo transcriptome assembler [21]. Thus, in several cases, error correction needs to be performed without a reference genome, relying instead on the reads themselves to identify and correct errors prior to assembling reads into contigs.

6.1.2 Error Correction of Nonuniform Sequencing Assays

Many recent studies deal with the subject of read sequencing error correction. However, most of the methods assume a *uniform* sequencing coverage [8, 22], which is usually obtained for genome sequencing of thousands of cells, each of which contains the same number of chromosomes. Sequencing of small or long RNAs [23], metagenomes or metatranscriptomes [3], or single-cell genomes lead to nonuniform sequencing coverage [14, 16]. We illustrate this difference in Fig. 6.1 where we show a genomic, a transcriptomic, and a metagenomic dataset. The latter two are sequencing experiments with nonuniform coverage and are devoid of a clear peak of the median or expected coverage in the dataset. Most of the error correction

[1] http://www.ncbi.nlm.nih.gov/genome.

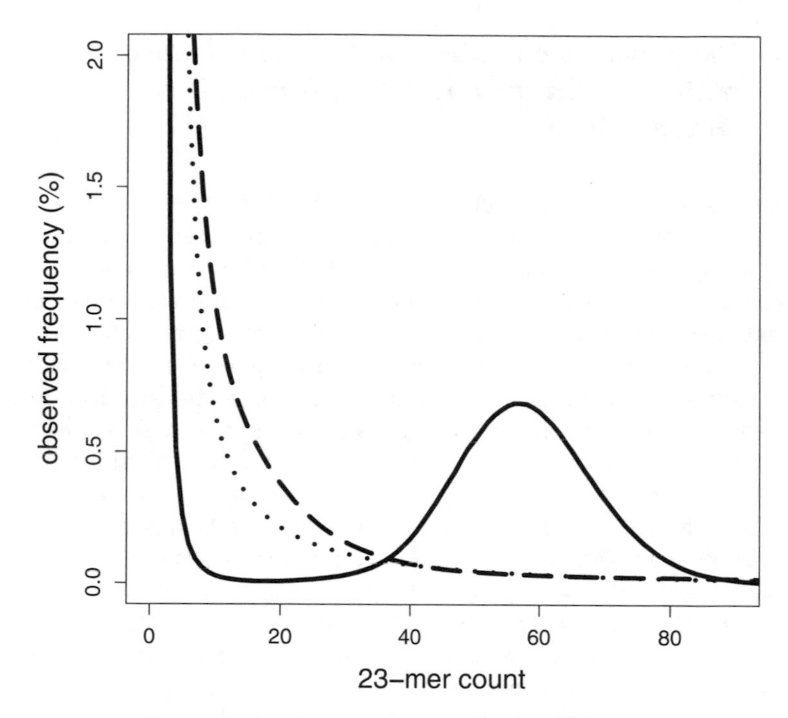

Fig. 6.1 Analysis of frequency distribution of 23-mers in human brain RNA (*dotted*), *E. coli* genome (*full line*), and human gut metagenome (*tiled line*) sequencing experiments. The number of 23-mers (*x*-axis) is compared to the frequency of distinct 23-mers that occur in the sequencing dataset (*y*-axis). Due to uniform coverage, the genome dataset shows a clear peak around the median sequencing coverage of 57. On the other hand, the RNA and metagenome experiments do not lead to such a clear peak but rather contain a large fraction of 23-mers that are only present 20 times or less in the dataset

methods that assume uniform coverage utilize a threshold-based approach to detect sequencing reads with errors. The idea is to find a minimum coverage level at which erroneous *k*-mers are unlikely to appear. For example, consider the *E. coli* genome sequencing experiment in Fig. 6.1. It shows that erroneous 23-mers can be separated from correct 23-mers by selecting a 23-mer count threshold (*x*-axis) of approximately 10. *k*-mers with higher coverage are likely real, whereas those with lower coverage are likely the result of sequencing errors. Unfortunately, for the nonuniform coverage data sets, such an approach would likely fail [14]. For example, in RNA-Seq data, read coverage is a function of expression, and therefore, to detect erroneous 23-mers from lowly expressed genes would require the use of a different (and unknown) threshold than erroneous 23-mers from highly expressed genes. As thresholds cannot be deduced from a coverage distribution plot as in Fig. 6.1, new approaches that use statistical methods have been proposed as we discuss below.

6.1.3 The Advantages of Machine Learning Methods in De Novo Error Correction of Nonuniform Coverage Data

Machine learning methods, which are often probabilistic and which combine statistical estimation, inference, and algorithmic efficiency are often used to deal with exactly this type of stochastic, noisy data. A number of different machine learning methods, ranging from clustering of short reads to the expectation maximization inference method to hidden Markov models have been suggested for the de novo error correction task. In most cases, these methods improve (sometimes quite a lot) upon deterministic methods when evaluated on simulated and real data. However, such improvements can come at a cost. As we discuss in more detail below, these methods can be either less memory efficient than their deterministic counterparts or slower (in terms of run time). These memory and run time requirements often result from the need to learn parameters for these probabilistic models which require either multiple iterations over (a subset of) the reads or from the need to store additional information regarding the relationship between reads. This tradeoff between efficiency and accuracy is an important issue to consider when deciding which method to use. Below, we discuss in detail a few representative examples of machine learning-based methods for de novo read error correction.

The rest of the chapter is organized as follows. In Sect. 6.2 algorithms that use the Hamming graph are introduced. In Sect. 6.3 an error correction algorithm that uses profile hidden Markov models is introduced. In Sect. 6.4 the performance of two algorithms is illustrated on a RNA-Seq dataset, and Sect. 6.5 ends with a discussion.

6.2 Hamming Graph-Based Error Correction Models

In addition to primarily relying on machine learning methods as discussed above, de novo, nonuniform, read error correction methods also often rely on a common data structure; the *Hamming graph*. In what follows, we will introduce the *Hamming graph*, which is either defined over reads or *k*-mers of reads.

6.2.1 Definitions

Consider m reads over the DNA alphabet $\Sigma = \{\text{A,C,G,T,N}\}$ with reads $\mathscr{R} = \{r_1, \ldots, r_m\}$ of length l sampled from one or several template sequences possibly containing sequencing errors. Define the Hamming distance between two sequences s_1 and s_2 as $dist(s_1, s_2)$. Define a graph $\mathscr{G} = (\mathscr{V}, \mathscr{E})$ where a node $n \in \mathscr{V}$ corresponds to a string $\in \Sigma^*$ denoted as $s(n)$. Two nodes $n_1, n_2 \in \mathscr{V}$ are connected

Table 6.1 Overview of methods that use the Hamming graph for read error correction using either full read sequences or k-mers of reads as nodes in the graph

Name	Datasets tested	Hamming graph	Citation
FreClu	Small RNA	Read	[18]
Recount	Small RNA, mRNA	Read	[26]
Hammer	Single-cell genome	k-mer	[14]
BayesHammer	Single-cell genome	k-mer	[16]
PREMIER	Genome	k-mer	[27]

by an edge $e \in \mathscr{E}$ if and only if $dist(s(n_1), s(n_2)) \leq T$, i.e., the Hamming distance between the strings is smaller than T.

The nodes in a *Hamming graph* $\mathscr{G}$ either represent the whole read, $s(n) \in \Sigma^l$, or k-mers (subsequences) of reads, $s(n) \in \Sigma^k$. In Table 6.1, we compare the methods that use the Hamming graph discussed here and specify if they use reads or k-mers.

6.2.2 *Hamming Graphs over Reads*

First we discuss two methods, *FreClu* and *Recount*, that work on the Hamming graph with nodes representing the complete read, i.e., $\mathscr{G} = (\mathscr{V}, \mathscr{E})$ where a node $n \in \mathscr{V}$ corresponds to $s(n) \in \Sigma^l$, $s(n) \in \mathscr{R}$. Building the Hamming graph over the entire read sequence is desirable when (1) the assumed number of errors is low and (2) the reads are short, as is the case for small RNA sequencing, for example.

6.2.2.1 Frequency-Based De Novo Short-Read Clustering (FreClu)

Qu et al. [18] introduced the *FreClu* method which is based on the idea of grouping or clustering reads such that each cluster represents a collection of erroneous short reads that can all be attributed to a parent (correct) read. Let r_1 and r_2 be two short reads and let f_1 and f_2 be their frequencies in the dataset.

We say that r_1 is a parent of r_2 (and so r_2 belongs to the r_1 cluster) if:

1. the Hamming distance between r_1 and r_2 is 1 and
2. a hypothesis test determines that the frequency observed for r_2 (f_2) can be explained as an error in reads coming from the same transcript as r_1.

The hypothesis test relies on f_1 as well as on the average quality value for Q in r_1. If we reject the null hypothesis, then we do not assign r_2 to r_1. Otherwise, r_2 is assigned to the r_1 cluster and base at position Q in r_2 is assumed to be an error which is corrected by changing it to the base at position Q in r_1.

The main challenge in the above procedure is how to perform such assignment process efficiently. Qu et al. described a $\mathscr{O}(ml)$ greedy algorithm for this task where

m is the total number of short reads and l is the length of each read. Their algorithm starts by hashing the set of 1-distance neighbors for each read and then sorting them according to their frequencies. Next, going from the bottom (lowest frequency) read r to the top, they perform the two tests for the most abundant read in the Hamming neighborhood of r and stop either when r is assigned to a parent or when no parents can be assigned in which case r is a parent of its own cluster. Thus, while the run time of the algorithm is very efficient (linear in the number of reads), memory requirements can be large because it depends on the neighborhood of each read which can be $3l$ large.

An application of the method to correct errors in data from small RNA from the HT-29 human colon adenocarcinoma cell line using the *Illumina* reads showed that the method is effective in reducing read errors. Using the reference human genes, Qu et al. showed that the percentage of nonaligned redundant reads decreased from 12% to 10%, and for nonredundant reads, it decreased from 38% to 32%. Perhaps more impressively, the fraction of nonredundant sequences that can be perfectly match *Recount*ed to the genome increased threefold, from 8% to 25% illustrating the power of the method.

6.2.2.2 *Recount*: Expectation Maximization-Based Read Error Corrections

While *FreClu* proved to be successful at reducing read errors, it requires large memory which may not be appropriate for all applications. To overcome this problem, Wijaya et al. [26] introduced *BayesHammer*, a probabilistic method that uses the *Expectation Maximization* (EM) algorithm to infer the correct counts of reads without performing genome alignments. Similar to *FreClu*, *Recount* assigns erroneous reads to other reads by increasing the count of the correct reads (effectively changing the erroneous bases that differ between erroneous and correct reads). However, unlike *FreClu*, *Recount* is not greedy and is also not deterministic.

Instead, to determine how to assign erroneous reads to correct reads the algorithm computes, for each short read sequence s of length l, the expected error at each base Q (by summing overall quality scores for Q in reads r with sequence s). For each pair of read sequences s_i and s_j, the algorithm computes the value $\alpha_{i,j}$ to denote the probability that sequence s_j is an erroneous read from sequence s_i. Using $\alpha_{i,j}$, *Recount* performs the EM iterations to determine the true counts for read r_i (denoted c_i) in the following way: in the E-step, the expectation for each read count, c_i is computed, and in the M-step, two parameters are learned for a Poisson distribution model of read counts. The first denotes the true fraction of reads from each possible sequence, and the second is the average number of correct reads, i.e., following error correction, λ. The E- and M-steps are iterated until convergence. Once the algorithm converges, new counts are estimated for each read sequence s_i, increasing the counts for correct reads and decreasing those for erroneous reads.

To improve efficiency (similar to *FreClu*), *Recount* only allows assignment of erroneous reads to reads that differ by 1 or 2 bases. However, unlike *FreClu*, it does not need to store a neighborhood for each read but rather computes the α parameters which leads to large reduction in memory requirements. On the other hand, while each iteration of the EM algorithm is linear in the number of reads and their length, unlike *FreClu* that only does one pass over all reads, the number of iterations of the E- and M-steps in *Recount* is not known in advance which can lead to longer runtimes.

In an analysis of error correction of mouse short reads expression data, *Recount* was shown to improve the number of mappable reads by almost 14%, whereas *FreClu* improved this number by 11.5%. As expected, *FreClu* was faster than *Recount* (by a factor of 1.35); however, *Recount* was almost 15 times more memory efficient than *FreClu* for this data.

6.2.3 Hamming Graphs over k-mers of Reads

In the previous subsection, the Hamming graph was built over the entire read sequence. However, as read length increases with advances in sequencing technology, the number of absolute errors per read increases. When Hamming graphs over reads are used, it becomes harder to recognize reads with errors.

Consider for example an erroneous read that overlaps with the beginning of a set of reads R_i and at the end with another abundant set of reads R_j. The methods described above cannot be used to correct this erroneous read because the Hamming distance to both sets is very large (the overlap is only half the read). However, it is likely that all three different sets of reads came from the same abundant sequence, and so we should be able to use R_i and R_j to correct the erroneous read. Note that the Hamming graph over k-mers may allow this if k is chosen small enough to recognize overlap with R_i and R_j. Therefore, methods that break the read into k-mers have been proposed to avoid the shortcomings for correcting long reads using Hamming graphs.

6.2.3.1 *Hammer*: Consensus Sequences in Connected Components

Medvedev et al. [14] proposed the *Hammer* algorithm for read error correction with Hamming graphs over k-mers, i.e., $\mathcal{G} = (\mathcal{V}, \mathcal{E})$ with a node $n \in \mathcal{V}$ corresponds to $s(n) \in \Sigma^k$, $s(n) \in \mathcal{R}$. Their algorithm is based on the following simple observation. A true k-mer u of a read r generates erroneous k-mers with sequencing error rate ϵ; all k-mers together are termed the generating set $G(u)$. If we build the Hamming graph $\mathcal{G}$, then all k-mers in $G(u)$ will form a connected component in $\mathcal{G}$. Of course, in reality we do not know the true k-mer u, so we have to compute the best possible generating k-mer from the connected components (CCs) of $\mathcal{G}$. Assuming that all k-mers in a CC are generated by the same k-mer u, the task becomes to find the k-mer that minimizes the distance to all other k-mers in the CC. Formulating that as a maximum likelihood inference, it turns out that the weighted consensus sequence of all k-mers in the *CC* represents the most likely generating k-mer, leading to the following simple algorithm:

***Hammer* algorithm**

1. Compute multiplicity of each k-mer $\in \mathscr{R}$.
2. Build Hamming graph $\mathscr{G}$.
3. Identify connected components in $\mathscr{G}$.
4. Compute the weighted consensus sequence for each component.
5. Correct k-mers in the component to the consensus.
6. Update and report corrected k-mers.

Medvedev et al. use a simple sorting algorithm to compute unique k-mers in $\mathscr{R}$ and their multiplicity in $\mathscr{O}(n_k \log n_k)$ time, where n_k is the number of k-mers in $\mathscr{R}$. Spaced seeds of the k-mers are used to determine pairwise overlaps between the k-mers, which are first sorted in $\mathscr{O}(K \log K)$ time, where K is the number of distinct k-mers in $\mathscr{R}$. Then, the *CCs* of the graph are computed by using a union-find data structure in $\mathscr{O}(K^2)$ time, although the authors note that as the components are small, this part of the algorithm runs much faster in practice. Finally, the consensus sequences of the clusters are computed by going over all distinct k-mers in $\mathscr{O}(K)$ time. We note here, that step 5 in the *Hammer* algorithm, is a bit more involved. For example, in order to deal with the problem that not all k-mers may be generated by the same k-mer (due to repeats, for example), the algorithm uses a threshold named `saveCutoff`, and all k-mers that appear more often than `saveCutoff` are never removed.

Medvedev et al. apply their algorithm to *E. coli* single-cell sequencing data with reads of length 100 bps. They show that the number of correct k-mers after correction with *Hammer* is higher than all other error correction methods tested. They observe that in the *E. coli* dataset, around 13,000 *CCs* contained more than one correct, generating k-mers, many of which could be rescued due to the use of the `saveCutoff` as explained above.

6.2.3.2 *BayesHammer*: Bayesian Subclusters of Connected Components

Nikolenko et al. [16] proposed an extension of the *Hammer* algorithm for read error correction with Hamming graphs over k-mers. Their main motivation for introducing *BayesHammer* was to tackle the problem that more than one generating, true k-mer can be part of a *CC* in the Hamming graph. As it is unknown how many generating k-mers are part of one *CC*, the authors suggest to use a clustering approach to find k-mer centers. They formulate a function $L(k_1|k_2)$ to quantify the likelihood that k-mer k_1 was generated by another k-mer k_2, assuming independent error probabilities. This function can then be used to express the likelihood

$L_c(C_1, \ldots, C_c)$ for a specific subclustering $C_1, \ldots, C_c$, where c is the number of subclusters of a component. Similar to the *Hammer* algorithm, the consensus k-mer of all k-mers represents the center of the subcluster. They use a modified k-means algorithm with the mentioned sub cluster likelihood function as objective for clustering of k-mers in a *CC*. The best number of subclusters is selected using the *Bayesian Information Criterion* (BIC). The complete algorithm looks as follows:

***BayesHammer* algorithm**

1. Compute the multiplicity of each k-mer $\in \mathscr{R}$.
2. Build Hamming graph $\mathscr{G}$.
3. Identify connected components in $\mathscr{G}$.
4. Compute Bayesian subclusters for each component.
5. Select high-quality *solid* k-mers from subcluster centers
6. Iteratively expand the set of *solid* k-mers.
7. Correct each read base with consensus over *solid* k-mers.
8. Update and report corrected reads

As seen in step 5 of the algorithm, the *BayesHammer* algorithm also uses a different approach for read error correction using the subclusters obtained in the previous steps. Briefly, each center k-mer of a subcluster is termed *solid* if its total quality exceeds a predefined threshold. Then, all k-mers of *solid* reads, i.e., reads that are completely covered with *solid* k-mers, are termed *solid* k-mers too. This procedure is done iteratively until no k-mer is turned *solid* anymore. Finally, all positions in a read are replaced with the consensus over *solid* k-mers and *solid* centers of all nonsolid k-mers overlapping the base.

The authors apply the *BayesHammer* algorithm to single-cell sequencing of *E. coli* and *S. aureus* and investigate the improvement of *BayesHammer* with Bayesian subclustering compared to *Hammer*. Interestingly, computing the subclusters has positive effects and increases the mappability of reads after correction as well as leading to the smallest set of nongenomic k-mers in the corrected read sets. In addition, they show that the de novo genome assembly from single-cell data with *Spades* [1] works best using *BayesHammer*, and produces the most contiguous assemblies with a small number of assembly errors.

6.2.3.3 *PREMIER*: HMM Error Correction of Individual Reads

Another approach, called *PREMIER*, which uses the Hamming graph over k-mers, is presented by Yin et al. [27]. Their method was initially designed and applied on multicell DNA sequence data. However, the method does not directly assume uniform coverage, and since the method is one of the first to use graphical models for the error correction task, we briefly mention it here.

PREMIER builds a *hidden Markov model* (HMM) for each read using the global frequency information for *k*-mers (across all reads). Each read is thought of as a continuous set of overlapping *k*-mers. The correct *k*-mers from which this read was obtained are represented by the (unobserved) states in the HMM. The basic idea is that the observed *k*-mers are emitted from the true *k*-mers using a probabilistic emission model. Thus, errors in the *k*-mers can be represented by this emission process, and by using the HMM inference process (the Viterbi algorithm), we can obtain the true states that generated each of the reads and use these to correct errors. The set of possible states is constrained to be only observed *k*-mers, which allows to drastically reduce the complexity of the model. Emission probabilities are also constrained by placing a cutoff on the potential difference between a state (unobserved *k*-mer) and the observed *k*-mer that is assumed to be emitted from that state. Since HMM learning is cubic in the number of states, which is too slow for the task of learning a model for each read, Lin et al. use a number of speedup methods. These methods are used to obtain an approximate solution to the forward-backward algorithm by severely limiting the number of potential transition probabilities the algorithm keeps track of for each state and path in the model. While the authors do not provide an explicit asymptotic run time analysis, it seems that the algorithm is likely linear in the length of the read (for each read).

The method was applied to relatively small datasets (one million and five million reads of sequencing data from *C. elegans* and *E. coli*, respectively) and was shown to improve upon other DNA-Seq error correction methods. Unfortunately, the authors do not discuss run time or memory usage, making it hard to determine how well this method can scale to much larger datasets.

6.3 *SEECER*: De Novo Error Correction Using Profile Hidden Markov Models

The HMM and other methods discussed above work by correcting individual reads one at a time. While such method can offer several advantages, it is restricted by the requirement to limit the number of errors allowed for each read (for example, Hamming distance 1) and the fact that in many cases, the context for the read or *k*-mer is lost which means that partially overlapping reads from the same genomic location are usually not used to improve read error correction. In addition, all previous methods can only correct for mismatches but cannot be used to correct for indels, since these change the location of all subsequent bases leading to a large Hamming distance.

To address these issues, we have recently developed the *SEquencing Error CorrEction in RNA-Seq* data (SEECER) method [10], which relies on *profile hidden Markov models* (PHMMs) for performing de novo read error correction. Here, PHMMs are used to describe the nucleotide distribution of all reads in a multiple sequence alignment. Unlike prior machine learning methods for this task, which are centered around individual reads, *SEECER* joins partially overlapping reads into contigs and uses these contigs to correct for mismatches and indels leading to both better usage of correct reads and increased ability to correct sequencing errors.

SEECER models contigs (a set of partially overlapping reads that represent a consensus region) with a PHMM recognizing substitutions, insertions and deletions. We start by selecting a random read from the set of reads that have not yet been assigned to any PHMM contig. Next, we extract (using a fast hashing of k-mers) all reads that overlap with the selected read in at least k nucleotides (with $k \ll r$, where the read length r, thus making hashing more efficient than *FreClu*). Because the subset of overlapping reads can be derived from alternatively spliced or repeated segments, we next identify the largest coherent subset of these reads (where coherent in this context refers to the entropy along the matched positions in the multiple sequence alignment) and use these as the seed reads for learning an initial PHMM contig. Next, we use the consensus sequence defined by the PHMM to extract more reads from our unassigned set (again, using the hashed k-mers). These additional reads are used to extend the PHMM in both directions. This process, retrieving new overlapping reads and extending the PHMM, repeats until no more reads overlap the current PHMM or the entropy at the edges of the PHMM exceeds a predefined threshold.

Unlike standard PHMMs which usually have a predefined number of states, *SEECER* needs to determine the number of states, which corresponds to the final length of the contig and is unknown in advance. In addition, because *SEECER* learns a PHMM for each transcript (and in many cases, multiple PHMMs for alternatively spliced transcripts), the total number of PHMMs *SEECER* learns is often hundreds of thousands. This means that the standard PHMM learning procedure, which is cubic in the length of the transcript, cannot be used (PHMM learning takes $\mathcal{O}(n^t)$ where n is the number of states and t is the length of the input sequences; in our case, both are linearly related to the length of the reconstructed PHMM transcript). Instead, *SEECER* uses a number of speedup procedures that reduce run time for learning a single transcript PHMM to $\mathcal{O}(ml)$ where m is the number of reads assigned to that PHMM and l is the length of each read leading to a linear run time in the number of total reads.

To achieve fast run time while maintaining accurate models, *SEECER* implements several strategies. These include initial subset selection to drastically reduce the seed set size of a PHMM contig, constrained models that only allows limited insertions and deletions based on the number of current expected errors [4], blockwise online learning of model parameters, and using entropy to determine when to terminate the learning process and output the final contig PHMM. Using these methods, it was shown that *SEECER* can be applied to large datasets in a reasonable time (145M paired-end, 101 bps reads in 40 h using a eight node cluster).

SEECER was tested using diverse human RNA-Seq datasets and was shown to improve both the percentage of mappable reads and the performance of downstream de novo assembly of these reads. We also compared *SEECER* against other deterministic de novo error correction methods. In all cases, the probabilistic, machine learning- based, approach of *SEECER* outperformed the other methods [10]. In addition to the improved alignment and assembly, using *SEECER* improves the ability to perform SNP calling. Using *SEECER* led to the largest number of SNP calls matching annotated SNPs in dbSNP [20] when compared to uncorrected data and to data corrected using other methods. To illustrate the ability of *SEECER* to perform de novo analysis of nonsequenced species, it was used to analyze RNA-Seq data for the sea cucumber [10]. The ability to accurately analyze de novo RNA-Seq data allowed the identification of both conserved and novel transcripts and provided important insights into sea cucumber development.

6.4 Comparison Between Approaches

In order to provide an insight into the correction of nonuniform datasets, we analyzed a RNA-Seq dataset (SRX016367, SRA database[2]) from the human reference tissue [2] with the *BayesHammer* and *SEECER* method introduced before. The dataset contains about 92 million single end reads of length 35 bps. We consider the effect of error correction on the performance of the de novo transcriptome assembler *Oases* [21]. *Oases* is a multi-k de novo transcriptome assembler that uses the de Bruijn graph for assembly. It was shown that methods based on de Bruijn graphs are susceptible to sequencing errors. Short reads are combined into isolated components in the de Bruijn graph and are used for prediction of full length transcripts.

We ran *SEECER* and *BayesHammer* with default parameters, *SEECER* uses $k = 17$ and *BayesHammer* $k = 21$. We note that the *Oases* assembler uses 21 as the smallest k-mer for assembly, using *Oases* default parameters. As *BayesHammer* discards low-quality reads after correction, we added all these discarded reads back to the file with corrected reads for better comparability with *SEECER*. Assembled transcripts were aligned against the human genome using *Blat* [9] and compared with gene annotation from *Ensembl* version 63. Error correction performance was measured with four metrics, nucleotide sensitivity and specificity, full length and 80% length reconstruction rate for *Ensembl* annotated transcripts, as previously introduced [21], see Table 6.2 legend for details.

In Table 6.2 we list the performance of *Oases* assembling the original data after correction with *BayesHammer* or *SEECER*. The correction with *SEECER* before de novo assembly results in a significant increase of full length and 80% transcripts assembled and higher nucleotide sensitivity and specificity of the predicted transcripts. Correction with the *BayesHammer* method leads to a slightly decreased performance compared to the original data, which may be due to the fact that *BayesHammer* was originally designed for single-cell genomic sequencing and the default parameters do not work well for RNA-Seq data.

[2]http://www.ncbi.nlm.nih.gov/sra.

Table 6.2 Assessment of error correction on the performance of de novo transcriptome assembly with *Oases*, on the original data or after using *BayesHammer* or *SEECER*

Data	# transcripts	Sensitivity	Specificity	100% length	80% length
Original	146,783	20.85	80.08	2776	16,789
BayesHammer	154,840	20.74	76.23	2368	15,024
SEECER	133,681	21.89	80.86	3291	18,157

The RNA-Seq dataset was derived from universal human reference tissue. The total number of assembled transcripts is shown in column 2. Column 3 denotes the nucleotide sensitivity, i.e., the percentage of base pairs of the reference annotation (*Ensembl* 63) that are reconstructed with the assembled transcripts. Similarly, column 4 denotes nucleotide specificity, i.e., the percentage of base pairs of the mapped assembled transcripts that overlap with reference annotation. Column 5 and 6 denote *Ensembl* 63 human transcripts recovered to 100% and 80% of their original length with assembled transcripts, respectively

6.5 Discussion

In this chapter, we reviewed different approaches for the correction of nonuniform sequencing datasets, including RNA-Seq, metagenomics, metatranscriptomics, or single-cell sequencing. Most methods reviewed in this paper utilize the Hamming graph over reads or k-mers and can only correct substitution errors by using nodes with a small Hamming distance to the corrected node (one or two). A notable exception to this strategy is the *SEECER* error correction method that indexes k-mers of reads to construct contig HMMs for read error correction and is able to correct indel and substitution errors.

There are several attributes that are shared between all methods discussed in this review. First, all try to address the fact that the frequency of reads and errors is nonuniform. Second, all methods use a probabilistic model, often based on a machine learning method, to detect and correct reads with errors. The common underlying idea behind these methods is that a read with low abundance which is very similar to a highly abundant read is very likely the result of a sequencing error.

While several similarities exist between these methods, there are also differences. Some, though not all, of the methods use the read quality values to assess the likelihood of error in a specific base pair, though relatively little work attempted to investigate the benefit of using such quality values compared to not using them. In addition, most of the methods only attempt to correct substitution errors, as they base their algorithms on the Hamming graph, a decision that is motivated by the large number of studies using *Illumina* sequencing data. However, more recent sequencing technologies show more bias toward indel errors, and so methods that can correct for such errors may become more important in the future. One such method is *SEECER* that was shown to perform well on an illustrative benchmark of error correction of RNA-Seq data before de novo assembly. However, we believe that there is plenty of room for improvement of nonuniform error correction methods to better handle indel errors and improve error correction accuracy and computational speed and memory usage.

Currently there are very few studies that apply methods tailored to nonuniform error correction and investigate the effect of error correction on downstream analyses. For example, we did not find any studies investigating these issues for metagenomics or metatranscriptomics de novo assembly or SNP calling. There is an urgent need for such studies to investigate the importance of nonuniform error correction methods to guide future research.

Acknowledgements We would like to thank Dilip Ariyur Durai for his help with the *Oases* benchmark.

References

1. Bankevich, A., Nurk, S., Antipov, D., Gurevich, A.A., Dvorkin, M., Kulikov, A.S., Lesin, V.M., Nikolenko, S.I., Pham, S., Prjibelski, A.D., Pyshkin, A.V., Sirotkin, A.V., Vyahhi, N., Tesler, G., Alekseyev, M.A., Pevzner, P.A.: SPAdes: a new genome assembly algorithm and its applications to single-cell sequencing. J. Comput. Biol. **19**(5), 455–477 (2012)
2. Bullard, J.H., Purdom, E., Hansen, K.D., Dudoit, S.: Evaluation of statistical methods for normalization and differential expression in mRNA-seq experiments. BMC Bioinform. **11**, 94 (2010)
3. Embree, M., Nagarajan, H., Movahedi, N., Chitsaz, H., Zengler, K.: Single-cell genome and metatranscriptome sequencing reveal metabolic interactions of an alkane-degrading methanogenic community. ISME J. **8**(4), 757–767 (2014)
4. Glenn, T.C.: Field guide to next-generation DNA sequencers. Mol. Ecol. Resour. **11**(5), 759–769 (2011)
5. Grabherr, M.G., Haas, B.J., Yassour, M., Levin, J.Z., Thompson, D.A., Amit, I., Adiconis, X., Fan, L., Raychowdhury, R., Zeng, Q., Chen, Z., Mauceli, E., Hacohen, N., Gnirke, A., Rhind, N., di Palma, F., Birren, B.W., Nusbaum, C., Lindblad-Toh, K., Friedman, N., Regev, A.: Full-length transcriptome assembly from RNA-seq data without a reference genome. Nat. Biotechnol. **29**(7), 644–652 (2011)
6. Hemme, C.L., Deng, Y., Gentry, T.J., Fields, M.W., Wu, L., Barua, S., Barry, K., Tringe, S.G., Watson, D.B., He, Z., Hazen, T.C., Tiedje, J.M., Rubin, E.M., Zhou, J.: Metagenomic insights into evolution of a heavy metal-contaminated groundwater microbial community. ISME J. **4**(5), 660–672 (2010)
7. Hinman, V.F., Nguyen, A.T., Davidson, E.H.: Expression and function of a starfish Otx ortholog, AmOtx: a conserved role for Otx proteins in endoderm development that predates divergence of the eleutherozoa. Mech. Dev. **120**(10), 1165–1176 (2003)
8. Kelley, D.R., Schatz, M.C., Salzberg, S.L.: Quake: quality-aware detection and correction of sequencing errors. Genome Biol. **11**(11), R116 (2010)
9. Kent, W.J.: Blat—the blast-like alignment tool. Genome Res. **12**(4), 656–664 (2002)
10. Le, H.-S., Schulz, M.H., McCauley, B.M., Hinman, V.F., Bar-Joseph, Z.: Probabilistic error correction for RNA sequencing. Nucleic Acids Res. **41**(10), e109 (2013)
11. Le Chatelier, E., Nielsen, T., Qin, J., Prifti, E., Hildebrand, F., Falony, G., Almeida, M., Arumugam, M., Batto, J.-M., Kennedy, S., Leonard, P., Li, J., Burgdorf, K., Grarup, N., Jorgensen, T., Brandslund, I., Nielsen, H.B., Juncker, A.S., Bertalan, M., Levenez, F., Pons, N., Rasmussen, S., Sunagawa, S., Tap, J., Tims, S., Zoetendal, E.G., Brunak, S., Clement, K., Dore, J., Kleerebezem, M., Kristiansen, K., Renault, P., Sicheritz-Ponten, T., de Vos, W.M., Zucker, J.-D., Raes, J., Hansen, T., MetaHIT consortium, Bork, P., Wang, J., Ehrlich, S.D., Pedersen, O., MetaHIT consortium additional members: Richness of human gut microbiome correlates with metabolic markers. Nature **500**(7464), 541–546 (2013)

12. Mardis, E.R.: Next-generation DNA sequencing methods. Annu. Rev. Genomics Hum. Genet. **9**, 387–402 (2008)
13. Marioni, J.C., Mason, C.E., Mane, S.M., Stephens, M., Gilad, Y.: RNA-seq: an assessment of technical reproducibility and comparison with gene expression arrays. Genome Res. **18**(9), 1509–1517 (2008)
14. Medvedev, P., Scott, E., Kakaradov, B., Pevzner, P.: Error correction of high-throughput sequencing datasets with non-uniform coverage. Bioinformatics (Oxford, England) **27**(13), i137–i141 (2011)
15. Mortazavi, A., Williams, B.A., McCue, K., Schaeffer, L., Wold, B.: Mapping and quantifying mammalian transcriptomes by RNA-seq. Nat. Methods **5**(7), 621–628 (2008)
16. Nikolenko, S., Korobeynikov, A., Alekseyev, M.: Bayeshammer: Bayesian clustering for error correction in single-cell sequencing. BMC Genomics **14**(Suppl. 1), S7 (2013)
17. Peng, Z., Cheng, Y., Tan, B.C.-M., Kang, L., Tian, Z., Zhu, Y., Zhang, W., Liang, Y., Hu, X., Tan, X., Guo, J., Dong, Z., Liang, Y., Bao, L., Wang, J.: Comprehensive analysis of RNA-seq data reveals extensive RNA editing in a human transcriptome. Nat. Biotechnol. **30**(3), 253–260 (2012)
18. Qu, W., Hashimoto, S.-I., Morishita, S.: Efficient frequency-based de novo short-read clustering for error trimming in next-generation sequencing. Genome Res. **19**(7), 1309–1315 (2009)
19. Richard, H., Schulz, M.H., Sultan, M., Nürnberger, A., Schrinner, S., Balzereit, D., Dagand, E., Rasche, A., Lehrach, H., Vingron, M., Haas, S.A., Yaspo, M.-L.: Prediction of alternative isoforms from exon expression levels in RNA-seq experiments. Nucleic Acids Res. **38**(10), e112 (2010)
20. Saccone, S.F., Quan, J., Mehta, G., Bolze, R., Thomas, P., Deelman, E., Tischfield, J.A., Rice, J.P.: New tools and methods for direct programmatic access to the dbSNP relational database. Nucleic Acids Res. **39**(Database issue), D901–D907 (2011)
21. Schulz, M.H., Zerbino, D.R., Vingron, M., Birney, E.: Oases: robust de novo RNA-seq assembly across the dynamic range of expression levels. Bioinformatics (Oxford, England) **28**(8), 1086–1092 (2012)
22. Schulz, M.H., Weese, D., Holtgrewe, M., Dimitrova, V., Niu, S., Reinert, K., Richard, H.: Fiona: a parallel and automatic strategy for read error correction. Bioinformatics **30**(17), i356–i363 (2014)
23. Sultan, M., Schulz, M.H., Richard, H., Magen, A., Klingenhoff, A., Scherf, M., Seifert, M., Borodina, T., Soldatov, A., Parkhomchuk, D., Schmidt, D., O'Keeffe, S., Haas, S., Vingron, M., Lehrach, H., Yaspo, M.-L.: A global view of gene activity and alternative splicing by deep sequencing of the human transcriptome. Science **321**(5891), 956–960 (2008)
24. Treangen, T., Koren, S., Sommer, D., Liu, B., Astrovskaya, I., Ondov, B., Darling, A., Phillippy, A., Pop, M.: Metamos: a modular and open source metagenomic assembly and analysis pipeline. Genome Biol. **14**(1), R2 (2013)
25. Wang, Z., Gerstein, M., Snyder, M.: RNA-seq: a revolutionary tool for transcriptomics. Nat. Rev. Genet. **10**(1), 57–63 (2009)
26. Wijaya, E., Frith, M.C., Suzuki, Y., Horton, P.: Recount: expectation maximization based error correction tool for next generation sequencing data. Genome Inform. **23**(1), 189–201 (2009). International Conference on Genome Informatics
27. Yin, X., Song, Z., Dorman, K., Ramamoorthy, A.: PREMIER Turbo: probabilistic error-correction using Markov inference in errored reads using the turbo principle. In: 2013 IEEE Global Conference on Signal and Information Processing, December, pp. 73–76. IEEE, New York (2013)
28. Zeller, G., Tap, J., Voigt, A.Y., Sunagawa, S., Kultima, J.R., Costea, P.I., Amiot, A., Böhm, J., Brunetti, F., Habermann, N., Hercog, R., Koch, M., Luciani, A., Mende, D.R., Schneider, M.A., Schrotz-King, P., Tournigand, C., Van Nhieu, J.T., Yamada, T., Zimmermann, J., Benes, V., Kloor, M., Ulrich, C.M., von Knebel Doeberitz, M., Sobhani, I., Bork, P.: Potential of fecal microbiota for early-stage detection of colorectal cancer. Mol. Syst. Biol. **10**(11), 766 (2014)

Chapter 7
DNA-Seq Error Correction Based on Substring Indices

David Weese, Marcel H. Schulz, and Hugues Richard

7.1 Introduction

Next-Generation Sequencing (NGS) has revolutionized genomics. NGS technologies produce millions of sequencing reads of a few hundred bases in length. In the following, we focus on NGS reads produced by genome sequencing of a clonal cell population, which has important applications like the de novo genome assembly of previously unknown genomes, for example, recently mutated parasites [22] or newly sequenced genomes [16]. Another important area is the resequencing of already assembled genomes and the analysis of *Single Nucleotide Polymorphisms* (SNPs) or larger structural variations, for example, to predict virulence factors in genetically diverse strains [3] or to reveal differences in populations [2, 37].

D. Weese (✉)
Department of Mathematics and Computer Science, Freie Universität Berlin, Berlin, Germany
e-mail: david.weese@fu-berlin.de

M.H. Schulz
Excellence Cluster for Multimodal Computing and Interaction, Saarland University, Saarbrücken, Germany

Computational Biology and Applied Algorithms, Max Planck Institute for Informatics, Saarbrücken, Germany
e-mail: mschulz@mmci.uni-saarland.de

H. Richard
Sorbonne Universités, UPMC Université Paris 06, UMR 7238, F-75005 Paris, France

Laboratory of Computational and Quantitative Biology, Institut Biologie Paris-Seine (LCQB-IBPS), F-75005 Paris, France
e-mail: hugues.richard@upmc.fr

M. Elloumi (ed.), *Algorithms for Next-Generation Sequencing Data*,
DOI 10.1007/978-3-319-59826-0_7

Due to the plethora of applications of genome sequencing, the correction of errors introduced by the sequencing machine—substitution errors as well as insertions or deletions (indels)—has recently attracted attention. It has been shown that error correction before analysis can improve de novo genome or transcriptome assembly performance [12, 17, 29, 30, 35], e.g., in the form of larger contigs and a reduced mismatch rate. Error correction prior to alignment improves SNP detection as well because it leads more accurate alignment and deeper read coverage at SNP locations [11, 12].

7.1.1 Sequencing Technologies

Let us explore now the existing sequencing technologies and what kind of errors may occur during sequencing. During the last few years, sequencing throughput increased dramatically with the introduction of so-called *High-Throughput Sequencing* (HTS), also known as *deep sequencing* or *Next-Generation Sequencing* (NGS), which nowadays allows the sequencing of billions of *base pairs* (bp) per day and produces millions of reads of length 100 bp and more. Since 2005, when *454 Life Sciences* released the first commercially available NGS machine, throughput has continued to increase and new technologies provide longer reads than currently available. Moreover, the sequencing costs are decreasing more rapidly than the costs for hard disk storage or Moore's law for computing costs [36].

Currently available NGS platforms include ABI SOLiD (*Life Tech. Corp.*), *Illumina* (*Illumina Inc.*), *Ion Torrent*™ (*Life Tech. Corp.*), *Single Molecule Real Time* (SMRT)™ (*Pacific Biosciences Inc.*), and *454* (*Roche Diagnostics Corp.*). See Table 7.1 for a comparison of their throughputs, run times, and expendable costs. Compared to older sequencing technologies by gel electrophoresis, a key improvement is its ability to cycle and image the incorporation of nucleotides or to detect the incorporation in real time. Removing the necessity of placing each sequence on a separate gel tremendously reduced the sequencing costs and has made it possible to miniaturize and parallelize sequencing. Common to all technologies is that the DNA is first fractionated into smaller double-stranded fragments, which

Table 7.1 Approximate run times, yields, read lengths, costs, and sequencing error rates of different high-throughput sequencing technologies by mid-2011 [7]

Technology	Instrument	Run time	Yield (Mb/run)	Read length (bp)	Costs ($/Mb)	Error rate (%)
Sanger	*3730xl* (capillary)	2 h	0.06	650	1500	0.1–1
Illumina	*HiSeq 2000*	8 days	200,000	2×100	0.10	≥ 0.1
SOLiD	SOLiD 4	12 days	71,400	$50 + 35$	0.11	> 0.06
454	*FLX Titanium*	10 h	500	400	12.4	1
SMRT™	*PacBio RS*	0.5–2 h	5–10	860–1100	11–180	16

are optionally amplified and then sequenced in parallel from one or both ends. The sequencing process is either cycled, where a single base (*Illumina*), a single primer (SOLiD), or a run of a single nucleotide (*454*) is incorporated in each, or in real time (SMRT, *Ion Torrent*), where the length, the intensity, or the color of a signal determines the length and nucleotide of the incorporated homopolymer run.

Cycled technologies amplify the DNA fragments within small clusters (*Illumina*, SOLiD, *454*) and require that each cluster contain thousands of identical molecules that are sequenced synchronously and whose combined signal is measured. Synchronization is realized either by using reversible terminator chemistry, which ensures that only one nucleotide is incorporated at a time (*Illumina*), by ligation of only one primer at a time (SOLiD) or by washing only one nucleotide over all fragments, which results in the incorporation of only one homopolymer run at a time (*454*).

One type of error of technologies with a prior amplification step is the so-called *dephasing*, where due to a synchronization loss in a cluster, e.g., due to missed incorporations, the signals of the current base and their neighbors in the DNA template interfere and lead to an increase of miscalls toward the end of the reads, see Fig. 7.1a. Another type of error is related to a limitation in signal resolution. Technologies that use the signal intensity (*454*) or length (SMRT, *Ion Torrent*) to determine the length of homopolymer runs cannot reliably detect the length of large runs due to a limited sensor resolution and an increase of noise, e.g., caused by variations of the DNA polymerase speed. Such technologies typically produce reads with insertions or deletions in homopolymer runs. A third type of error causes deletions of bases in technologies that omit a prior template amplification (SMRT, *Ion Torrent*) due to weak and undetected signals.

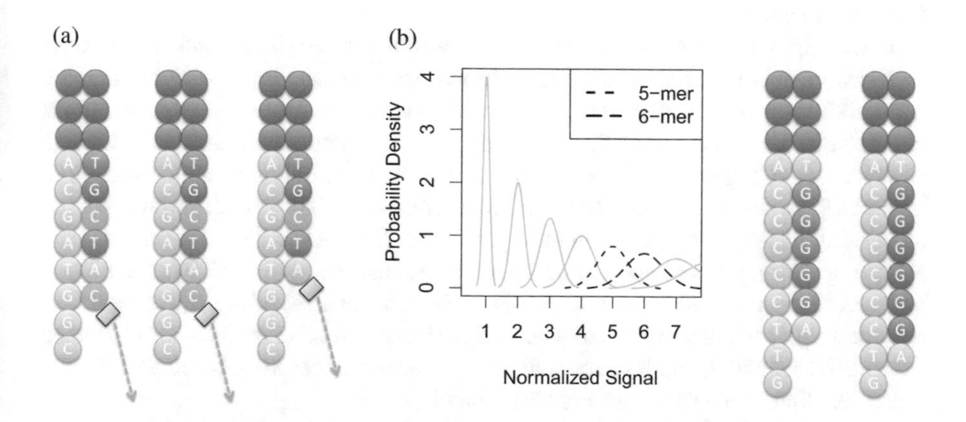

Fig. 7.1 (**a**) Missed nucleotide incorporations or failed terminator cleavage contribute to a loss of synchronization, also called *dephasing*. (**b**) In *454* sequencing, a whole homopolymer run is incorporated at a time. The intensity of the emitted signal follows a normal distribution with mean and deviation proportional to the homopolymer length [19]

In summary, it can be said that the most prevalent error types differ depending on the technology. *Illumina* technology produces mostly substitution errors [23], whereas *454* sequencers cannot accurately determine the length of longer runs of the same nucleotide and tend to produce insertions [34]. More recent technologies such as reads produced by the *Ion Torrent Personal Genome Machine* and *Pacific Biosciences RS* were shown to have a higher rate of indel errors and substitution errors [27].

7.1.2 Substring Indices

The advances in sequencing technology and the continuously increasing sequencing throughput demand for novel approaches and efficient data structures specifically designed for the analysis of mass data. One such data structure is the substring index that represents all substrings or substrings up to a certain length contained in a given text.

One of the first substring indices is the *suffix tree* proposed by Weiner [39]. It represents all substrings of a text of length n in $\mathcal{O}(n)$ memory, can be constructed in $\mathcal{O}(n)$ time[1] [39], and supports exact string searches in optimal time. In the following years, practically faster linear-time construction algorithms were proposed that use less memory [20] or read the text in a sequential scan [38]. However, its memory consumption of roughly $20n$ bytes makes it inapplicable to large analyses of whole genomes. Nowadays, many alternative substring indices are available that behave like a suffix tree but at a smaller memory or running time footprint. The most commonly used indices are the *suffix array* [18], the *enhanced suffix array* [1], and the *FM-index* [6].

In its original definition, the *suffix tree* is a rooted tree whose edges are labeled with strings over an alphabet Σ. Each internal node is branching, i.e., it has at least two children, and outgoing edges begin with distinct characters. It is called *suffix* tree because each path from the root node to a leaf spells out a suffix of the text and vice versa. Figure 7.2 shows an example of a suffix tree for a single sequence `banana$` and the generalization to multiple sequences.[2] The suffix tree compresses repetitions in the text into a common path from the root, which not only allows searching a string in a single top-down traversal but also finding all exact overlaps between sequence reads with a Depth-First Search traversal of the suffix tree. The so-called *suffix trie* is a suffix tree where edges labeled with more than one character are split into a path of nodes such that the connecting edges are labeled with single characters that spell out the original edge label.

Another much simpler substring index is the *k-mer index* which stores all substrings of a fixed length k. It can be constructed much faster than suffix tree

[1] Assuming a constant-sized alphabet.

[2] The artificial `$` characters must be appended to each sequence to let all suffixes end in leaves.

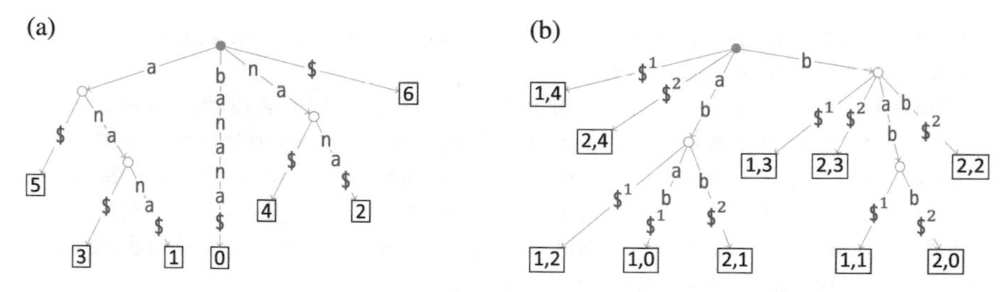

Fig. 7.2 (**a**) The suffix tree of `banana$`. (**b**) The generalized suffix tree of the two sequences `abab$`^{1} and `babb$`^{2}. Leaves are labeled with string positions i or (j, i) representing suffixes (of the j'th sequence) starting at position i

indices and typically consists of two tables. The first table stores all possible begin positions of k-mers and arranges them so that occurrences of the same k-mer are stored continuously. The second table maps each k-mer to its continuous block in the first table. This can be realized with a *hash map* or a *lookup table* that has an entry for each possible k-mer. Wherever sufficient, k-mer indices are the preferred choice due to their simplicity and efficacy. They can be used for approximate string search, exact seeding, or k-mer statistics.

7.1.3 Existing Error Correction Approaches

All read error correction methods can be divided into three essential tasks that are performed sequentially: (1) read overlap computation, (2) error detection in reads, and (3) actual correction of the error. To perform those tasks, different types of data structure are usually used. Accordingly, Yang et al. [41] classified the existing methods into the three following categories: k-spectrum based, *multiple sequence alignment* (MSA) based and suffix tree/array based.

k-spectrum approaches use a simpler form of the k-mer index that only stores the number of occurrences per k-mer. k-mers with an abundance below a threshold t are considered erroneous. Using *Hamming distance*[3] to other k-mers, an erroneous k-mer is replaced with the k-mer that has the smallest Hamming distance and is not erroneous. Recent approaches additionally incorporate base quality values [12, 40], select the best abundance threshold t using mixture models [4, 5, 12], or replace absolute thresholds by a likelihood formulation [21]. However, the general disadvantage of the k-spectrum approach is that only base substitution errors can be corrected, a severe limitation for NGS technologies like *454* or *Ion Torrent* sequencing. In addition, most of these approaches need to be run with different

[3]The *Hamming distance* between two sequences of same length is the number of positions at which the letters are different.

parameters for a dataset because k and often other parameters have a strong influence on error correction performance.

MSA methods use a k-mer index to retrieve reads with a common k-mer and use the k-mer as an anchor for an initial MSA, which they refine using *dynamic programming* [11, 29]. Errors are corrected using a *majority vote* [29] or a *maximum a posteriori estimation* [11] over the MSA columns. *Coral* [29] and SEECER [14] are the only MSA methods to correct indel errors which on the other hand require a more costly dynamic programming algorithm.

Suffix tree and suffix array approaches use a variable seed length for read overlapping and error detection [10, 28, 31–33]. They consider for each erroneous read a set of correcting reads such that all reads share a $(k-1)$-mer left of the error and the set of correcting reads share a k-mer that ends with the correct base which outvotes the erroneous base. To efficiently find erroneous reads and correcting candidates, they either traverse a generalized suffix tree of all reads [28, 32, 33], in which erroneous and correcting reads occur as children of branching nodes with string depth $k-1$ or a generalized suffix array [10, 33], where each $(k-1)$-mer is a suffix array interval that includes erroneous and correct reads as subintervals. Another approach is to process each read individually and search its suffixes in several top-down traversals [31]. In this case, it is not necessary to record all correction candidates until the end of the traversal but after processing the read, however, at the expense of a higher running time and memory consumption, as the full suffix tree must be kept in memory and cannot be partitioned. The general suffix tree traversal approach can be extended to correct indel errors [28, 33] as well as sequences in color space [28]. However, many suffix tree/array methods lack a clear formulation of error correction [10, 28, 32], i.e., when the same error is encountered multiple times through different seed lengths in the tree, it is not defined which correction has to be applied and in which order multiple errors in a read should be corrected.

In the following, we detail the different suffix tree/array methods and their algorithmic steps as they show the greatest flexibility to correct reads with indels and reads of variable length which are common in *IonTorrent* or *454* datasets. The rest of this chapter is organized as follows: Sect. 7.2 details the statistical properties of k-mers counts and the method used to detect erroneous k-mers; Sect. 7.3 details how current methods use suffix tree and suffix arrays to compute the overlap between reads, and we introduce in Sect. 7.4 a new method that corrects error using multiple suffixes. Finally, Sect. 7.5 compares the performance of the various published methods, and we conclude with a few perspectives in the last section.

7.2 Statistical Detection of Sequencing Errors

The main ingredient to detect reads containing errors relies on the observation that sequencing errors will introduce sequences which are not present in the genome, see Fig. 7.3a. Given the depth of the sequencing, every genomic position is expected

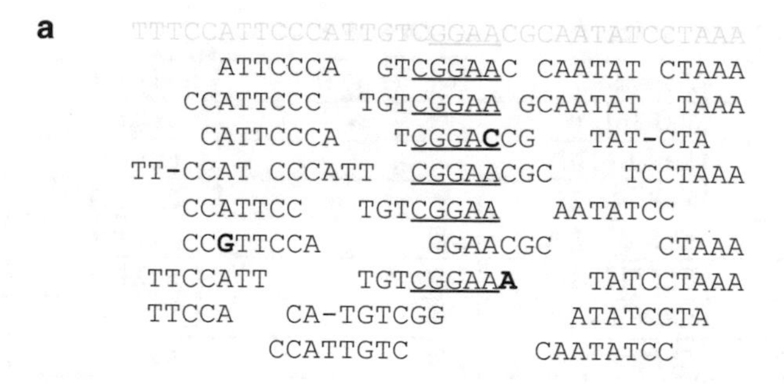

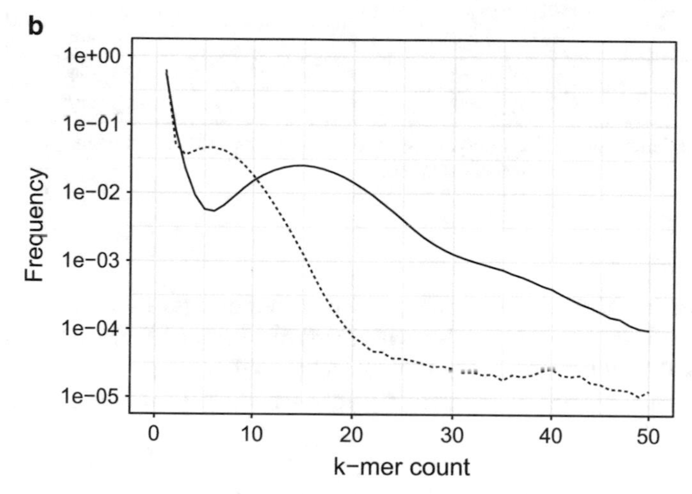

Fig. 7.3 Coverage properties of sequencing errors. (**a**) Due to the redundancy of DNA-Seq data, each string in the genome (*grey*) is expected to be covered by multiple reads. For instance, the string CGGAA is covered by five reads. Sequencing errors (in *boldface*) create substrings which are usually not present in the genome (for instance, CGGAC). However, sequence repeats will obscure this effect, as the string CCATT, repeated twice, and the leftmost read TTCCCAT will be misplaced despite the error. (**b**) Histogram of all substring frequency of length 14 (*plain*) and 30 (*dashed*) in the dataset SRR611140 (see Table 7.2). Most of the very low-frequency *k*-mers are expected to contain sequencing errors—in particular, singletons. Different *k*-mer lengths have different histograms and will not identify the same errors

to be covered by multiple reads, and sequencing errors should thus have a lower coverage, see Fig. 7.3b. Statistical error detection then primarily consists in setting an abundance value under which a string is deemed erroneous. This strategy was first initiated in the correction module of the EULER read assembler [26] which, after selecting the least abundant substrings (usually singletons), corrected them using a nearest-neighbor strategy. The approach has been extended with a more general modeling framework, using models for the sampling of genomic positions [5, 12, 32]. We will first present this strategy in the next two sections and then describe how error detection works for suffix trees and how parameters can be set.

Table 7.2 The experimental read datasets used for evaluation

Organism	Accession	Avg. length (bp)	Read count	Coverage
B. pertussis	ERR161541[b]	142	2,464,690	85×
D. melanogaster	SRX016210[a]	544	4,692,486	18×
E. coli K-12	SRR000868[a]	253	230,517	13×
E. coli K-12	ERR039477[b]	92	390,976	8×
E. coli K-12	SRR611140[b]	162	4,669,065	163×
E. coli K-12	SRR620425[b]	170	4,237,734	156×
E. coli O104:H4	SRR254209[b]	178	977,971	32×
P. falciparum	ERR161543[b]	154	1,959,564	13×
S. aureus	ERR236069[b]	228	1,338,465	109×
S. aureus	SRR070596[a]	514	185,384	34×
S. cerevisae	SRX039441[a]	274	690,237	16×

The average length of all reads (avg. length), the number of reads (read count), and the average genome coverage of all a datasets are shown
[a] *454*
[b] *Ion Torrent*

First we introduce a few notations: we consider strings of a fixed length k over a set $\mathcal{R}$ of m reads sampled over the two strands of a genome $\mathcal{G}$ of length n. The reads have (possibly) different lengths $(\ell_i)_{i=1}^{m}$, and we will always consider that the reads are such that $k < \min(\ell_i)$ and $\max(\ell_i) \ll n$ which is always the case for genome sequencing.

7.2.1 Detecting Erroneous Reads

In order to detect a k-mer with a sequencing error, the strategy is to classify it according to its coverage. If we denote as Y_k the number of occurrences of a k-mer in the population of reads $\mathcal{R}$, its coverage will depend on the number of errors z, which were made while sequencing it. Independently of any assumption regarding the placement of the reads, Y_k can be modelled as a mixture distribution depending on z (and possibly other covariates, see below). Classification is then made according to the value of "$Y_k \mid z = 0$" (no error in the k mer) or "$Y_k \mid z > 0$" (at least one error).

The first publications on this subject proposed to control the proportion of genuine reads detected as erroneous under the model, namely $P(Y_k = c \mid z = 0)$ also called type-I error or specificity [28, 32]. The other strategy [5, 10, 12, 33] finds the count c where specificity is balanced with the proportion of errors that can be detected (e.g. the sensitivity). For instance, Fiona [33] classifies k-mers given their log odds ratio (positive value for erroneous k-mers):

$$\log \frac{P(Y_k = c \mid z > 0)}{P(Y_k = c \mid z = 0)} + w \tag{7.1}$$

The constant w will have an effect on the proportion of erroneous reads detected. To match the setup of a naive Bayes classifier, one reasonable use would be to set it to the log odds ratio of the probability that the k-mer has errors compared to the probability that the k-mer has no error: $w = \log(1 - (1 - \varepsilon)^k)/(1 - \varepsilon)^k$. As we will show in the following section, under a uniform model for reads placement, the only parameters needed to determine the threshold value are the genome length n and the average error rate ε.

7.2.2 Properties of DNA-Seq Sequencing Errors

The fragments before sequencing and, thus, the k-mers composing them are drawn according to binomial sampling along each position in $\mathcal{G}$. If we assume the sampling to be uniform along positions, that any word of length k is unique in $\mathcal{G}$, then the expected count λ_k of a k-mer before sequencing is:

$$\lambda_k = \sum_{i=1}^{m} \frac{\ell_i - k + 1}{n - \ell_i + 1} \tag{7.2}$$

As n is usually large, X_k can be approximated by a Poisson distribution of rate λ_k. The assumption that a word is unique in $\mathcal{G}$ is important, as repeats in $\mathcal{G}$ will have a higher λ_k and would necessitate to add specific terms to the mixture. We restrict ourselves to this hypothesis within this context, although repeats can be integrated in the distribution of Y_k (see below). Errors derived from repeats are in any case difficult to infer based exclusively on their coverage, and it is therefore important to treat and identify words that can originate from repetitive regions.

If we assume a uniform error rate of ε at each base (thus a probability of $\varepsilon/3$ for each substitution), we can derive the expected count for a given k-mer, given its number of sequencing errors, i.e., the distribution of $Y_k \,|z$. The coverage of a k-mer possessing i sequencing errors is distributed according to a Poisson distribution with an expectation of

$$\mu_i = \lambda_k \cdot (\varepsilon/3)^i (1 - \varepsilon)^{k-i} \tag{7.3}$$

Note that this formulation does not incorporate cases where errors would accumulate on other reads in the neighborhood of the k-mer. This effect can be neglected given the relatively low error rate of current sequencers (<5%).

The distribution of Y_k can be obtained by summing over the possible number of errors, which results in a mixture of Poisson distributions with rates μ_i. We denote the proportion of reads with exactly i errors as:

$$\pi_i = \binom{k}{i} \varepsilon^i (1 - \varepsilon)^{k-i} \tag{7.4}$$

which we call the mixture coefficients. It follows the formulation for Y_k:

$$P(Y_k = c) = \sum_{i=0}^{k} \pi_i \cdot P(Y_k = c \mid z = i) = \sum_{i=0}^{k} \pi_i \cdot \frac{e^{-\mu_i} \mu_i^c}{\mu_i!} \tag{7.5}$$

Note that without the $i = 0$ term, this formulation denotes the distribution for reads with at least one error. In practice, when error rates are around 1–5%, one only needs to compute the distribution of Y_k up to $i = 2$.

However, the hypothesis of uniform sampling of genomic positions is rarely met, and eukaryotic genomes reportedly have a significant fraction of repeated content. Therefore, in a more data- driven approach, Chaisson and Pevzner [5] first proposed to empirically adjust a mixture distribution, with correct and erroneous k-mers modelled by *Normal* and *Poisson* distributions, respectively. This model was later extended in *Quake* [12] by adding a covariate r to account for the higher multiplicity of some of the sequences. r is supposed to follow a *Zeta distribution*, a distribution with a long tail that can account for the common characteristic of repeats in genomes. Erroneous k-mers are modelled by a *Gamma distribution*. This mixture model has a total of six free parameters which can be estimated by numerical optimization or an EM procedure, given the histogram of word counts. However, despite being more flexible and adapted to each dataset, this strategy presents two disadvantages. First, the distributions used are meant to model real values and not counts. Second, it is necessary to obtain the histogram of all k-mers before doing error correction, which comes at a significant computational cost.

7.2.3 Detecting Errors with Suffix Trees

Due to the flexibility of the suffix tree data structures, error detection could in principle be performed for any seed length (up to the length of the reads). However, given genome complexity and characteristics of the sequencing experiment (depth, read length), only a limited range of values for k are relevant. Certainly, the smallest seed length $k_{\min}$ cannot be too small because in this case, repetitive sequences greatly complicate error correction. Conversely, longer seeds will have a lower coverage, resulting in a lack of sensitivity of erroneous reads (as shown in Fig. 7.3b for two k-mer sizes).

A first way of fixing k_{min}, proposed in *Shrec* [32], is to consider a random sequence of length n and then to control the expected proportion of repeats. When all bases have the same probability, $1/4$, a repeat will occur with probability $2n/4^k$. Fixing the maximal expected proportion of repeats to α (for instance 0.1%), $k_{min} = \log(2n/\alpha)$. This parameter is easy to set but ignores the interplay between error rate and seed length for being able to place a correct seed within a read.

In a fashion similar to what was done for seed design in sequence alignment [25], the authors of HiTEC [10] proposed to determine the best value for k_{min} by balancing the sensitivity of the seeds and their accuracy, given the error rate and the read lengths. This formulation was later extended in *Fiona* [33] to account for heterogeneous read lengths. They compute for each possible seed length two quantities, proxies for sensitivity and specificity, corresponding to two types of reads that cannot be corrected:

- Uncorrectable reads (U_k) are the expected number of reads where errors are distributed in such a way that they will cover all possible seeds of length k (for instance, a read of length 10 with 3 errors cannot be corrected using a seed of length 8).
- Destructible reads (D_k) are the expected number of reads where a sequencing error in the seed misdirects the genomic location it should be corrected to. For instance, in Fig. 7.3a, suppose that a read coming from location `TCCTAA` on the right is sequenced with an error on the second position as `TCCCAA`. The 5-bp seed `TCCCA` will then be corrected according to the left position on the genome with sequence `TCCCAT`.

Then, one way to select k_{min} is to choose k such that $U_k + D_k$ is minimal. The maximum value k_{max} can be set to ensure that a minimum expected coverage is obtained.

7.3 Index-Based Overlapping

While all existing index-based approaches for error correction have a common strategy to detect errors using seeds with a relatively low frequency, they differ in the way they determine the overlapping bases between erroneous and correct reads. In the following, we present and compare the different approaches proposed for the computation of the overlap.

The first and simplest approach, used in HiTEC [10], only considers the seed itself for the overlapping information. This corresponds to searching branches in a suffix trie at a fixed string depth and correct errors right of branching nodes. In practice, HiTEC proceeds by examining a suffix array for blocks of suffixes that begin with a common seed. It partitions each block according to the first base observed right of the seed into subsets of reads and flags each subset as correct or erroneous according to its coverage. If there is only one correct subset, it is used to correct all reads from the erroneous subset on the differing base. For more than one correct subset, erroneous reads have to overlap on more than two bases after the mismatching position to be corrected.

In a more accurate approach, *Shrec* [32] and *HybridShrec* [28] verify whether the potentially erroneous read can be overlapped right of the seed by at least a correct one. To this end, they skip the first base right of the common seed and descend in two parallel subtrees of the generalized suffix trie of the reads. The overlap is computed by recursively examining whether the correct subtree contains a path down the leaves that can be matched to the erroneous read up to m mismatches ($m = 0$ for *HybridShrec*, $m = 1$ for *Shrec*). In case of a success, the erroneous read will be corrected using the base skipped in the correct subtree. *HybridShrec* extends this approach to correct single-base indels a well.

Fiona [33] searches a generalized suffix trie as well to find branches that separate erroneous from correct reads. However, instead of overlapping only the suffixes right of the seed it computes anchored overlap alignments between the whole read sequences. This approach reduces the number of false positive corrections and allows to tolerate arbitrary error rates in the overlap without additional computational costs when compared with the exhaustive search in both subtrees. To improve the running time compared to classic dynamic programming-based overlap alignment algorithms, *Fiona* uses a banded variant of Myers' bit-vector algorithm [24].

7.3.1 Practical Considerations

Although the above-mentioned approaches traverse a generalized suffix trie, it is not necessary to construct and keep the whole tree in memory. In order to reduce their memory footprints and for parallelization, *Shrec* and *HybridShrec* enumerate all possible 5-mers and construct separate subtrees of the suffix trie in parallel to only those suffixes that start with the same 5-mer up to the deepest traversal depth k_{max}. This way, each thread traverses a separate subtree and only consumes memory required to represent that subtree. However, both tools store subtree nodes explicitly as objects, and their memory consumption grows quadratically in the total number of base pairs (in the worst-case, they construct and traverse each subtree up to its leaves). This makes them inapplicable to data volumes produced by current high-throughput sequencers.

Fiona uses a more memory-efficient representation of the suffix tries by storing only the contained suffixes in lexicographical order; hence, its memory consumption grows linearly with the number of total base pairs. Suffix trie nodes correspond to intervals in the sorted array and are not explicitly stored. Instead, child nodes of the currently visited node are determined using a binary search for the five possible outgoing edges (`A`, `C`, `G`, `T`, and `N`). To go up in the tree, the intervals of all nodes on the path to the root are maintained on a stack. For better load balancing and a lower memory footprint, *Fiona* builds subtrees that share the same 10-mer prefix. Before a thread can traverse the subtree of a 10-mer, all suffixes beginning with that 10-mer must be collected and lexicographically sorted. The necessary string comparison algorithm is interrupted after the first k_{max} characters, as nodes at a deeper level will not be traversed.

7.4 Correcting Reads with Information from Multiple Suffixes

After errors are detected and possible corrections are identified as explained above, the final task becomes to decide which correction is the true one. For example, consider two read suffixes **ACCG** and **ACG** and assume that the latter one was detected to have an error at position 3 (the **G**). Considering read 1 as the correct read there are two possibilities how to correct read 2: either an insertion of a **C** at position 3 of the suffix or a substitution of the **G** by a **C** at position 3. Therefore, algorithms need to have rules to select the best correction for each error.

In addition, as substring indices record all suffixes (and their reverse complements) of the reads, the same error appears several times in the tree. An example is shown in Fig. 7.4, where an error is depicted at three different depths in the tree. Obviously, this property is one of the main advantages of substring-based approaches for error correction as the error can be found multiple ways, and thus, false negatives can be avoided. However, when the error is found at multiple length suffixes, it must not be true that the same corrections are found. The figure highlights this fact as for $k = 4$, **G** or **T** are possible, whereas for $k = 6$, only **T** is a possible correction. Thus, depending on how the algorithms perform the search for errors in the index, the same error may be corrected differently as we explain below.

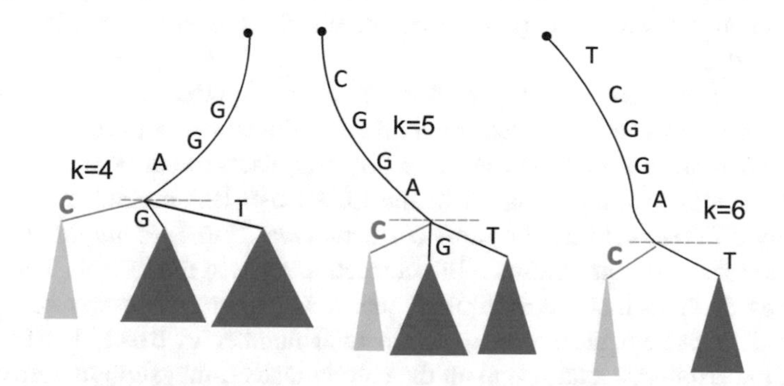

Fig. 7.4 Sequencing errors add branches to the suffix tree over reads. An error in a suffix at position k will create a branch at depth k. Here, we show a sequence error in the string **TCGGAT**, where the last position is sometimes sequenced as a C and thus creates three branches at $k = 4, 5, 6$. Further, there is a repeat copy at branches for $k = 4, 5$ which has a G instead of T

7.4.1 First Come, First Corrected Approaches

The first algorithm to suggest using suffix indices for correction was introduced in the *Shrec* method [32]. As explained above, *Shrec* builds subtrees of the full suffix trie for traversal and finding of erroneous nodes. Whenever errors are encountered, they are corrected immediately. That means the next subtree build already contains corrected read suffixes. This has the advantage that in one round of correction, i.e., one complete traversal of the tree, more than one error per read can be corrected. The disadvantage is that the first occurrence of the error may have led to a false positive correction. This type of *first come, first corrected* paradigm was adopted by most of the other substring-based methods, including HiTEC [10] and *HybridShrec* [28].

7.4.2 Integrating Information from Multiple Length Suffixes

Thus far, there have been two different approaches to integrate information from multiple length suffixes. The first was introduced by Savel et al. [31] and termed the PLURIBUS algorithm. Different to other substring-based methods, in their work, read errors are not found by tree traversal but by searching for all suffixes of a read. In this way, whenever a search path of a read suffix contains an erroneous node in the tree, the algorithm collects possible corrections for that read. After searching all suffixes, the correction that was found most often is selected as the best possible correction. This heuristic approach uses the fact that errors are encountered at different suffixes. The disadvantage of this approach is that the search is quadratic in the read length and takes $\mathcal{O}(L \cdot \ell^2)$, where L denotes the total length of all reads. Further, the memory required for building the suffix tree over all reads is prohibitive for larger datasets.

A second approach was suggested in the *Fiona* algorithm. In order to collect possible corrections for each read, *Fiona* holds a linked list of possible corrections for each position in a read. During tree traversal, each found possible correction for an erroneous position is saved in the linked list. But instead of counting in how many different suffixes the same correction was found, *Fiona* defines a score that summarizes the contribution of all correcting reads at the branch. This score is inspired by methods that use multiple sequence alignments for correction [14, 29]. As such, it considers the $score(c)$ as the total number of bases that are shared between the erroneous read r_e and all the correct reads r_i that support correction c:

$$score(c) = \sum_{i \in correct} overlap(r_e, r_i)\,, \tag{7.6}$$

where $overlap(r_e, r_i)$ denotes the number of bases shared between two reads and the sum goes over all reads in the correct branch supporting correction c.

After traversal of the tree, *Fiona* has collected these scores for each possible mismatch, insertion, or deletion correction for an error. Then, for each read, the best correction for a read position is chosen as the one with maximal score, avoiding indel corrections in the same read that are closer than seed length. Compared to PLURIBUS, *Fiona*'s approach has the advantage that the information about all corrections is stored in the linked list, which takes less space than storing the whole substring index in memory. Therefore, *Fiona* is able to correct large datasets from complex organisms like humans.

It was shown that the approach to integrate information from multiple length suffixes in *Fiona* outperforms the first come, first corrected approach, in particular for indel-prone datasets [33].

7.5 Performance Evaluation

Using 11 different datasets, seven from the *IonTorrent* and four from the *454* sequencer, we previously compared many error correction methods [33]. Table 7.2 lists the datasets with general characteristics and their accession numbers that were used for comparison. We compare the following read correction programs that can correct indel errors: *Fiona* [33], *Allpaths*-LG [8], *Coral* [29], and *Hybrid-Shrec* [28]. *Allpaths*-LG is an error correction program that was primarily designed for the correction of *Illumina* data but is also able to correct small indel errors. It uses a hash table for read positions of k-mers in the reads as index and looks at stacked sequence reads to find and correct errors. A similar index is used by *Coral*, but further, an optimal multiple sequence alignment for overlapping reads is computed to properly account for indel errors.

Because *Coral* and *Hybrid-Shrec* did not perform well with default parameters, we optimized the parameters in all the experiments reported; therefore, both methods are denoted with a star (*). For *HybridShrec*, we optimized the range of k-mers as well as the correction threshold, and for *Coral*, we optimized the sequencing error rate parameter. For both methods we took the parametrization that achieved the highest gain value for each dataset.

7.5.1 Performance Measures

For the evaluation of read correction quality, the metric *gain*, also known as accuracy, has been established in [40, 41] as a measure that combines sensitivity and precision, and we use it with edit distance to capture indel errors. Uncorrected reads are aligned against the genome with BWA-SW [15] using default parameters, and these alignments are considered as the true origin of the read. After correction, changes in the read in comparison to the read origin are used to compute the *gain* $= (b - a)/b$, where a and b denote the number of errors observed after and before read correction. The `compute_gain` tool was used to produce the gain values and read error rates (number of edit distance errors divided by the total number of bases considered) [33]. The gain is negative if more errors are introduced than corrected.

7.5.1.1 Error Correction Performance

Four different *454* datasets with different coverage values were collected, see Table 7.2. The per-base error rate ranges from 0.6% to 1.76% for all datasets. We observe that *HybridShrec* yields negative gain values on some datasets, thus producing more errors than correcting errors, as reported previously by Yang et al. [41]. *Fiona* shows the highest gain for all but the *S. aureus* dataset, where *Coral** works better. However, *Fiona* runs roughly 10 times faster than *Coral** on this dataset.

Next, seven different *Ion Torrent* datasets are analyzed, see Table 7.2. On six out of seven the *Fiona* method leads to the smallest error rate and best gain value, often with a significant improvement in error correction capability and generally more stable results than the other methods tested. The exception is the *E. coli* (156×) dataset, where *Fiona* and *Coral** have comparable gain values (Table 7.3).

Figure 7.5 shows a comparison of runtime and memory consumption for two different datasets for all methods, all methods are run with multiple threads. It is interesting to see how the methods based on different index types differ in their performance. For example, *Fiona* uses a partial suffix array which always outperforms the partial suffix trie in terms of run time and memory usage. The *k*-spectrum-based method *Allpaths*-LG is always the fastest with the lowest memory

Table 7.3 This table shows the base error rate (e-rate, in percent) before and after read correction as well as the gain metric

		Original	*Allpaths*-LG		*Coral**		*Fiona*		*HybridShrec**	
Dataset		e-rate	e-rate	Gain	e-rate	Gain	e-rate	Gain	e-rate	Gain
D. melanogaster	18×	1.17	1.07	8.87	0.55	53.30	**0.42**	**64.62**	0.73	38.17
E. coli K-12	13×	1.06	0.74	30.68	0.54	49.42	**0.25**	**76.88**	0.64	40.05
S. aureus	34×	1.76	1.34	23.85	**0.44**	**74.90**	0.53	69.87	1.40	20.50
S. cerevisae	16×	0.95	0.78	18.45	0.92	2.99	**0.61**	**36.04**	0.73	23.11
B. pertussis	85×	3.71	2.22	40.13	2.57	30.60	**1.01**	**72.83**	4.07	−9.68
E. coli K-12	8×	0.62	0.28	54.46	0.30	51.86	**0.06**	**90.52**	0.36	41.81
E. coli K-12	163×	1.46	1.23	15.99	0.38	73.72	**0.27**	**81.24**	1.46	0.00
E. coli K-12	156×	1.11	0.75	31.98	**0.28**	**74.70**	0.29	74.06	1.11	0.00
E. coli O104:H4	32×	5.19	3.09	40.53	3.44	33.82	**1.59**	**69.33**	4.31	16.76
P. falciparum	13×	5.06	3.97	21.39	3.80	24.94	**2.33**	**54.12**	4.63	8.50
S. aureus	109×	3.32	2.83	14.89	1.44	56.91	**1.17**	**64.94**	3.31	0.32

For each dataset, the method achieving the highest gain is highlighted in bold. The results are separated by sequencing technology, the *454* results are the top box and the *IonTorrent* results are the lower box

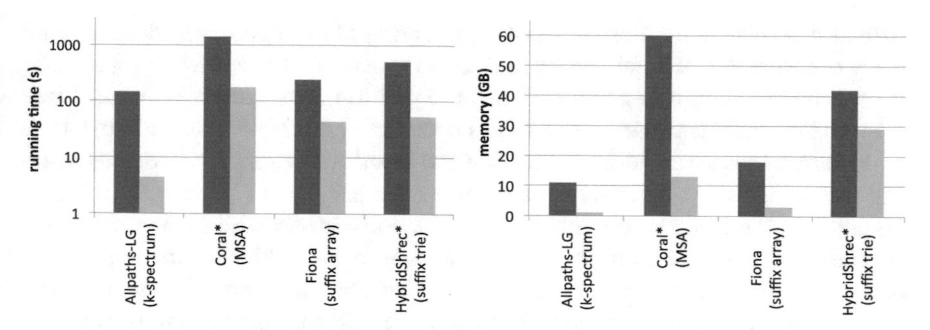

Fig. 7.5 Analysis of runtime in seconds (*left*) and memory consumption in GB (*right*) on two datasets to illustrate the differences between the methods. The bars denote the *D. melanogaster* (*dark grey*) and *S. aureus* dataset (*light grey*), respectively

but also shows huge losses in the gain value as compared to *Fiona*, for example. However, the *Coral* method uses more memory than the partial suffix array in *Fiona*, despite the simpler index structure. Overall, one can conclude that *Fiona* shows the best correction performance with a scalable resource consumption.

7.6 Conclusion and Perspectives

The exponential increase of DNA sequencing throughput in the last years resulted in an ever- higher accessibility of whole genome sequencing or resequencing datasets. As a consequence, methods for the automatic correction of sequencing errors had shown a regain of interest. However, most of the tools are still limited in their design to a specific sequencing technology and assay. For instance, most of the available methods correct only substitution errors—in practice only applicable to *Illumina* technology. Another practical limitation relates to genome ploidy (the copy number of each chromosome). Many tools do not specifically integrate the presence of heterozygous SNPs in diploid genomes, which is an important concern in genetic studies. Moreover, the more complex cases of polyploid genomes, very common in plants, has still not been addressed.

In addition, the genomic assays considered until now are sequencing the genome of one single clonal population. This may become the exception rather than the rule in the future, as various assays are considering a mixture of genomes. This is the case for metagenome sequencing, with the DNA content of a microbial community being sequenced, or cancer genomics, where the material being sequenced is usually a mixture of somatic and tumor cells (with extremely variable genomic content). Additionally, when a small quantity of starting material is available, single cell sequencing generates a highly nonuniform coverage distribution and thus invalidates the uniform coverage assumption that is used to detect erroneous reads (see Sect. 7.2.2). While the methods presented in this chapter could in principle be extended to account for those types of assays, many of the algorithmic details do not carry over directly. Nonuniformity in datasets is often addressed with probabilistic models on the Hamilton graph of k-mers to perform error correction instead (see Chap. 6 by Schulz and Bar-Joseph in this book for a presentation on the subject).

Finally, to successfully assemble long and repetitive genomes, various sequencing technologies are typically used on the same sample. For instance, it is now common to sequence long reads from *Pacific Biosciences* with higher error rates together with shorter and more accurate *Illumina* reads at higher throughput. This category of experiments slightly changes the problem of error correction, as one could in principle combine the information from diverse sequencing runs in one go. Existing read error correction tools only use one technology as a template to correct the one with the higher error rate, either using the profile from pairwise alignments [13] or k-spectrum techniques [9]. Merging all reads in one global correction scheme would necessitate to extend the formalism for the classification of erroneous substrings, as well as the criteria for correction of an error to account for multiple libraries at the same time. While this can be done relatively easily, some challenges remain to be able to index and access those informations efficiently.

References

1. Abouelhoda, M.I., Kurtz, S., Ohlebusch, E.: Replacing suffix trees with enhanced suffix arrays. J. Discrete Algorithms **2**(1), 53–86 (2004)
2. Auton, A., Fledel-Alon, A., Pfeifer, S., Venn, O., Ségurel, L., Street, T., Leffler, E.M., Bowden, R., Aneas, I., Broxholme, J., Humburg, P., Iqbal, Z., Lunter, G., Maller, J., Hernandez, R.D., Melton, C., Venkat, A., Nobrega, M.A., Bontrop, R., Myers, S., Donnelly, P., Przeworski, M., McVean, G.: A fine-scale chimpanzee genetic map from population sequencing. Science **336**(6078), 193–198 (2012)
3. Bart, R., Cohn, M., Kassen, A., McCallum, E.J., Shybut, M., Petriello, A., Krasileva, K., Dahlbeck, D., Medina, C., Alicai, T., Kumar, L., Moreira, L.M., Neto, J.R., Verdier, V., Santana, M.A., Kositcharoenkul, N., Vanderschuren, H., Gruissem, W., Bernal, A., Staskawicz, B.J.: High-throughput genomic sequencing of cassava bacterial blight strains identifies conserved effectors to target for durable resistance. Proc. Natl. Acad. Sci. U. S. A. **109**, E1972–1979 (2012)
4. Butler, J., MacCallum, I., Kleber, M., Shlyakhter, I.A., Belmonte, M.K., Lander, E.S., Nusbaum, C., Jaffe, D.B.: ALLPATHS: de novo assembly of whole-genome shotgun microreads. Genome Res. **18**(5), 810–820 (2008)
5. Chaisson, M.J., Pevzner, P.A.: Short read fragment assembly of bacterial genomes. Genome Res. **18**(2), 324–330 (2008)
6. Ferragina, P., Manzini, G.: Opportunistic data structures with applications. In: FOCS'00, pp. 390–398 (2000)
7. Glenn, T.C.: Field guide to next-generation DNA sequencers. Mol. Ecol. Resour. **11**(5), 759–769 (2011)
8. Gnerre, S., MacCallum, I., Przybylski, D., Ribeiro, F.J., Burton, J.N., Walker, B.J., Sharpe, T., Hall, G., Shea, T.P., Sykes, S., Berlin, A.M., Aird, D., Costello, M., Daza, R., Williams, L., Nicol, R., Gnirke, A., Nusbaum, C., Lander, E.S., Jaffe, D.B.: High-quality draft assemblies of mammalian genomes from massively parallel sequence data. Proc. Natl. Acad. Sci. U. S. A. **108**(4), 1513–1518 (2010)

9. Greenfield, P., Duesing, K., Papanicolaou, A., Bauer, D.C.: Blue: correcting sequencing errors using consensus and context. Bioinformatics **30**(19), 2723–2732 (2014)
10. Ilie, L., Fazayeli, F., Ilie, S.: HiTEC: accurate error correction in high-throughput sequencing data. Bioinformatics **27**(3), 295–302 (2011)
11. Kao, W., Chan, A., Song, Y.: ECHO: a reference-free short-read error correction algorithm. Genome Res. **21**(7), 1181 (2011)
12. Kelley, D.R., Schatz, M.C., Salzberg, S.L.: Quake: quality-aware detection and correction of sequencing errors. Genome Biol. **11**(11), R116 (2010)
13. Koren, S., Schatz, M.C., Walenz, B.P., Martin, J., Howard, J.T., Ganapathy, G., Wang, Z., Rasko, D.A., McCombie, W.R., Jarvis, E.D., Phillippy, A.M.: Hybrid error correction and de novo assembly of single-molecule sequencing reads. Nat. Biotechnol. **30**(7), 693–700 (2012)
14. Le, H.S., Schulz, M.H., McCauley, B.M., Hinman, V.F., Bar-Joseph, Z.: Probabilistic error correction for RNA sequencing. Nucleic Acids Res. **41**(10), e109 (2013)
15. Li, H., Durbin, R.: Fast and accurate long-read alignment with burrows-wheeler transform. Bioinformatics **26**(5), 589–595 (2010)
16. Locke, D.P., Hillier, L.W., Warren, W.C., Worley, K.C., Nazareth, L.V., Muzny, D.M., Yang, S.P., Wang, Z., Chinwalla, A.T., Minx, P., Mitreva, M., Cook, L., Delehaunty, K.D., Fronick, C., Schmidt, H., Fulton, L.A., Fulton, R.S., Nelson, J.O., Magrini, V., Pohl, C., Graves, T.A., Markovic, C., Cree, A., Dinh, H.H., Hume, J., Kovar, C.L., Fowler, G.R., Lunter, G., Meader, S., Heger, A., Ponting, C.P., Marques-Bonet, T., Alkan, C., Chen, L., Cheng, Z., Kidd, J.M., Eichler, E.E., White, S., Searle, S., Vilella, A.J., Chen, Y., Flicek, P., Ma, J., Raney, B., Suh, B., Burhans, R., Herrero, J., Haussler, D., Faria, R., Fernando, O., Darre, F., Farre, D., Gazave, E., Oliva, M., Navarro, A., Roberto, R., Capozzi, O., Archidiacono, N., Valle, G.D., Purgato, S., Rocchi, M., Konkel, M.K., Walker, J.A., Ullmer, B., Batzer, M.A., Smit, A.F.A., Hubley, R., Casola, C., Schrider, D.R., Hahn, M.W., Quesada, V., Puente, X.S., Ordonez, G.R., Lopez-Otin, C., Vinar, T., Brejova, B., Ratan, A., Harris, R.S., Miller, W., Kosiol, C., Lawson, H.A., Taliwal, V., Martins, A.L., Siepel, A., RoyChoudhury, A., Ma, X., Degenhardt, J., Bustamante, C.D., Gutenkunst, R.N., Mailund, T., Dutheil, J.Y., Hobolth, A., Schierup, M.H., Ryder, O.A., Yoshinaga, Y., de Jong, P.J., Weinstock, G.M., Rogers, J., Mardis, E.R., Gibbs, R.A., Wilson, R.K.: Comparative and demographic analysis of orangutan genomes. Nature **469**, 529–533 (2011)
17. MacManes, M.D., Eisen, M.B.: Improving transcriptome assembly through error correction of high-throughput sequence reads. PeerJ **1**, e113 (2013)
18. Manber, U., Myers, E.W.: Suffix arrays: a new method for on-line string searches. In: SODA'90, pp. 319–327. SIAM, Philadelphia (1990)
19. Margulies, M., Egholm, M., Altman, W.E., Attiya, S., Bader, J.S., Bemben, L.A., Berka, J., Braverman, M.S., Chen, Y.J., Chen, Z., Dewell, S.B., Du, L., Fierro, J.M., Gomes, X.V., Godwin, B.C., He, W., Helgesen, S., Ho, C.H., Irzyk, G.P., Jando, S.C., Alenquer, M.L.I., Jarvie, T.P., Jirage, K.B., Kim, J.B., Knight, J.R., Lanza, J.R., Leamon, J.H., Lefkowitz, S.M., Lei, M., Li, J., Lohman, K.L., Lu, H., Makhijani, V.B., Mcdade, K.E., Mckenna, M.P., Myers, E.W., Nickerson, E., Nobile, J.R., Plant, R., Puc, B.P., Ronan, M.T., Roth, G.T., Sarkis, G.J., Simons, J.F., Simpson, J.W., Srinivasan, M., Tartaro, K.R., Tomasz, A., Vogt, K.A., Volkmer, G.A., Wang, S.H., Wang, Y., Weiner, M.P., Yu, P., Begley, R.F., Rothberg, J.M.: Genome sequencing in microfabricated high-density picolitre reactors. Nature **437**, 376–380 (2005)
20. McCreight, E.M.: A space-economical suffix tree construction algorithm. J. ACM **23**(2), 262–272 (1976)
21. Medvedev, P., Scott, E., Kakaradov, B., Pevzner, P.: Error correction of high-throughput sequencing datasets with non-uniform coverage. Bioinformatics **27**(13), i137–i141 (2011)
22. Mellmann, A., Harmsen, D., Cummings, C.A., Zentz, E.B., Leopold, S.R., Rico, A., Prior, K., Szczepanowski, R., Ji, Y., Zhang, W., McLaughlin, S.F., Henkhaus, J.K., Leopold, B., Bielaszewska, M., Prager, R., Brzoska, P.M., Moore, R.L., Guenther, S., Rothberg, J.M., Karch, H.: Prospective genomic characterization of the German enterohemorrhagic Escherichia coli o104:h4 outbreak by rapid next generation sequencing technology. PLoS ONE **6**(7), e22751 (2011)

23. Minoche, A.E., Dohm, J.C., Himmelbauer, H.: Evaluation of genomic high-throughput sequencing data generated on Illumina HiSeq and genome analyzer systems. Genome Biol. **12**(11), R112 (2011)
24. Myers, E.W.: A fast bit-vector algorithm for approximate string matching based on dynamic programming. J. ACM **46**(3), 395–415 (1999)
25. Pevzner, P., Waterman, M.: Multiple filtration and approximate pattern matching. Algorithmica **13**(1–2), 135–154 (1995)
26. Pevzner, P.A., Tang, H., Waterman, M.S.: An Eulerian path approach to DNA fragment assembly. Proc. Natl. Acad. Sci. U. S. A. **98**(17), 9748–9753 (2001)
27. Quail, M.A., Smith, M., Coupland, P., Otto, T.D., Harris, S.R., Connor, T.R., Bertoni, A., Swerdlow, H.P., Gu, Y.: A tale of three next generation sequencing platforms: comparison of Ion Torrent, Pacific Biosciences and Illumina MiSeq sequencers. BMC Genomics **13**(1), 341 (2012)
28. Salmela, L.: Correction of sequencing errors in a mixed set of reads. Bioinformatics **26**(10), 1284–1290 (2010)
29. Salmela, L., Schröder, J.: Correcting errors in short reads by multiple alignments. Bioinformatics **27**(11), 1455–1461 (2011)
30. Salzberg, S.L., Phillippy, A.M., Zimin, A., Puiu, D., Magoc, T., Koren, S., Treangen, T.J., Schatz, M.C., Delcher, A.L., Roberts, M., Marçais, G., Pop, M., Yorke, J.A.: GAGE: a critical evaluation of genome assemblies and assembly algorithms. Genome Res. **22**(3), 557–567 (2012)
31. Savel, D.M., LaFramboise, T., Grama, A., Koyutürk, M.: Suffix-tree based error correction of NGS reads using multiple manifestations of an error. In: Proceedings of the International Conference on Bioinformatics, Computational Biology and Biomedical Informatics, BCB'13, pp. 351:351–351:358. ACM, New York (2013)
32. Schröder, J., Schröder, H., Puglisi, S.J., Sinha, R., Schmidt, B.: SHREC: a short-read error correction method. Bioinformatics **25**(17), 2157–2163 (2009)
33. Schulz, M.H., Weese, D., Holtgrewe, M., Dimitrova, V., Niu, S., Reinert, K., Richard, H.: Fiona: a parallel and automatic strategy for read error correction. Bioinformatics **30**(17), i356–i363 (2014)
34. Shendure, J., Ji, H.: Next-generation DNA sequencing. Nat. Biotechnol. **26**(10), 1135–1145 (2008)
35. Simpson, J.T., Durbin, R.: Efficient de novo assembly of large genomes using compressed data structures. Genome Res. **22**(3), 549–556 (2012)
36. Stein, L.: The case for cloud computing in genome informatics. Genome Biol. **11**(5), 207 (2010)
37. The 1000 Genomes Project Consortium: A map of human genome variation from population-scale sequencing. Nature **467**(7319), 1061–1073 (2010)
38. Ukkonen, E.: On-line construction of suffix trees. Algorithmica **14**(3), 249–260 (1995)
39. Weiner, P.: Linear pattern matching algorithms. In: Proceedings of the 14th Symposium on Switching and Automata Theory, SWAT'73, pp. 1–11. IEEE Computer Society, Washington, DC (1973)
40. Yang, X., Dorman, K.S., Aluru, S.: Reptile: representative tiling for short read error correction. Bioinformatics **26**(20), 2526–2533 (2010)
41. Yang, X., Chockalingam, S.P., Aluru, S.: A survey of error-correction methods for next-generation sequencing. Brief. Bioinform. **14**(1), 56–66 (2013)

Chapter 8
Error Correction in Methylation Profiling From NGS Bisulfite Protocols

Guillermo Barturen, José L. Oliver, and Michael Hackenberg

8.1 Introduction

DNA methylation presents the most characteristic traits of epigenetic marks [1]. Basically, it consists of a covalent bond between a methyl group and the cytosine's fifth carbon, which will result in methylcytosine, called the "fifth base of the DNA" by some authors [2]. During the last years, efforts have been made to elucidate how methylation works and in which biological processes it is involved [3]. However, despite the increasing knowledge about its working mechanism and functions, DNA methylation is still a broadly unknown epigenetic regulatory mark. Since the discovery of DNA methylation [4] and its implication in the regulation of gene expression [5, 6], many different methods have been proposed for its detection and quantification; these have been reviewed and compared in recent reviews [2, 7].

Recently, the advent of high-throughput sequencing techniques together with the bisulfite treatment of the DNA [8] allows whole-genome methylation profiling at a single cytosine resolution in a relatively short time. Consequently, *Whole-Genome Bisulfite Sequencing* (WGBS) has become the primary method for studying DNA methylation. However, unlike other methylation assays, such as, for example, microarrays, the WGBS data sets have not had a straightforward processing

G. Barturen (✉)
Centro de Genómica e Investigaciones Oncológicas, Pfizer-Universidad de Granada-Junta de Andalucía, Granada 18016, Spain
e-mail: guillermo.barturen@genyo.es

J.L. Oliver • M. Hackenberg
Dpto. de Genética, Facultad de Ciencias, Universidad de Granada, Campus de Fuentenueva s/n, 18071 Granada, Spain
e-mail: oliver@ugr.es; mlhack@go.ugr.es

M. Elloumi (ed.), *Algorithms for Next-Generation Sequencing Data*,
DOI 10.1007/978-3-319-59826-0_8

protocol, and many quality controls should be included to ensure the reliability of the methylation ratios. The WGBS protocol is composed of two major separated steps: the alignment of the bisulfite-treated reads to a reference genome and the profiling of the methylation levels. The alignment to reference genomes of bisulfite-treated reads cannot be addressed as with non-treated reads because of the reduction of the sequence complexity caused by the bisulfite conversion. Currently, many tools exist to align these reads, such as *BSMAP/RRBSMAP* [9], *BS-Seeker* [10, 11], *mrsFAST* [12], *Bismark* [13], *MethylCoder* [14] or *NGSmethPipe* [15], among others. The alignment is an important step during the methylation calling process, and researchers must choose the most appropriate aligner for their experimental protocol. However, this chapter will not focus on that issue, as there are good reviews and comparisons about this topic [16–18]. Regarding the methylation ratio quantification, due to its nature, methylation can be measured in different ways, which have also been reviewed [19]. In addition, both steps inherit potential important error sources from either the *High-Throughput Sequencing* (HTS) technology or the bisulfite conversion, which must also be taken into account and properly managed to avoid undesirable biases in the final results. In this chapter, all the known additional bias sources are exhaustively reviewed and critically evaluated. In Sect. 8.3, potential sequence contamination biases are reviewed for different methylation profiling techniques. In Sect. 8.4, coverage and sequence composition biases are evaluated and some recommendations are given to deal with them. In Sect. 8.5, we discuss how redundant sequencing calls might affect the methylation values and recommend some available software to manage and remove this redundant information. In Sect. 8.6, some recommendations are given to remove bad sequencing bases. In Sect. 8.7, we reviewed some approaches to assess and deal with bisulfite conversion failures. In Sect. 8.8, we go over some biases that can only be assessed during or after the methylation value estimation. Finally, according to our knowledge, an extensive review of the error sources that can bias the methylation level measurement and the different algorithms that have been proposed to deal with them does not exist. Therefore, in this chapter, all known WGBS error sources are extensively reviewed and critically evaluated to suggest a few of the best practices to deal with all sources of bias in WGBS assays.

8.2 Bisulfite Sequencing Basis

The determination of the methylation status by means of HTS experiments is based on the conversion of unmethylated cytosines, but not methylcytosines, to uracils during sodium bisulfite treatment [8]. Then, the PCR amplification step of the treated DNA fragments prior to the sequencing converts these uracils to thymines. Once the reads have been correctly aligned against a reference genome, the methylation status of all covered reference cytosines can be easily estimated as the cytosine proportion is sequenced at each position ($C/(C + T)$), obtaining a methylation ratio or β value. Bisulfite HTS methods present important advantages compared to other genome-wide methodologies (reviewed in [7, 20]) because of the full genome-wide coverage, known methylation sequence context, strand-specific methylation levels and lack of CpG density biases. However, the results still must be processed carefully to avoid undesirable biases from either the HTS methodology or the bisulfite treatment protocol, which might lead to a misinterpretation of the results.

8.3 Sequence Contamination

The sequence contamination in high-throughput sequencing HTS experiments is defined as all the artificial nucleotides that could be added to both ends of the reads and could introduce important biases to the results. Different types of sequences, with different purposes, are added to the sequences of interest during the library construction. Bisulfite methodologies could show some artificial sequences common to every HTS experiment (adapters or barcodes) and others specific to bisulfite-treated libraries (BS-Seq tags or those derived from end repairs).

8.3.1 Adapter and Barcodes

The adapters are oligonucleotides added during the library construction of all HTS techniques and are composed of a primer and a binding sequence to the flow cell. Generally, these sequences should not be part of the reads. However, when the number of sequencing cycles exceeds the DNA fragment length, variable lengths of the adapter can appear at the 3′ end of the reads (see read 2 in Fig. 8.1). Barcodes

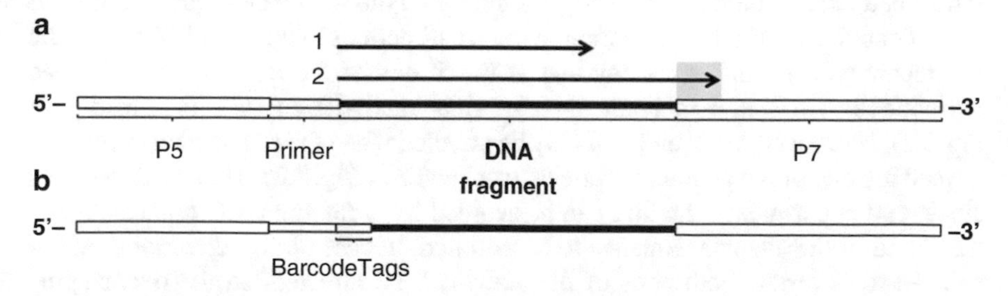

Fig. 8.1 Artificial sequences in WGBS libraries. (**a**) Typical read schema of artificial sequences added to WGBS libraries. (**b**) Reads schema for barcodes or tags added to libraries. Read 1 sequencing cycles do not exceed the DNA fragment length and thus will not contain adapter sequences, while in read 2 part of the adapter will be sequenced at its 3′ end

can be added to sequence multiple libraries together within the same sequencing lane (multiplexing); however, they are usually not used in bisulfite sequencing experiments. Barcodes are located at the 5′ end of the reads (see Fig. 8.1).

Both adapters and barcodes must be detected and trimmed before aligning the reads against the reference genome. Otherwise, their presence can either reduce the mapping efficiency [21] or even bias the final results.

8.3.2 BS-Seq Tags

The BS-Seq protocol [22] applies another type of barcode (also ligated at the 5′ end of the reads and usually called tags), which, instead of allow sequencing multiple libraries together, is used to distinguish between reads sequenced from the bisulfite-converted strand and its reverse complementary strand (see Fig. 8.1). However, they must be found and discarded before the alignment step using a proper trimming software, because most of the alignment programs do not manage them properly. To our knowledge, only *BS-Seeker* [10, 11] manages those tags properly. *BS-Seeker* detects the tags and tries to align each read against its theoretical orientation, thus reducing the search space from four possible strands to two (the first version of the *BS-Seeker* paper presents a detailed explanation of the process [10]).

8.3.3 RRBS Ends-Repair

The most common restriction enzyme used for *RRBS* methodologies [23] is the MspI. This endonuclease cleaves the DNA, leaving sticky ends formed by an overhanging CG, which must be repaired before adding the adapter. The nucleotides introduced during the repair will align perfectly on the reference sequence, but will not reflect the original methylation state. In directional single-end libraries, the end repair will not introduce any bias at the 5′ end of the read as the sequenced nucleotides at this end maintain their original methylation state (see read 1 in Fig. 8.2). Moreover, the 3′ end will only be affected when the sequencing cycle goes beyond the end of the genomic fragment (see read 2 in Fig. 8.2). Then, in directional single-end experiments, the bias can be avoided by including a CG dinucleotide at the 5′ end of the adapter sequence to be trimmed. In case of non-directional and/or paired-end libraries, both ends of the reads can be affected, so the overhanging number of nucleotides left by the endonuclease must be trimmed from both the 3′ and 5′ ends of the reads (see 3 reads in Fig. 8.2).

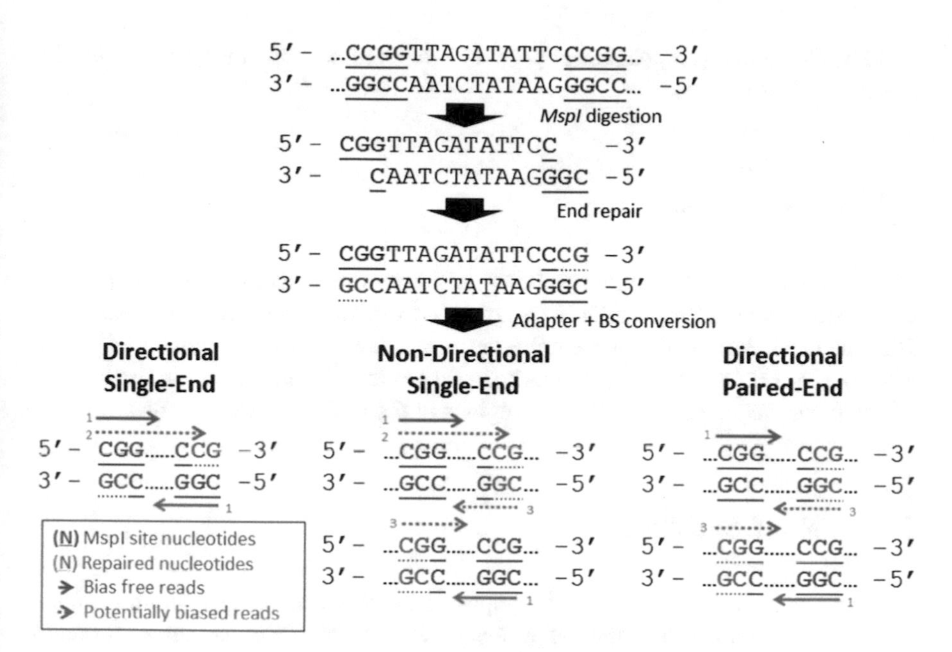

Fig. 8.2 Artificial methylation in the *RRBS* end-repair step. The schema shows all the unmethylation cytosines that could be artificially added in the *RRBS* experiments during the end repair step. Read 1 does not sequence any artificially added cytosines, while reads 2 and 3 present unmethylated cytosines at the 3′ and 5′ end, respectively

8.3.4 Whole-Genome Ends-Repair

The *Illumina* WGBS protocol includes an end-repair step after the sonication of the genome. To maintain the double-stranded DNA, the potential overhanging ends are filled up with complementary unmethylated nucleotides, which might result in an underestimation of the methylation at those positions. This kind of bias at the end of the reads could be included under the term *M-bias* [24], as it should be treated differently from other sequence contaminations; *M-bias* detection and management will be reviewed below (see Sect. 8.7.3).

Even if all the sequence contaminations reviewed can introduce biases in methylation calling steps, the methylation biases depend on the presence of thymines and/or cytosines in the added artificial sequence. Nevertheless, this is not the case with the nucleotides coming from end repair (Sects. 8.3.3 and 8.3.4), which must be taken into special consideration. These nucleotides maintain the sequence information but not the original methylation status, and they then might be aligned perfectly in a methylation context, introducing a large bias in the methylation levels of these positions.

In addition to the control steps included in some of the alignment and methylation calling programs, there are some suitable tools to identify and clip these sequences, for example, *FastQC* [25] to check the sequence contamination present in the sample and *FASTX-Toolkit* [26], *Cutadapt* [27] or *Trimmomatic* [28], among others, to trim the artificial sequences.

8.4 Coverage Bias

One of todays' HTS technology bases is the random location and independent composition sequencing of DNA fragments from the genome. In this way, the HTS method should obtain an unbiased representation of the sequenced DNA. However, the HTS methodology has been shown to introduce biases in the number of fragments sequenced for different regions of the genome (coverage bias).

8.4.1 Read End Bias

Beyond the well-described 5′ end compositional bias caused by random hexamer priming on *Illumina* transcriptome protocols [29], this bias was also shown for reads obtained by means of protocols such as DNA-Seq, Chip-Seq [30] or even WGBS, which do not incorporate the random hexamer priming. Thus, the bias probably has other additional sources, such as, for example, the non-random fragmentation of the genome [31] or a potential preference of some read end nucleotides during the adapter ligation. This would mean that the sequenced reads might preferentially start at some locations of the genome, and then a sampling bias could be found, in the worst case losing the methylation information from the disfavored regions.

8.4.2 GC Composition

Another common issue affecting HTS *Illumina* libraries is the GC bias, mainly caused by the amplification during library preparation [32], which reduces the coverage at extremely AT- and GC-rich regions [33]. By definition, protocols without library amplification, such as PBAT [34], will not present a GC bias. However, at least currently, they are not broadly used.

Both sources of coverage bias lead to a non-random distribution of the reads along the genome, which would mainly affect quantitative signal approaches (copy number estimation, RNA-Seq or Chip-Seq). However, on bisulfite sequencing analysis, the coverage bias would also reduce the sensitivity of further analysis (as differential methylation identification), especially on low coverage data sets (because of the potential loss of cytosine information). The only way to reduce its effects on library amplification WGBS protocols, beyond improving the experimental protocols, is sequencing with the appropriate coverage for the further methylation analysis to be done. Choosing an appropriate sequencing coverage and number of replicas for each sample in the study is the first technical decision when designing a bisulfite sequencing experiment. Increased coverage may lead to wasted resources, while low coverage sequencing could affect the methylation profiles. Recently, it has been suggested that the sequencing coverage per replica in single CpG methylation analysis should be between 5× and 15×, ideally reaching 30–40× sum coverage for all replicas in the sample. Moreover, samples with the same total coverage but higher numbers of replicas have been shown to present better results [35]. Then, choosing the appropriate coverage and number of replicas will not only reduce the potential coverage bias shown by *Illumina* sequencers, but will also increase the reliability of the results.

8.5 Redundant Calls

Another relevant bias of the random and compositional independent sequencing of the genome comes from the fact that sometimes the information from the same DNA fragment could be read two or more times. Redundant sequencing would overestimate the real coverage of some regions and overweigh the methylation information from these fragments during the downstream analysis, which might cause an incorrect inference at positions with allele-specific methylation (genetic imprinting), sequence variation, hemi-methylation, sequencing errors and bisulfite failure or those that are heterogeneous over the cell population. This bias can appear from either duplicate reads or overlapping mate reads.

8.5.1 Duplicated Reads

The most common source of reads sequenced more than once in HTS protocols arises from duplications during the PCR amplification event (PCR duplicates) [36]. PCR duplications occur when two copies of the same original read occupy different beads or primer lawns in the flow cell. Optical duplicates are another source of artificial sequence duplication, which occurs when a cluster on the flow cell is erroneously split into two or more clusters.

The percentage of PCR duplicates should be controlled during the experimental protocol, for example, starting with enough DNA or not including many short reads in the library. However, due to the experimental procedure, duplication percentages of around 5% cannot be avoided. Therefore, some computational methods have been proposed to deal with them. The PCR duplicates frequently are detected by means of their 5′ coordinates, choosing the best read in terms of sequencing quality (*SAMtools* [37] or *Picard-tools* [38]). Although these tools are widely used, at high read coverage (≥40×), they might incorrectly discard many reads as the probability to obtain several non-duplicated reads starting at the same position increases with the coverage. Furthermore, they do not take into account that reads with the

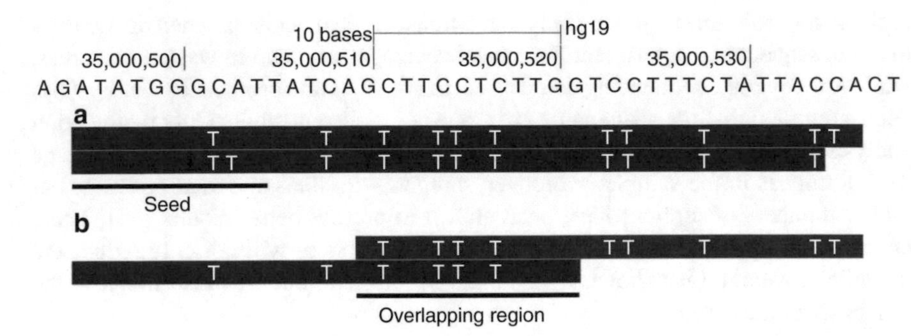

Fig. 8.3 Single and paired read potential redundant calls. Example of potential single-end (**a**) and paired-end (**b**) read PCR duplications. The figure is represented using UCSC *Genome Browser* [41]

same start coordinate could represent different alleles or come from different cells (see Single-End Reads in Fig. 8.3). Recently, to reduce the number of incorrectly discarded duplications on bisulfite-treated protocols, a new method was proposed in *MethylExtract* [39], which combines the positional information and sequence composition of the reads.

Regarding the optical duplicates, by definition, they are discarded among the PCR duplicates (as all of them present the same 5′ position), but they could be detected in a more specific way. In RRBS experiments [23], where discarding the PCR duplicates will result in a large data loss, and in PBAT experiments [34], where theoretically PCR duplicates should not appear, discarding only the optical duplicates might be relevant. Currently, *Picard-tools* [38] are able to detect and quantify this type of duplication using the tile and image coordinate information contained in the *FastQ* identifiers [40]. To our knowledge, currently no existing tools allow discarding optical duplicates independently from the PCR ones.

8.5.2 Overlapping 3′ Ends

Paired-end mates contain sequence and methylation information from the same DNA strand. Therefore, when the sum of sequencing cycles from both mates exceeds the DNA length fragment, the 3′ ends of both mates will overlap and contain duplicated information (see Paired-End Reads in Fig. 8.3). To avoid scoring the information twice, programs such as *Bismark* [13] or *MethylExtract* [39] discard the overlapping information from the second pair, which usually shows lower quality.

8.6 Base Sequencing Quality

The sequencing quality is another issue common to all HTS protocols, and the bisulfite-treated protocols are not an exception. Each base call has a PHRED quality score encoded as a single ASCII character within the *FastQ* file [40], which gives an estimation of the probability of the sequencing errors. Since the advent of the HTS techniques, this sequencing quality score has been used to improve the read alignment as well as downstream analysis [39, 42–44].

8.6.1 3′ End Decay

In bisulfite protocols, *Illumina* is the sequencing platform par excellence. However, due to its sequencing by the synthesis strategy, the sequencing quality decays toward the 3′ end of the reads, which is a potential well-characterized bias [45, 46]. These low-quality positions at the 3′ end of the reads might result in a reduced number of reads mapped against the genome or even an increased incorrect number of alignments that will lead to incorrect methylation values [21]. Moreover, the presence of these low-quality positions could also represent an obstacle to detecting and trimming the adapter sequences, which in turn will increase the alignment and methylation calling errors. Therefore, the 3′ end of the reads should be quality trimmed before the adapter trimming step. Recently, an extensive evaluation of quality-trimming methods has been performed for some types of HTS experiments [47]. Although its effect on methylation calling has not been evaluated, it should be similar to the SNP calling results, as for both approaches the single base information is crucial for the downstream analysis.

8.6.2 Low-Quality Positions

Apart from the 3′ end quality decay, some other low-quality positions can exist along the reads. These low-quality positions are usually, but not exclusively, associated to specific sequencing contexts [48] and must be managed carefully to obtain the most reliable results. Recently, it has been shown that discarding positions with low-quality scores can greatly increase the accuracy of the methylation calls in bisulfite sequencing experiments [39], thus stressing the importance of the sequencing quality information also in the methylation calling. Currently, only *MethylExtract* discards low-quality positions for methylation calling, which probably leads to the higher specificity compared to *Bis-SNP* [49]. A commonly accepted previous step to manage single base sequencing quality scores, at least for SNP calling experiments, is the base-quality recalibration. Currently, the *Bis-SNP* package [49] includes a modified *GATK* recalibration algorithm [50] to work on bisulfite-treated data sets.

However, the recalibration depends directly on a well-curated SNP annotation, which is not always available. This, together with continuously improving accuracy between the expected and observed error rates on *Illumina* sequencing platforms [48], calls into doubt the necessity of this step for genomes that are not well annotated. According to our knowledge, there are no studies on the effect of the quality recalibration, but they would be necessary to form an opinion.

8.7 Bisulfite Bias

The bisulfite sequencing methods are based on the premise that the bisulfite treatment of single-stranded DNA converts unmethylated cytosines into uracils (converted to thymines after PCR amplification) while leaving methylcytosines unconverted [8]. Then, after sequencing and alignment against a reference genome, the original methylation state can be recovered. However, the bisulfite conversion has a certain error rate, and usually reads with an incomplete bisulfite conversion can be found, which can lead to an overestimation of the methylation levels. Detecting and discarding the incompletely converted reads are critical steps to infer the correct methylation levels. Two types of incomplete read conversions have been observed in the unmethylated spiked phage genome: partially unconverted and global unconverted [11].

8.7.1 Global Unconverted Reads

Most of the unconverted cytosines in a CpG context (around 85%) have been found in global unconverted reads, which has been proposed to be the result of spontaneously secondary structure formation [11]; therefore, the bisulfite cannot access the sequence. Generally, a low percentage of non-CpG contexts is methylated in the genomes, so most of these contexts should be converted to thymines during the treatment. Therefore, Lister et al. proposed a method where reads with $\geq$3 unconverted non-CpG cytosines were considered incompletely converted [51]. Currently, *MethylExtract* [39] and *BS-Seeker2* [11] include methods based on the same idea. *MethylExtract* has an option to include either a percentage or an absolute number threshold of unconverted non-CpG contexts, while *BS-Seeker2* requires both thresholds to be satisfied. Although the method seems to be effective in discarding global unconverted reads, special care must be taken when the samples are known to have increased non-CpG methylated contexts (for example, plant or mammal embryonic stem cells). In these cases, stricter thresholds should be set or the detection of unconverted reads should be deactivated.

8.7.2 Partially Unconverted Reads

The partially unconverted reads cannot be detected by the method described in the last section, because most of the cytosines in these reads are correctly converted. The unconverted cytosines on partially unconverted reads have been suggested to be caused by random conversion failures or confused with T to C sequencing errors [11]. Some of these conversion failures can be removed during the *M-bias* correction, particularly those coming from 3′ end decay sequencing errors. However, unconverted random cytosines cannot be detected or trimmed. So, the methylation level reliability should be estimated after the methylation calling step and will be reviewed in Sect. 8.8.2.

8.7.3 M-bias

M-bias [24] refers to an uneven proportion of cytosines and thymines at both ends of the reads compared with the theoretically independent distribution of the cytosine conversion ratios along them. This bias can be easily observed plotting the percentage of the cytosine conversion as a function of the position (see Fig. 8.4).

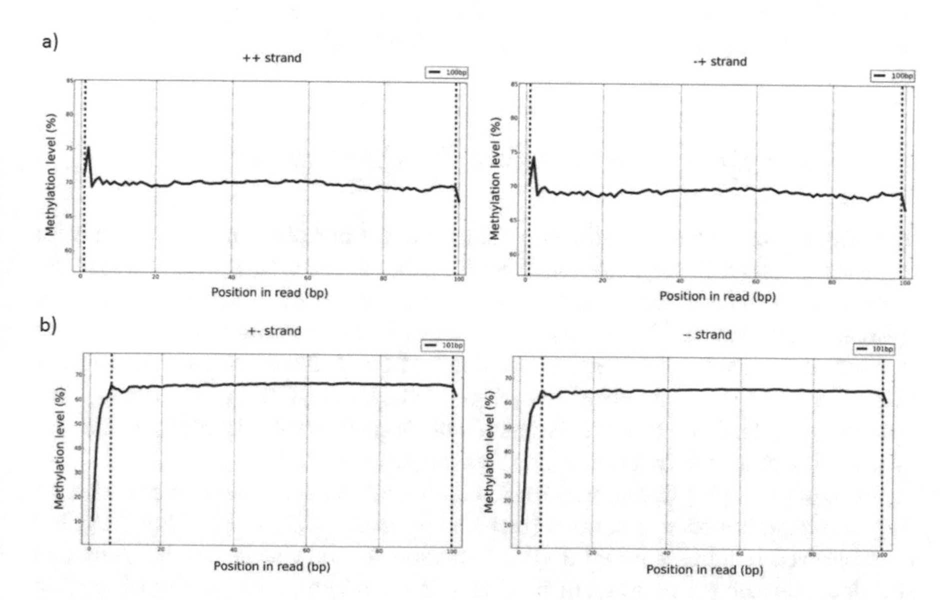

Fig. 8.4 *M-bias* plots for the single-end read SRR202019 data set from Hon et al. [52] (**a**) and a second pair of paired-end read SRR949208 data sets from Ziller et al. [53] (**b**). (**a**) show a slight bias at the 3′ end, while (**b**) presents the bias at the 5′ end. *M-bias* plots are represented using *BSeQC* [54] after trimming the adapter using *CutAdapt* [27] and aligning the reads using *Bismark* default parameters [13]

The most common observation is an increased proportion of unconverted cytosines at the 5′ end, which might be caused by the re-annealing of sequences adjacent to the methylated adapters in bisulfite conversion [49]. Although 5′ end re-annealing is the most common source of *M-bias* and leads to an overestimation of the methylation levels, there are other sources that might also produce biases at both ends of the reads and over- or underestimation of the methylation levels. For example, the overhanging end repair (see Sects. 8.3.3 and 8.3.4) might provoke underestimations of the methylation levels at either the 5′ or/and 3′ end of the reads, or both adapter and erroneous sequencing might lead to an over- or underestimation at the 3′ end.

Currently, *BSeQC* [54] is the most complete software to view and correct the *M-bias* on bisulfite sequencing protocols. It automatically cleans the *M-bias* from both ends of the reads using a statistical cutoff and takes into account that reads from different strands and with different lengths present different biases, whereas *BSmooth* [24] only detects and shows the bias and *Bis-SNP* [49] only corrects the 5′ end overestimation.

8.8 Methylation Levels

Although most of the bias that might affect the methylation estimation are corrected or discarded before calculating the methylation levels for each cytosine, some bias can only be managed during or after the methylation calling.

8.8.1 Methylation and Sequence Variation Calling

The methylation levels are estimated comparing the bisulfite converted reads with the genome reference sequence (see Sect. 8.2), and then a C to T substitution in the sample can be interpreted as an unmethylated cytosine, when actually there is no cytosine in the sample. This natural sequence variation is an important error source in bisulfite sequencing data sets, as two thirds of *Single Nucleotide Polymorphisms* (SNPs) in human genome occur at CpG contexts, having C/T or G/A alleles. This error not only leads to an underestimation of the methylation levels, but also to a miscounting of the methylation contexts in the genome.

Some authors [53] have opted for discarding all the genome positions where a polymorphism has been described (included in *dbSNP* [55] or *HapMap* [56]), but this approach requires curated SNP information for the species in the study and still does not solve the problem in a sample-specific way (it overlooks specific sample variants and discards more contexts than necessary). Another potential solution resides in the fact that the bisulfite only converts single-stranded cytosines to thymines, while the guanines on their reverse complementary strands will remain unaffected [57]. The information on the unconverted strand can then be used to infer the sample sequence allele. Therefore, a variant calling algorithm could use the unconverted strand information during the methylation estimation step to check the sequence context present in the sample and solve this issue in a sample-specific way. Currently, only *Bis-SNP* [49] and *MethylExtract* [39] include modified variant calling algorithms to work on bisulfite sequencing data sets, based on *GATK* [58] and *VarScan* [59], respectively.

8.8.2 *Bisulfite Conversion Efficiency*

Despite the important biases checked before the methylation level estimation, it is not possible to detect, and therefore discard, the random bisulfite failures. The only way to take these potential failures into account in further analyses is assessing the reliability of the methylation levels statistically based on the bisulfite conversion efficiency. It is widely accepted that the best way to measure the bisulfite conversion rate is to spike an unmethylated genome into the experiment (in plants the chloroplast genome can be used). In mammals, for example, some authors have argued that non-CpG contexts [60] or cytosines on the mitochondrial genome [49] can be used for this purpose because of their low methylation rate. However, using them might introduce an additional bias, as these cytosines could present functional and non-random methylation levels [61–63]. Once the bisulfite conversion rate has been estimated, binomial statistics can be used to calculate the probability of a cytosine being really methylated and not being a conversion failure artifact [19, 51]. However, other biologically relevant methylation levels besides methylated or un-methylated do exist, such as intermediate levels (around 0.5) reflecting allele-specific methylation or partially methylated levels associated with distal regulatory regions [64]. To assess the reliability of any methylation states, an error interval should be chosen, and, by means of the binomial distribution, the probability of the methylation level relying on the chosen interval can be estimated using an additional script provided by *MethylExtract* [39].

8.9 Conclusion

Nowadays, bisulfite HTS is the gold-standard technique for the study of DNA methylation as it is the most powerful technique for methylation profiling in terms of specificity and speed. However, its raw data sets must be managed carefully to obtain the most reliable results because of the great number of potential biases that can arise from either the HTS technique or bisulfite conversion.

Once the libraries have been sequenced with appropriate coverage, the first step before the alignment must be properly finding and trimming all the artificial sequences added to the $5'$ and $3'$ ends of the reads, which also should include a $3'$ end quality trimming. This step improves the alignment and discards the potential bias introduced by the artificial methylation of the added sequences.

After the alignment step, duplicated reads and those potentially not affected by the bisulfite conversion must be discarded (remember that duplicates removing is not required in all protocols). The next issue to deal with is the bisulfite conversion biases that can be found at both ends of the reads (*M-bias*); if *M-bias* is detected in your data set, the entire library should be trimmed accordingly to the *M-bias* plots. These steps are the major modifications that must be done to your libraries before starting the methylation calling. During the methylation calling, there are other important considerations: the reference sequence must be corrected using an appropriate variant caller, and redundant calls from $3'$ overlapping pair-end reads, low-quality bases and nucleotides affected by end repair must be discarded. Finally, after the methylation calling and despite all the quality steps, statistically assessing the methylation conversion ratio for each cytosine covered by the data set is recommended.

To our knowledge, no other complete reviews of the known bisulfite high-throughput sequencing biases like the one included in this chapter are available. For each error source, how it may affect the methylation ratios and what methods are proposed for dealing with it (usually already implemented in the available software) are described. Moreover, and because there is no single program that covers all known error sources, in conclusion, a methylation analysis workflow has been proposed, which can be shown as a summary of best practices that should be followed to obtain the best reliable methylation ratios from bisulfite HTS experiments.

References

1. Bonasio, R., Tu, S., Reinberg, D.: Molecular signals of epigenetic states. Science **330**(6004), 612–616 (2010)
2. Lister, R., Ecker, J.R.: Finding the fifth base: genome-wide sequencing of cytosine methylation. Genome Res. **19**(6), 959–966 (2009)
3. Jones, P.A.: Functions of DNA methylation: islands, start sites, gene bodies and beyond. Nat. Rev. Genet. **13**(7), 484–492 (2012)
4. Hotchkiss, R.D.: The quantitative separation of purines, pyrimidines, and nucleosides by paper chromatography. J. Biol. Chem. **175**(1), 315–332 (1948)
5. Riggs, A.D.: X inactivation, differentiation, and DNA methylation. Cytogenet. Cell Genet. **14**(1), 9–25 (1975)
6. Holliday, R., Pugh, J.E.: DNA modification mechanisms and gene activity during development. Science **187**(4173), 226–232 (1975)
7. Laird, P.W.: Principles and challenges of genomewide DNA methylation analysis. Nat. Rev. Genet. **11**(3), 191–203 (2010)
8. Frommer, M., McDonald, L.E., Millar, D.S., Collis, C.M., Watt, F., Grigg, G.W., Molloy, P.L., Paul, C.L.: A genomic sequencing protocol that yields a positive display of 5-methylcytosine residues in individual DNA strands. Proc. Natl. Acad. Sci. U. S. A. **89**(5), 1827–1831 (1992)
9. Xi, Y., Li, W.: Bsmap: whole genome bisulfite sequence mapping program. BMC Bioinf. **10**, 232 (2009)
10. Chen, P.Y., Cokus, S.J., Pellegrini, M.: Bs seeker: precise mapping for bisulfite sequencing. BMC Bioinf. **11**, 203 (2010)

11. Guo, W., Fiziev, P., Yan, W., Cokus, S., Sun, X., Zhang, M.Q., Chen, P.Y., Pellegrini, M.: Bs-seeker2: a versatile aligning pipeline for bisulfite sequencing data. BMC Genomics **14**, 774 (2013)
12. Hach, F., Hormozdiari, F., Alkan, C., Hormozdiari, F., Birol, I., Eichler, E.E., Sahinalp, S.C.: mrsFAST: a cache-oblivious algorithm for short-read mapping. Nat. Methods **7**(8), 576–577 (2010)
13. Krueger, F., Andrews, S.R.: Bismark: a flexible aligner and methylation caller for Bisulfite-Seq applications. Bioinformatics **27**(11), 1571–1572 (2011)
14. Pedersen, B., Hsieh, T.F., Ibarra, C., Fischer, R.L.: Methylcoder: software pipeline for bisulfite-treated sequences. Bioinformatics **27**(17), 2435–2436 (2011)
15. Hackenberg, M., Barturen, G., Oliver, J.L.: In: Tatarinova, T. (ed.) DNA Methylation Profiling from High-Throughput Sequencing Data, DNA Methylation - From Genomics to Technology, InTech (2012). doi:10.5772/34825
16. Chatterjee, A., Stockwell, P.A., Rodger, E.J., Morison, I.M.: Comparison of alignment software for genome-wide bisulphite sequence data. Nucleic Acids Res. **40**(10), e79 (2012)
17. Frith, M.C., Mori, R., Asai, K.: A mostly traditional approach improves alignment of bisulfite-converted DNA. Nucleic Acids Res. **40**(13), e100 (2012)
18. Kunde-Ramamoorthy, G., Coarfa, C., Laritsky, E., Kessler, N.J., Harris, R.A., Xu, M., Chen, R., Shen, L., Milosavljevic, A., Waterland, R.A.: Comparison and quantitative verification of mapping algorithms for whole-genome bisulfite sequencing. Nucleic Acids Res. **42**(6), e43 (2014)
19. Schultz, M.D., Schmitz, R.J., Ecker, J.R.: 'leveling' the playing field for analyses of single-base resolution DNA methylomes. Trends Genet. **28**(12), 583–585 (2012)
20. Beck, S., Rakyan, V.K.: The methylome: approaches for global DNA methylation profiling. Trends Genet. **24**(5), 231–237 (2008)
21. Krueger, F., Kreck, B., Franke, A., Andrews, S.R.: DNA methylome analysis using short bisulfite sequencing data. Nat. Methods **9**(2), 145–151 (2012)
22. Cokus, S.J., Feng, S., Zhang, X., Chen, Z., Merriman, B., Haudenschild, C.D., Pradhan, S., Nelson, S.F., Pellegrini, M., Jacobsen, S.E.: Shotgun bisulphite sequencing of the arabidopsis genome reveals DNA methylation patterning. Nature **452**(7184), 215–219 (2008)
23. Meissner, A., Gnirke, A., Bell, G.W., Ramsahoye, B., Lander, E.S., Jaenisch, R.: Reduced representation bisulfite sequencing for comparative high-resolution DNA methylation analysis. Nucleic Acids Res. **33**(18), 5868–5877 (2005)
24. Hansen, K.D., Langmead, B., Irizarry, R.A.: Bsmooth: from whole genome bisulfite sequencing reads to differentially methylated regions. Genome Biol. **13**(10), R83 (2012)
25. Andrews, S.: FastQC: a quality control application for fastq data (2010). Available online at: http://www.bioinformatics.babraham.ac.uk/projects/fastqc/
26. Hannon: Fastx-toolkit (2009)
27. Martin, M.: Cutadapt removes adapter sequences from high-throughput sequencing reads. EMBnet. J. **17**(1), 10–12 (2011)
28. Bolger, A.M., Lohse, M., Usadel, B.: Trimmomatic: a flexible trimmer for Illumina sequence data. Bioinformatics **30**(15), 2114–2120 (2014)
29. Hansen, K.D., Brenner, S.E., Dudoit, S.: Biases in Illumina transcriptome sequencing caused by random hexamer priming. Nucleic Acids Res. **38**(12), e131 (2010)
30. Schwartz, S., Oren, R., Ast, G.: Detection and removal of biases in the analysis of next-generation sequencing reads. PLoS One **6**(1), e16685 (2011)
31. Poptsova, M.S., Il'icheva, I.A., Nechipurenko, D.Y., Panchenko, L.A., Khodikov, M.V., Oparina, N.Y., Polozov, R.V., Nechipurenko, Y.D., Grokhovsky, S.L.: Non-random DNA fragmentation in next-generation sequencing. Sci. Rep. **4**, 4532 (2014)
32. Aird, D., Ross, M.G., Chen, W.S., Danielsson, M., Fennell, T., Russ, C., Jaffe, D.B., Nusbaum, C., Gnirke, A.: Analyzing and minimizing PCR amplification bias in Illumina sequencing libraries. Genome Biol. **12**(2), R18 (2011)
33. Benjamini, Y., Speed, T.P.: Summarizing and correcting the GC content bias in high-throughput sequencing. Nucleic Acids Res. **40**(10), e72 (2012)

34. Miura, F., Enomoto, Y., Dairiki, R., Ito, T.: Amplification-free whole-genome bisulfite sequencing by post-bisulfite adaptor tagging. Nucleic Acids Res. **40**(17), e136 (2012)
35. Ziller, M.J., Hansen, K.D., Meissner, A., Aryee, M.J.: Coverage recommendations for methylation analysis by whole-genome bisulfite sequencing. Nat. Methods **12**(3), 230–232 (2015)
36. Kozarewa, I., Ning, Z., Quail, M.A., Sanders, M.J., Berriman, M., Turner, D.J.: Amplification-free Illumina sequencing-library preparation facilitates improved mapping and assembly of (g+c)-biased genomes. Nat. Methods **6**(4), 291–295 (2009)
37. Li, H., Handsaker, B., Wysoker, A., Fennell, T., Ruan, J., Homer, N., Marth, G., Abecasis, G., Durbin, R., Subgroup Genome Project Data Processing: The sequence alignment/map format and samtools. Bioinformatics **25**(16), 2078–2079 (2009)
38. Broad-Institute: A set of tools for working with next generation sequencing data in the BAM. Available online at: http://broadinstitute.github.io/picard/
39. Barturen, G., Rueda, A., Oliver, J.L., Hackenberg, M.: MethylExtract: high-quality methylation maps and SNV calling from whole genome bisulfite sequencing data. F1000Res **2**, 217 (2013)
40. Cock, P.J., Fields, C.J., Goto, N., Heuer, M.L., Rice, P.M.: The sanger FASTQ file format for sequences with quality scores, and the Solexa/Illumina FASTQ variants. Nucleic Acids Res. **38**(6), 1767–1771 (2010)
41. James Kent, W., Sugnet, C.W., Furey, T.S., Roskin, K.M., Pringle, T.H., Zahler, A.M., Haussler, D.: The human genome browser at UCSC. Genome Res. **12**(6), 996–1006 (2002)
42. Langmead, B., Trapnell, C., Pop, M., Salzberg, S.L.: Ultrafast and memory-efficient alignment of short DNA sequences to the human genome. Genome Biol. **10**(3), R25 (2009)
43. Li, H.: Improving SNP discovery by base alignment quality. Bioinformatics **27**(8), 1157–1158 (2011)
44. Langmead, B., Salzberg, S.L.: Fast gapped-read alignment with Bowtie 2. Nat. Methods **9**(4), 357–359 (2012)
45. Fuller, C.W., Middendorf, L.R., Benner, S.A., Church, G.M., Harris, T., Huang, X., Jovanovich, S.B., Nelson, J.R., Schloss, J.A., Schwartz, D.C., Vezenov, D.V.: The challenges of sequencing by synthesis. Nat. Biotechnol. **27**(11), 1013–1023 (2009)
46. Taub, M.A., Corrada Bravo, H., Irizarry, R.A.: Overcoming bias and systematic errors in next generation sequencing data. Genome Med. **2**(12), 87 (2010)
47. Del Fabbro, C., Scalabrin, S., Morgante, M., Giorgi, F.M.: An extensive evaluation of read trimming effects on Illumina NGS data analysis. PLoS One **8**(12), e85024 (2013)
48. Minoche, A.E., Dohm, J.C., Himmelbauer, H.: Evaluation of genomic high-throughput sequencing data generated on Illumina HiSeq and genome analyzer systems. Genome Biol. **12**(11), R112 (2011)
49. Liu, Y., Siegmund, K.D., Laird, P.W., Berman, B.P.: Bis-SNP: combined DNA methylation and SNP calling for Bisulfite-seq data. Genome Biol. **13**(7), R61 (2012)
50. DePristo, M.A., Banks, E., Poplin, R., Garimella, K.V., Maguire, J.R., Hartl, C., Philippakis, A.A., del Angel, G., Rivas, M.A., Hanna, M., McKenna, A., Fennell, T.J., Kernytsky, A.M., Sivachenko, A.Y., Cibulskis, K., Gabriel, S.B., Altshuler, D., Daly, M.J.: A framework for variation discovery and genotyping using next-generation DNA sequencing data. Nat. Genet. **43**(5), 491–498 (2011)
51. Lister, R., Pelizzola, M., Dowen, R.H., Hawkins, R.D., Hon, G., Tonti-Filippini, J., Nery, J.R., Lee, L., Ye, Z., Ngo, Q.M., Edsall, L., Antosiewicz-Bourget, J., Stewart, R., Ruotti, V., Millar, A.H., Thomson, J.A., Ren, B., Ecker, J.R.: Human DNA methylomes at base resolution show widespread epigenomic differences. Nature **462**(7271), 315–322 (2009)
52. Hon, G.C., Hawkins, R.D., Caballero, O.L., Lo, C., Lister, R., Pelizzola, M., Valsesia, A., Ye, Z., Kuan, S., Edsall, L.E., et al.: Global DNA hypomethylation coupled to repressive chromatin domain formation and gene silencing in breast cancer. Genet. Res. **22**(2), 246–258 (2012)
53. Ziller, M.J., Gu, H., Muller, F., Donaghey, J., Tsai, L.T., Kohlbacher, O., De Jager, P.L., Rosen, E.D., Bennett, D.A., Bernstein, B.E., Gnirke, A., Meissner, A.: Charting a dynamic DNA methylation landscape of the human genome. Nature **500**(7463), 477–481 (2013)

54. Lin, X., Sun, D., Rodriguez, B., Zhao, Q., Sun, H., Zhang, Y., Li, W.: Bseqc: quality control of bisulfite sequencing experiments. Bioinformatics **29**(24), 3227–3229 (2013)
55. Sherry, S.T., Ward, M.H., Kholodov, M., Baker, J., Phan, L., Smigielski, E.M., Sirotkin,K.: dbSNP: the NCBI database of genetic variation. Nucleic Acids Res. **29**(1), 308–311 (2001)
56. Consortium Genomes Project, Abecasis, G.R., Altshuler, D., Auton, A., Brooks, L.D., Durbin, R.M., Gibbs, R.A., Hurles, M.E., McVean, G.A.: A map of human genome variation from population-scale sequencing. Nature **467**(7319), 1061–1073 (2010)
57. Weisenberger, D.J., Campan, M., Long, T.I., Kim, M., Woods, C., Fiala, E., Ehrlich, M., Laird, P.W.: Analysis of repetitive element DNA methylation by methylight. Nucleic Acids Res. **33**(21), 6823–6836 (2005)
58. McKenna, A., Hanna, M., Banks, E., Sivachenko, A., Cibulskis, K., Kernytsky, A., Garimella, K., Altshuler, D., Gabriel, S., Daly, M., DePristo, M.A.: The genome analysis toolkit: a mapreduce framework for analyzing next-generation DNA sequencing data. Genome Res. **20**(9), 1297–1303 (2010)
59. Koboldt, D.C., Chen, K., Wylie, T., Larson, D.E., McLellan, M.D., Mardis, E.R., Weinstock, G.M., Wilson, R.K., Ding, L.: Varscan: variant detection in massively parallel sequencing of individual and pooled samples. Bioinformatics **25**(17), 2283–2285 (2009)
60. Seisenberger, S., Andrews, S., Krueger, F., Arand, J., Walter, J., Santos, F., Popp, C., Thienpont, B., Dean, W., Reik, W.: The dynamics of genome-wide DNA methylation reprogramming in mouse primordial germ cells. Mol. Cell **48**(6), 849–862 (2012)
61. Iacobazzi, V., Castegna, A., Infantino, V., Andria, G.: Mitochondrial DNA methylation as a next-generation biomarker and diagnostic tool. Mol. Genet. Metab. **110**(1–2), 25–34 (2013)
62. Guo, J.U., Su, Y., Shin, J.H., Shin, J., Li, H., Xie, B., Zhong, C., Hu, S., Le, T., Fan, G., Zhu, H., Chang, Q., Gao, Y., Ming, G.L., Song, H.: Distribution, recognition and regulation of non-CpG methylation in the adult mammalian brain. Nat. Neurosci. **17**(2), 215–222 (2014)
63. Guo, W., Chung, W.Y., Qian, M., Pellegrini, M., Zhang, M.Q.: Characterizing the strand-specific distribution of non-CpG methylation in human pluripotent cells. Nucleic Acids Res. **42**(5), 3009–3016 (2014)
64. Stadler, M.B., Murr, R., Burger, L., Ivanek, R., Lienert, F., Scholer, A., van Nimwegen, E., Wirbelauer, C., Oakeley, E.J., Gaidatzis, D., Tiwari, V.K., Schubeler, D.: DNA-binding factors shape the mouse methylome at distal regulatory regions. Nature **480**(7378), 490–495 (2011)

Part III
Alignment of NGS Data

Chapter 9
Comparative Assessment of Alignment Algorithms for NGS Data: Features, Considerations, Implementations, and Future

Carol Shen, Tony Shen, and Jimmy Lin

9.1 Introduction

Due to the nature of massively parallel sequencing use of shorter reads, the algorithms developed for alignment have been crucial to the widespread adoption of *Next-Generation Sequencing* (NGS). There has been great progress in the development of a variety of different algorithms for different purposes. Researchers are now able to use sensitive and efficient alignment algorithms for a wide variety of applications, including genome-wide variation studies [1], quantitative RNA-seq expression analyses [2], the study of secondary RNA structure [3], microRNA discovery [4], identification of protein-binding sites using ChIP-sequencing [5], recognizing histone modification patterns for epigenetic studies [6], simultaneous alignment of multiple genomes for comparative genomics [7], and the assembly of de novo genomes and transcriptomes [8]. In clinical settings, alignment to reference genomes has led to rapid pathogen discovery [9], identification of causative mutations for rare genetic diseases [10–12], detection of chromosomal abnormalities in tumor genomes [13], and many other advances which similarly depend on rapid and cost-effective genome-wide sequencing.

As previously mentioned, the ability of NGS to perform massively parallel sequencing depends on the alignment of short sequence reads to a reference genome. Prior to the development of NGS, Sanger sequencing had been the gold standard for automated sequencing, utilizing electrophoretic separation to accurately identify linear sequences of randomly terminated nucleotide sequences. However, limited

C. Shen (✉) • T. Shen
School of Medicine, Washington University in St. Louis, St. Louis, MO, USA
e-mail: clshen@wustl.edu; tsshen@wustl.edu

J. Lin
Rare Genomics Institute, St. Louis, MO, USA
e-mail: jimmy.lin@raregenomics.org

M. Elloumi (ed.), *Algorithms for Next-Generation Sequencing Data*,
DOI 10.1007/978-3-319-59826-0_9

by its linear workflow, an automated Sanger machine can only read small amounts of DNA of 500 bp to 1 kb in length within a reasonable time frame [14]. NGS overcomes this bottleneck by fragmenting the target DNA into short fragments, sequencing the short sequences in parallel, and aligning them to a reference genome. The processes of alignment, data storage, and analysis contribute to the immense computing resource requirement of NGS programs. As a result, new alignment tools are constantly and rapidly developed in order to satisfy the lofty requirements of resources and accuracy of NGS technologies.

Alignment algorithms must be able to map reads accurately despite sequence variations and errors inherent in the reference genome, e.g., repetitive regions and polymorphisms. The selection of an inappropriate aligner which fails to account for such variations and errors would lead to inaccurate interpretations of biological outcomes. Therefore, the ability to evaluate the features, performance, and accuracy of aligners in the context of various biological and clinical applications is of paramount importance [14, 15].

The rest of this chapter is organized as follows: In Sect. 9.2, we illustrate the importance of NGS and the development of new alignment algorithms. In Sect. 9.3, we present an overview of algorithms and discuss the technical and theoretical implications of some example aligners. In Sect. 9.4, we describe the biological and clinical applications of aligners and evaluate criteria for selection. Finally, in Sect. 9.5, we discuss future developments of NGS algorithms.

9.2 Importance of NGS and Aligner Development

NGS refers to sequencing technologies designed to detect massive numbers of reads simultaneously, followed by the alignment of short reads to a reference genome.

9.2.1 Overview

NGS technologies are able to produce massive throughput by means of a parallelized workflow. The massive throughput of NGS is made possible by first fragmenting starting genomic materials into shorter read lengths of about 30–400 *base-pairs* (bp). The reads are then converted into a sequencing library, which is then either sequenced or amplified. The library is then cycled through a series of automated, standardized chemical reactions to determine multiple template sequences simultaneously [14]. This parallelized sequencing technology allows NGS to overcome the limitations of discrete DNA fragment separation faced by Sanger sequencing. Popular sequencing platforms such as *Roche 454*, *Illumina Hi-Seq*, ABI SOLiD, and *Ion Torrent* produce data on the order of *Giga base-pairs* (Gbp) per machine day [16], which is at least four orders of magnitude greater than the throughput of a Sanger machine. It is estimated that NGS has decreased the cost of sequencing more than 100,000-fold in the past decade [17].

9.2.2 Implications and Limitations of Short-Read Alignment

NGS technology relies on the fragmentation of the DNA template into shorter fragments. The most widely used NGS platforms (*Roche 454*, *Illumina*, ABI SOLiD, *Ion Torrent*) generate shorter read lengths (30–400 bp) compared to Sanger sequencing (500 bp–1 *Kilo base-pairs* (Kbp)). The subsequent mapping or alignment of DNA fragments to a reference genome is therefore required to assemble the target sequence. Any reference genome, assembled from a mosaic of representative DNA sequences, likely contains inherent sequence variations and sequencing errors, such as repetitive regions; structural variations, e.g. insertions, deletions, and translocations; and polymorphisms [15]. New alignment algorithms are constantly developed and modified to better predict these variations and to increase the accuracy of mapping. In addition to the challenges of aligning short reads to a reference genome, short read lengths also make *de novo* assembly of a genome very difficult, such that the original sequence is almost impossible to reconstruct. Therefore, despite its advantages of speed and efficiency, short-read alignment represents the rate-limiting step in progressing toward the widespread adoption of NGS [18].

9.2.3 Implications and Limitations of Long-Read Alignment

Currently, sequencing platforms are being developed to produce longer read lengths in order to overcome some of the challenges described above. Particularly, longer read lengths allow for easier de novo genome assembly and more sensitive detection of structural variations and repetitive regions. Since longer reads are more likely to contain structural variations and repetitive regions than short reads, long-read aligners must be more permissive toward alignment gaps and inexact matches. The algorithms used by long-read aligners, which are hash-table-based, are therefore fundamentally different from those preferred by short-read aligners, such as *Burrows-Wheeler-Transform* (BWT)-based programs [18]. The different classes of algorithms and their implementation will be further discussed in this chapter.

9.2.4 Paired-End and Mate-Paired Sequencing

Some sequencing technologies produce paired-end or mate-paired reads [18], which are additional strategies to increase sensitivity for structural variations. In paired-end sequencing, both forward and reverse reads of a linear segment of DNA are paired together in order to map both ends of the sequence. This is achieved by first independently aligning the two reads belonging to a pair, then identifying the pairs of hits with the correct orientation and the right distance apart [19]. Since sequencing errors and structural variations have to be accounted for in both reads in a pair, not all aligners can achieve paired-end alignment. An aligner in paired-end sequencing must allow for mismatches and/or gaps in alignment. In mate-paired sequencing, DNA fragments are circularized, enabling both ends to be mapped at once. When alignment at both ends fails, the error is detected and fixed. Paired-end alignment is shown to have higher sensitivity and specificity than single-end alignment [18].

9.2.5 Overview of Aligners for Different Platforms

Currently, there are three NGS sequencing platforms used to a large extent: *Roche 454*, *Illumina*, ABI SOLiD, and *Ion Torrent* [15]. These platforms vary in their engineering and sequencing chemistry; therefore, aligners with different properties are required to map the different types of reads produced by these platforms.

Roche 454 sequencers are able to produce longer reads more than 500 bp in length. Because of this capability, it has wide applications in de novo sequencing and resequencing, as well as transcriptome sequencing and ultra-deep sequencing [20]. An additional advantage of this platform is that the reads generated have high 5′ base quality [18]. On the other hand, reads generated by *Roche 454* instruments are more prone to contain insertions and deletions due to their longer length. Some popular aligners used on this platform are RMAP, SHRiMP2 [21], BWA [22], SOAP2 [19], *Bowtie* [23], and SSAHA2 [15].

Illumina Hi-Seq sequencers amplify DNA on the surface of micro-fluidic chips, providing templates for the sequential addition of fluorescent reversible terminator deoxyribonucleotides. The chips are imaged subsequent to each nucleotide addition and analyzed to generate a sequence [24]. *Illumina Hi-Seq* sequencers historically could produce hundreds of millions of shorter reads (25–50 bp) [25] but are now moving toward longer reads (250–300 bp). Shorter read lengths allow for higher speeds and lower costs of sequencing [26]. The high 5′ base quality of reads and the low indel error rate are also advantages [18]. On the other hand, the shorter read lengths of *Illumina* are unlikely to encompass all structural variations and repetitions, leading to a higher sequencing error of mismatches. Examples of popular aligners used on this platform include RMAP, *SeqMap*, MAQ, SHRiMP2, BWA, SOAP2, and *Bowtie* [15].

ABI SOLiD sequencers generate up to 50 bp read lengths and use a unique ligation-mediated sequencing strategy, which increases its sensitivity for homopolymers [27]. In addition, they generate two-base encoding data, which allow discrimination between sequencing errors and polymorphisms. On the other hand, ABI SOLiD sequencers are comparatively slower than other sequencers. Instead of using a letter-based encoding system, ABI SOLiD processes data in a color-space. The computational challenge of aligning color-based data to a letter-based reference genome necessitates computational strategies to overcome the problem of translation. Four dyes are used to encode for the 16 nucleotide pairs, so each color codes for a set of four nucleotide pairs. The colors are each given a reference number, and the algorithm proceeds as a finite state automaton: Each shift from one letter to the next gives off a particular emission, which is interpreted as color. If the first nucleotide in the pair is known, the next nucleotide can be inferred given the color [1]. Utilizing color-space necessitates the extra step of translating reads into letter-space before alignment. A solution is to translate the reference genome into color-space and align the colors [28]. Sequencing nucleotides in pairs and assigning each a color also means that if one of the bases is misidentified due to a sequencing error, the rest of the letters are mistranslated. This can be useful in distinguishing between sequencing errors and true structural variations or polymorphisms. However, ABI SOLiD is not sensitive at recognizing two adjacent SNPs. One solution is to utilize a color-space Smith-Waterman algorithm [1] allowing for sequencing errors. ABI SOLiD Aligners used widely on this platform include RMAP, MAQ, SHRiMP2, BWA, *Bowtie*, and SSAHA2 [15].

Ion Torrent utilizes semi-conductor technology in detecting the release of protons by the incorporation of nucleotides during synthesis [29]. DNA fragments are linked to microbeads via adapter sequences and clonally amplified. The beads are loaded onto a silicon chip for sequencing. Each nucleotide base is introduced sequentially, and a release of protons signals the addition of one or more of that nucleotide at that particular location in the sequence. *Ion Torrent* cannot accurately detect the number of bases in homopolymers more than 8 bases long [30]. The read length generated by *Ion Torrent* machines is about 200 bp. Examples of popular aligners are *Novoalign* [31], SMALT, *SRmapper*, *Bowtie2* [32], MOSAIK, *segemehl* [33], and SHRiMP2 [34].

9.3 Technical and Theoretical Implications of Alignment

Alignment algorithms are derived from two fundamental algorithms, namely the hash-table-based algorithm and the BWT algorithm [18].

9.3.1 *Hash-Table-Based Algorithms*

The idea of hash-table indexing originates in the *Basic Local Alignment Search Tool* (BLAST) [35], which searches for regions of similarity between biological sequences (nucleotides or proteins). Hash-table-based algorithms use a strategy named seed-and-extend. First, during the seed step, the program performs a homology search, localizing the k-mer substrings shared by both reads and the reference genome through hash tables. Then, the program extends these local alignments of seeds using a slower and more accurate dynamic programming algorithm, e.g., the Smith-Waterman algorithm [36]. Aligners classified under the hash-table-based method include *SeqMap*, PASS, MAQ, GASSST, RMAP, PErM, *RazerS, microread Fast Alignment Search Tool (mrFAST), microread (substitutions only) Fast Alignment Search Tool* (mrsFAST), *GenomeMapper*, and BOAT [15].

9.3.1.1 Spaced Seeds

Different alignment algorithms have different strategies for seed detection. A seed allowing mismatches at certain positions is called a spaced seed. The BLAST algorithm performs seed alignment with 11 consecutive matches between read and reference sequences. However, spaced seed requires matches only at certain pre-selected positions, allowing for the possibility of sequence errors and structural variations in the read sequence. The result is improved speed and sensitivity of the homology searches [37]. Examples of aligners that perform spaced seeds include PerM, SHRiMP2, *RazerS*, BOAT, and GASSST [15].

9.3.1.2 The Pigeonhole Principle

The pigeonhole principle is an extension of the spaced seed principle. The general principle states that if there are more items than the number of containers available for holding them, then at least one container will hold more than one item. Using this principle, if there is an alignment of $k + n$ bases with at most n mismatches, $k/2$ bases must be exact matches. It follows that given the type of inexact alignment, the number of guaranteed exact matches can be quantified. For alignment using the pigeonhole principle, after the initial seeding of inexact alignment, the expected exact matches within the inexact alignment can be extended to produce longer matches. Examples of alignment algorithms that use the pigeonhole principle include SeqMap, MAQ, RMAP [38], and SOAP2 [15].

9.3.1.3 Q-Gram Filter

One problem with consecutive seeds, spaced seeds, and seeds that use the *pigeonhole principle,* however, is that they are not permissive toward gaps within the seed [15]. The g-gram filter is based on the observation that in a w-long read with at most k differences (mismatches and gaps), the read and the w-long reference sequence share at least $(w + 1)–(k + 1)q$ common subsequences of length q [18, 39]. Non-candidate filtration is then performed based on a seed-and-extend alignment of clusters of short, overlapping q-grams. This method is useful for accelerating long-read alignment. Examples of aligners that use the q-gram filter include SHRiMP2 [21] and *RazerS* [40].

9.3.2 Burrows-Wheeler-Transform-Based Algorithms

Instead of aligning seeds of reads against substrings from the reference genome, alignment algorithms based on the BWT align entire reads against the reference genome. In order to do this, all the suffixes of reference genome sequences are stored in a database based on certain representations of data structure, such as the *prefix-suffix tree*, *suffix array*, and *FM-index* [18]. A prefix-suffix tree data structure allows the alignment of multiple identical copies of a segment in the reference genome all at once since they are all collapsed in a single path in the tree. The benefit of this strategy is that it efficiently solves alignment to multiple identical copies in the reference genome sequence, instead of aligning repetitive sequences individually as in a *hash table index* [18]. The *FM-index* was developed by Ferregina and Manzini as a reversible data compression algorithm which allowed for reduced memory occupation of data structures [41]. At 0.5–2 bytes of memory per nucleotide, the *FM-index* of the entire human genome occupies 2–8 GB of memory [18], which allows it to be stored on any personal computer. Aligners classified under the *FM-index* include SOAP2 [19], BWA [22], *Bowtie* [32], and CUSHAW [42].

9.3.3 Combination of Hash-Table-Based and Index-Based Approaches

Recently, novel approaches to reducing the memory footprint and improving processing speeds have involved developing algorithms which are a combination of both hash-table-based and index strategies [43]. One study [43] describes a novel alignment strategy named HIVE-Hexagon, which first non-redundifies short read data sets then stores such reads and reference sets in a computational cloud. Then, parallelized reference sets are compiled into a hash table, and each read undergoes inexact seeding and extension. If an inexact alignment doesn't meet score requirements, it is filtered out. Finally, an adaptation of the Smith-Waterman algorithm [36] is computed for the candidate reads that had the best matches in the alignment stage. This method reduces the time requirement of alignment by reducing the number of computations necessary without sacrificing accuracy.

9.4 Biological and Clinical Applications of Alignment

9.4.1 DNA

NGS of genomes is made possible by aligning short reads to a reference genome, allowing for genome-wide association studies and the identification of structural variants. Aligners applied to DNA sequencing should be able to map reads to the reference genome accurately and efficiently in terms of computational cost. They should ideally be sensitive to structural variations in the reference genome, such as insertions, deletions, and polymorphisms. Below are some examples of popular aligners for DNA sequencing.

9.4.1.1 Bowtie

Bowtie is an aligner which utilizes the *FM-index* based on BWT. *Bowtie* was developed in 2009 as a response to the high computational cost of aligning short reads to a large genome. As a result of the BWT indexing of the reference genome, alignment with *Bowtie* incurs a very small computational cost compared to that of aligners previously developed, which relied largely on hash tables. *Bowtie* aligns 35-bp reads at over 25 million reads per CPU hour and has a memory footprint of only about 1.3 GB for the human genome [23]. The increase in speed does come at a cost of accuracy, especially in the context of resequencing. However, the aligner's backtracking algorithm that allows inexact matches and favors high-quality alignments help increase its sensitivity.

Although *Bowtie* is able to perform efficient ungapped alignment of short reads, it fails to align reads spanning gaps efficiently. In 2012, the authors developed an extension of *Bowtie*, *Bowtie 2*, to allow for gapped alignment. *Bowtie 2* first undergoes an ungapped seeding stage based on the *FM-index*, followed by a gapped extension stage that uses dynamic programming [32].

9.4.1.2 BWA

The *Burrows-Wheeler Alignment* (BWA) tool was also developed in 2009 as a response to the enormous volume of short reads generated by sequencing technologies at the time. It is an aligner which is based on backward search with BWT, enabling rapid alignment at low computational costs. BWA is able to perform gapped alignment and paired-end mapping and operates on both base-space reads and color-space reads. BWA has a memory footprint of 2.3 GB for single-end mapping and 3 GB for paired-end mapping [22]. Since BWA requires full reads to be aligned, it is less able to account for structural variations or sequencing errors in the reference genome, making it less reliable for long-read alignment.

In order to address the inability of BWA to align long reads accurately and to respond to the longer reads produced by sequencing platforms, the authors developed a new algorithm, BWA's Smith-Waterman Alignment (BWA-SW) in 2010. BWA-SW is able to align reads of up to 1 MB against the reference genome, using 4 GB of memory. In contrast to hashing-based algorithms used exclusively to align long reads at the time, e.g., BLAT and SSAHA2, BWA-SW uses dynamic programming to find seeds between two *FM indices*, allowing for gaps and mismatches. It then performs quality-conscious extension, eliminating unnecessary extension of repetitive sequences [44].

9.4.1.3 SOAP 2

SOAP2 [19] was developed in 2009 as an improvement on its previous version SOAP, which utilized hash-table indexing of the reference genome instead of BWT. SOAP2 is able to perform both single-end and paired-end mapping with inexact matches. SOAP2 has an additional feature that allows additional mismatches in the low-quality 3′ region of reads. It is able to align reads up to 1024 bp in length. The memory usage of SOAP2 is 5.4 GB, which does not increase regardless of read count and reference genome size due to its BWT-based backtracking algorithm.

9.4.1.4 Novoalign

Novoalign [31] is a short-read alignment tool that supports single-end, paired-end, and mate-pair reads. It utilizes hash-table-based indexing combined with dynamic programming based on the Needleman-Wunsch algorithm. It is able to perform gapped alignment and allows for inexact matching, resulting in a high sensitivity for SNPs, indels, and other structural variations. While slower than BWT-based backtracking algorithms, it has a high level of accuracy by being aware of base quality at all levels of alignment. *NovoalignCS* is a version of *Novoalign* developed for color-space reads [31].

9.4.1.5 SHRiMP2

SHRiMP2 was developed in 2011 as an improvement on its earlier version, *Short Read Mapping Program* (SHRiMP). SHRiMP2 indexes the genome using spaced seeds. Each read is then projected to the reference genome using the spaced seeds. *Candidate Mapping Locations* (CMLs) are generated, requiring multiple matches between the reads and the reference genome. The CMLs are then investigated by the Smith-Waterman string matching algorithm. SHRiMP2 supports both color-space and base-space reads. Indexing of the entire human genome using SHRiMP2 requires 48 GB; however, SHRiMP2 provides tools to break the genome into segments to reduce RAM usage. SHRiMP2 is highly sensitive to polymorphisms and indels [21].

9.4.2 RNA

9.4.2.1 RNA-seq

Alignment of RNA reads can be useful for identifying differences in gene expression and better understanding certain biological pathways, such as tumorigenesis, development, and metabolism. The unique challenge of RNA alignment is that the aligner must be able to place spliced reads spanning introns and to correctly determine exon-intron boundaries. The criteria for evaluation of RNA aligners is thus accuracy, mismatch and gap sensitivity, exon junction discovery, and suitability of alignments for transcript reconstruction [2]. One study showed that FM-index-based aligners perform better than hash-based aligners when taking these criteria into account [45].

Aligners such as GSNAP [46], GSTRUCT, *MapSplice* [47], and STAR [48] are effective RNA-seq aligners. They use different techniques in identifying exon junctions and aligning RNA reads to a reference. GSTRUCT and *MapSplice* are able to infer exon junctions from the initial seeding based on the presence of gapped alignment then perform realignment to the reference genome in order to account for exon-intron boundaries. GSNAP detects splice junctions in reads as short as 14 bp by using probabilistic models or a database of known splice sites. STAR allows direct non-contiguous alignment of RNA reads to the reference genome, followed by stitching together clusters of the aligned seeds to build an entire read. MapSplice is the most conservative regarding exon junction calls, whereas GSNAP and GSTRUCT tend to detect more false junctions [2].

9.4.2.2 Secondary Structure

Non-coding RNAs regulate many processes in gene expression, including alternative splicing, transcription, translation, and mRNA localization. Genome-wide comparative RNA structurome analyses can result in the discovery of novel regulatory pathways. However, traditional methods for 3-dimensional RNA structure determination, e.g., X-ray crystallography, *Nuclear Magnetic Reasonance* (NMR), and *cryo-Electron Microscopy* (cryo-EM), are inappropriately costly for genome-wide applications. NGS has made the genome-wide experimental determination of RNA secondary structure possible by means of technologies such as PARS [49], *FragSeq* [50], and SHAPE-seq [51]. Alignment of RNA reads can then be performed to structuromes created by these methods for genome-wide comparative analyses of RNA secondary structure.

There are very few existing programs that can accurately and efficiently perform RNA structure-to-structure alignment. Among existing alignment strategies for RNA secondary structures are tree editing algorithms. In these algorithms, RNA secondary structures are represented as tree structures, and the structural similarity between tree structures can be represented by the edit-distance [52]. The tree alignment algorithm is a special case of the tree editing algorithm. Instead of computing edit-distance to compare similarities between trees, tree alignment algorithms compute the tree alignment distance [53]. *RNAforester* is an example of an aligner based on the tree alignment algorithm [54]. ERA is an example of an aligner which performs general edit-distance alignment of RNA secondary structure using a novel sparse dynamic programming technique [3].

9.4.2.3 MicroRNA

MicroRNAs (miRNAs) are known to be important regulators of many cellular processes, including development, metabolism, immunology, and tumorigenesis. They regulate RNA expression through selective hybridization and cleavage of mRNA or suppression of translation. miRNA sequencing and subsequent alignment to reference RNA data sets allow characterization and detection of known and novel miRNAs and comparative cross-species analyses. miRNA software tools use a variety of alignment algorithms and have varying functionality depending on the aligner they use. *miRExpress* and DSAP, which both use a Smith-Waterman algorithm, have high sensitivity performances when detecting known miRNAs. mirTools utilizes SOAP2 and has fast computational speed and a low memory requirement. MIReRNA and *miRDeep*, which utilize BLAST, as well as mirTools, are accurate and sensitive software for predicting novel miRNAs in humans [55].

9.4.3 Bisulfite Sequencing

Bisulfite sequencing with NGS enables genome-wide measurement of DNA methylation, which provides important information about epigenetics and genome-wide expression trends. Before alignment, unmethylated cytosine bases in the read sequences are converted to thymines, and the guanines complementary to those cytosines are converted to adenosines. Aligners of bisulfite sequencing data account for these conversions by creating a C-to-T reference, in which all Cs are converted to Ts and a G-to-A reference, in which all Gs are converted to As. In the first round of alignment to the C-to-T reference, C-T mismatches are regarded as methylated Cs. In the next round of alignment, a similar procedure is performed for the G-to-A conversion.

Five bisulfite aligners are widely used: *Bismark* [56], BSMAP [57], *Pash* [58], *BatMeth* [59], and *BS Seeker* [60]. *Bismark*, BSMAP, *Path,* and *BS Seeker* all have high coverage of genome-wide CpG sites and have reliable estimates of percentage methylation. *Bismark* has the advantage of speed, high genomic coverage, and accuracy. *Pash* is slower but has high sensitivity to structural variations [61]. Recently, GNUMAP-bs has been developed to incorporate probabilistic alignment, which accounts for uncertainty from sequencing error and imperfect bisulfite conversion in order to improve accuracy and read coverage [6].

9.5 Future Directions

9.5.1 Computational Performance

Improving the computational efficiency of alignment represents an important challenge as NGS usage becomes more widespread. Since its introduction, NGS throughput has improved 5-fold per year, while Moore's law estimates a doubling of computational performance every 18–24 months [62]. In order to handle this increasing volume of data, computational architectures need to increase efficiency, reduce operating costs, and be scalable.

Many investigators have turned to *cloud computing* to address these challenges. *Cloud computing* refers to the network-coordinated distribution of computational load over multiple machines simultaneously. The ability to outsource computing resources to cloud servers (e.g., *Amazon Web Services*) reduces costs and increases accessibility for researchers without access to dedicated computing centers. As a proof of concept, Maji et al. modified *TopHat*, a splice junction mapper using the *Bowtie* aligner, by changing certain serial computational steps into parallel ones. *Pipelined Version of TopHat* (PVT), designed to be implemented on a cloud architecture, reduced execution time by 41% for paired-end reads [63]. Researchers at the University of Minnesota implemented a cloud-based, web-based computing infrastructure designed to handle increasing volumes of NGS data. Initial testing demonstrated reduced the cost per sequencing sample without sacrificing accuracy [64]. Similarly, Reid et al. developed *Mercury* as a scalable cloud-based bioinformatics pipeline for the purpose of increasing efficiency [65].

Improved computational performance may also be achieved through the pairing of algorithms with dedicated hardware. The *Dynamic Read Analysis of GENomes* (DRAGEN) processor by *Edico Genomics* has been designed specifically for NGS computations [66]. Custom processors could be engineered to sacrifice the flexibility of a general-purpose computer in order to optimize efficiency for sequencing applications. While current bioinformatics programs are written to run on conventional computing hardware, future algorithms may be designed to maximize the hardware capabilities of dedicated processors such as DRAGEN.

9.5.2 Clinical Applications

In conjunction with a robust bioinformatics pipeline, the development of efficient alignment algorithms will generate novel clinical tools. Clinicians at the University of California-San Francisco developed *Sequence-based Ultrarapid Pathogen Identification* (SURPI) as a tool for public health surveillance, epidemiological investigation, and diagnosis for infectious diseases [9]. Relying on the speed of the BW aligner SNAP, SURPI is able to analyze pathogen sequences in less than 5 h. The time scale of this analysis is short enough to be clinically actionable [9].

9.5.3 End-User Friendliness

Improvements in end-user friendliness are necessary for the continued widespread adoption of NGS in research and clinical settings. The *Mayo Clinic* has created an integrated RNA-seq program that performs sequence alignment among other key computational tasks in an effort to standardize research workflows [67]. Several companies also offer integrated bioinformatics services, including *Seven Bridges Genomics*, *Appistry*, *BINA Technologies*, *DNA Nexus*, and *Flatiron Health*.

9.5.4 Long Read Lengths

Read lengths will continue to increase as DNA sequencing technologies mature [18]. *Third-generation sequencing systems* are projected to produce reads lengths on the order of thousands to tens of thousands. Single-molecule real-time sequencing (SMRT, *Pacific Biosciences*), available since 2011, produces reads in the range of 5000–6000 bp [68, 69]. *Nanopore sequencing* (*Oxford Nanopore*), though yet to see widespread use, has produced read lengths of 4500 bp [70, 71].

The ability to generate longer read lengths will change the role of alignment in the bioinformatics pipeline. In some applications, alignment to a reference genome may no longer be necessary. SMRT has been used to sequence microbial genomes de novo [68]. Long read lengths will also enable sequencing of regions rich in tandem repeats which are unresolvable by short reads. Ummat and Bashir developed an aligner specifically designed to map tandem repeats from long reads obtained from SMRT [72]. As read lengths increase, there will likely be reduced pressure to optimize short-read algorithms. However, the principles and strategies gained from building short-read aligners will inform the development of novel strategies to accommodate advances in sequencing technology.

References

1. Dalca, A.V., Brudno, M.: Genome variation discovery with high-throughput sequencing data. Brief. Bioinform. **11**(1), 3–14 (2010)
2. Engstrom, P.G., et al.: Systematic evaluation of spliced alignment programs for RNA-seq data. Nat. Methods. **10**(12), 1185–1191 (2013)
3. Zhong, C., Zhang, S.: Efficient alignment of RNA secondary structures using sparse dynamic programming. BMC Bioinformatics. **14**, 269 (2013)
4. Sun, Z., et al.: CAP-miRSeq: a comprehensive analysis pipeline for microRNA sequencing data. BMC Genomics. **15**, 423 (2014)
5. Johnson, D.S., et al.: Genome-wide mapping of in vivo protein-DNA interactions. Science. **316**(5830), 1497–1502 (2007)
6. Hong, C., et al.: Probabilistic alignment leads to improved accuracy and read coverage for bisulfite sequencing data. BMC Bioinformatics. **14**, 337 (2013)
7. Kim, J., Ma, J.: PSAR-align: improving multiple sequence alignment using probabilistic sampling. Bioinformatics. **30**(7), 1010–1012 (2014)
8. Li, R., et al.: De novo assembly of human genomes with massively parallel short read sequencing. Genome Res. **20**(2), 265–272 (2010)
9. Naccache, S.N., et al.: A cloud-compatible bioinformatics pipeline for ultrarapid pathogen identification from next-generation sequencing of clinical samples. Genome Res. **24**(7), 1180–1192 (2014)

10. Ng, B.G., et al.: Mosaicism of the UDP-galactose transporter SLC35A2 causes a congenital disorder of glycosylation. Am. J. Hum. Genet. **92**(4), 632–636 (2013)
11. Green, R.C., et al.: Exploring concordance and discordance for return of incidental findings from clinical sequencing. Genet. Med. **14**(4), 405–410 (2012)
12. Goh, V., et al.: Next-generation sequencing facilitates the diagnosis in a child with twinkle mutations causing cholestatic liver failure. J. Pediatr. Gastroenterol. Nutr. **54**(2), 291–294 (2012)
13. Schroder, J., et al.: Socrates: identification of genomic rearrangements in tumour genomes by re-aligning soft clipped reads. Bioinformatics. **30**(8), 1064–1072 (2014)
14. Rizzo, J.M., Buck, M.J.: Key principles and clinical applications of "next-generation" DNA sequencing. Cancer Prev. Res. (Phila.) **5**(7), 887–900 (2012)
15. Shang, J., et al.: Evaluation and comparison of multiple aligners for next-generation sequencing data analysis. Biomed. Res. Int. **2014**, 16 (2014)
16. Metzker, M.L.: Sequencing technologies—the next generation. Nat. Rev. Genet. **11**(1), 31–46 (2010)
17. Lander, E.S.: Initial impact of the sequencing of the human genome. Nature. **470**(7333), 187–197 (2011)
18. Li, H., Homer, N.: A survey of sequence alignment algorithms for next-generation sequencing. Brief. Bioinform. **11**(5), 473–483 (2010)
19. Li, R., et al.: SOAP2: an improved ultrafast tool for short read alignment. Bioinformatics. **25**(15), 1966–1967 (2009)
20. Margulies, M., et al.: Genome sequencing in microfabricated high-density picolitre reactors. Nature. **437**(7057), 376–380 (2005)
21. David, M., et al.: SHRiMP2: Sensitive yet Practical Short Read Mapping. Bioinformatics. **27**(7), 1011–1012 (2011)
22. Li, H., Durbin, R.: Fast and accurate short read alignment with Burrows–Wheeler transform. Bioinformatics. **25**(14), 1754–1760 (2009)
23. Langmead, B., Trapnell, C., Pop, M., Salzberg, S.: Ultrafast and memory-efficient alignment of short DNA sequences to the human genome. Genome Biol. **10**(3), R25 (2009)
24. Bentley, D.R., et al.: Accurate whole human genome sequencing using reversible terminator chemistry. Nature. **456**(7218), 53–59 (2008)
25. Smith, A.D., Xuan, Z., Zhang, M.Q.: Using quality scores and longer reads improves accuracy of Solexa read mapping. BMC Bioinformatics. **9**(128), 128 (2008)
26. Hoffmann, S., et al.: Fast mapping of short sequences with mismatches, insertions and deletions using index structures. PLoS Comput. Biol. **5**(9), e1000502 (2009)
27. Ondov, B.D., et al.: Efficient mapping of applied biosystems SOLiD sequence data to a reference genome for functional genomic applications. Bioinformatics. **24**(23), 2776–2777 (2008)
28. Kim, D., et al.: TopHat2: accurate alignment of transcriptomes in the presence of insertions, deletions and gene fusions. Genome Biol. **14**(4), R36 (2013)
29. Rothberg, J.M., et al.: An integrated semiconductor device enabling non-optical genome sequencing. Nature. **475**(7356), 348–352 (2011)
30. Quail, M.A., et al.: A tale of three next generation sequencing platforms: comparison of Ion Torrent, Pacific Biosciences and Illumina MiSeq sequencers. BMC Genomics. **13**, 341 (2012)
31. Novocraft Technologies: Novoalign 30 June 2014. Available from: http://www.novocraft.com/main/index.php (2014). Accessed 20 September 2014
32. Langmead, B., Salzberg, S.L.: Fast gapped-read alignment with Bowtie 2. Nat. Methods. **9**(4), 357–359 (2012)
33. Otto, C., Stadler, P.F., Hoffmann, S.: Lacking alignments? The next-generation sequencing mapper segemehl revisited. Bioinformatics. **30**(13), 1837–1843 (2014)
34. Caboche, S., et al.: Comparison of mapping algorithms used in high-throughput sequencing: application to Ion Torrent data. BMC Genomics. **15**, 264 (2014)
35. Altschul, S.F., Gish, W., Miller, W., Myers, E.W., Lipman, D.J.: Basic local alignment search tool. J. Mol. Biol. **215**(3), 8 (1990)

36. Smith, T.F., Waterman, M.S.: Identification of common molecular subsequences. J. Mol. Biol. **147**(1), 195–197 (1981)
37. Ma, B., Tromp, J., Li, M.: PatternHunter: faster and more sensitive homology search. Bioinformatics. **18**(3), 440–445 (2002)
38. Ruffalo, M., LaFramboise, T., Koyutürk, M.: Comparative analysis of algorithms for next-generation sequencing read alignment. Bioinformatics. **27**(20), 2790–2796 (2011)
39. Cao, X., Cheng, L.S., Tung, A.K.H.: Indexing DNA sequences using q-Grams. DASFAA, Lecture Notes in Computer Science, vol. 3453: p. 13 (2005)
40. Weese, D., et al.: RazerS—fast read mapping with sensitivity control. Genome Res. **19**(9), 1646–1654 (2009)
41. Ferragina, P., Manzini, G.: Opportunistic data structures with applications. Proceedings of the 41st symposium on foundations of computer science, Redondo Beach, CA, USA, p. 9. (2000)
42. Liu, Y., Schmidt, B., Maskell, D.L.: CUSHAW: a CUDA compatible short read aligner to large genomes based on the Burrows–Wheeler transform. Bioinformatics. **28**(14), 1830–1837 (2012)
43. Santana-Quintero, L., et al.: HIVE-hexagon: high-performance, parallelized sequence alignment for next-generation sequencing data analysis. PLoS One. **9**(6), e99033 (2014)
44. Li, H., Durbin, R.: Fast and accurate long-read alignment with Burrows–Wheeler transform. Bioinformatics. **26**(5), 589–595 (2010)
45. Lindner, R., Friedel, C.C.: A comprehensive evaluation of alignment algorithms in the context of RNA-Seq. PLoS One. **7**(12), e52403 (2012)
46. Wu, T.D., Nacu, S.: Fast and SNP-tolerant detection of complex variants and splicing in short reads. Bioinformatics. **26**(7), 873–881 (2010)
47. Wang, K., et al.: MapSplice: accurate mapping of RNA-seq reads for splice junction discovery. Nucleic Acids Res. **38**(18), e178 (2010)
48. Dobin, A., et al.: STAR: ultrafast universal RNA-seq aligner. Bioinformatics. **29**(1), 15–21 (2013)
49. Kertesz, M., et al.: Genome-wide measurement of RNA secondary structure in yeast. Nature. **467**(7311), 103–107 (2010)
50. Underwood, J.G., et al.: FragSeq: transcriptome-wide RNA structure probing using high-throughput sequencing. Nat. Methods. **7**(12), 995–1001 (2010)
51. Lucks, J.B., et al.: Multiplexed RNA structure characterization with selective 2'-hydroxyl acylation analyzed by primer extension sequencing (SHAPE-Seq). Proc. Natl. Acad. Sci. U. S. A. **108**(27), 11063–11068 (2011)
52. Zhang, K., Shasha, D.: Simple fast algorithms for the editing distance between trees and related problems. SIAM J. Comput. **18**, 1245–1262 (1989)
53. Jiang, T., Wang, L., Zhang, K.: Alignment of trees–an alternative to tree edit. Theor. Comput. Sci. **143**, 137–148 (1995)
54. Hochsmann, M., Toller, T., Giergerich, R., Kurtz, S.: Local similarity in RNA secondary structures. In: Proceedings of the 2nd IEEE Computer Society Bioinformatics Conference, Washington DC, (2003). pp. 159–168
55. Li, Y., et al.: Performance comparison and evaluation of software tools for microRNA deep-sequencing data analysis. Nucleic Acids Res. **40**(10), 4298–4305 (2012)
56. Krueger, F., Andrews, S.R.: Bismark: a flexible aligner and methylation caller for Bisulfite-Seq applications. Bioinformatics. **27**(11), 1571–1572 (2011)
57. Xi, Y., Li, W.: BSMAP: whole genome bisulfite sequence MAPping program. BMC Bioinformatics. **10**, 232 (2009)
58. Coarfa, C., et al.: Pash 3.0: A versatile software package for read mapping and integrative analysis of genomic and epigenomic variation using massively parallel DNA sequencing. BMC Bioinformatics. **11**, 572 (2010)
59. Lim, J.Q., et al.: BatMeth: improved mapper for bisulfite sequencing reads on DNA methylation. Genome Biol. **13**(10), R82 (2012)
60. Chen, P.Y., Cokus, S.J., Pellegrini, M.: BS Seeker: precise mapping for bisulfite sequencing. BMC Bioinformatics. **11**, 203 (2010)

61. Kunde-Ramamoorthy, G., et al.: Comparison and quantitative verification of mapping algorithms for whole-genome bisulfite sequencing. Nucleic Acids Res. **42**(6), e43 (2014)
62. Schatz, M.C., Langmead, B., Salzberg, S.L.: Cloud computing and the DNA data race. Nat. Biotechnol. **28**(7), 691–693 (2010)
63. Maji, R.K., et al.: PVT: an efficient computational procedure to speed up next-generation sequence analysis. BMC Bioinformatics. **15**, 167 (2014)
64. Onsongo, G., et al.: Implementation of cloud based next generation sequencing data analysis in a clinical laboratory. BMC Res. Notes. **7**, 314 (2014)
65. Reid, J.G., et al.: Launching genomics into the cloud: deployment of Mercury, a next generation sequence analysis pipeline. BMC Bioinformatics. **15**(1), 30 (2014)
66. Oldach, L.: Edico genome makes first sale of NGS processor. In: Bio-IT World, Cambridge Healthtech Institute, 2014
67. Kalari, K.R., et al.: MAP-RSeq: Mayo Analysis Pipeline for RNA sequencing. BMC Bioinformatics. **15**(1), 224 (2014)
68. Chin, C.-S., et al.: Nonhybrid, finished microbial genome assemblies from long-read SMRT sequencing data. Nat. Methods. **10**(6), 563–569 (2013)
69. English, A.C., et al.: Mind the Gap: Upgrading Genomes with Pacific Biosciences RS Long-Read Sequencing Technology. PLoS One. **7**(11), e47768 (2012)
70. Branton, D., et al.: The potential and challenges of nanopore sequencing. Nat. Biotechnol. **26**(10), 1146–1153 (2008)
71. Laszlo, A.H., et al.: Decoding long nanopore sequencing reads of natural DNA. Nat. Biotechnol. **32**(8), 829–833 (2014)
72. Ummat, A., Bashir, A.: Resolving complex tandem repeats with long reads. Bioinformatics. **30**(24), 3491–3498 (2014)

Chapter 10
CUSHAW Suite: Parallel and Efficient Algorithms for NGS Read Alignment

Yongchao Liu and Bertil Schmidt

10.1 Introduction

A variety of *Next-Generation Sequencing* (NGS) technologies have been developed, based on fundamentally different methods from the conventional Sanger sequencing approach. The emergence and application of such new technologies have triggered numerous ground-breaking discoveries and ignited a revolution in the field of genomic science. Propelled by ever-increasing throughput and decreasing cost, NGS technologies have been continually displacing Sanger sequencing in many biological applications such as de novo genome assembly [1–4], resequencing [5, 6], and metagenomics [7–9]. In some of these applications, we usually require aligning NGS reads to reference genomes at fast speed and high accuracy. This has driven a substantial amount of efforts into the research and development of NGS read aligners.

Existing aligners can be generally classified into two generations in terms of functionality. The first-generation aligners are usually designed and optimized for very short reads whose lengths are typically less than 100 *base pairs* (bps). These aligners tend to postulate very small deviations between the short reads and the reference genome and therefore typically allow only mismatches in the alignments. This assumption does simplify the algorithmic design but may lose some significant alignments with insertions or deletions (indels). Some first-generation aligners provide support for gaps, but the maximum allowable number of gaps is also very

Y. Liu (✉)
School of Computational Science & Engineering, Georgia Institute of Technology, Atlanta, GA, USA
e-mail: yliu@cc.gatech.edu

B. Schmidt
Institut für Informatik, Johannes Gutenberg Universität Mainz, Mainz, Germany
e-mail: bertil.schmidt@uni-mainz.de

M. Elloumi (ed.), *Algorithms for Next-Generation Sequencing Data*,
DOI 10.1007/978-3-319-59826-0_10

limited, typically one gap, for the sake of alignment throughput. RMAP [10], MAQ [11], BFAST [12], *Bowtie*[13], BWA [14], CUSHAW [15], and SOAP2 [16] are example first-generation aligners. Propelled by the advances in NGS, read length continually grows larger. Compared to shorter reads, these longer reads usually have higher sequencing error rates and, meanwhile, tend to have more true mismatches and indels to the reference genome. This challenges the first-generation aligners in terms of alignment quality, speed, or even both and calls for aligners that support fully gapped alignment with more mismatches and indels allowed. In this context, a few second-generation aligners have been developed, including BWA-SW [17], GASSST [18], *Bowtie2* [19], CUSHAW2 [20], CUSHAW2-GPU [21], CUSHAW3 [22, 23], GEM [24], *SeqAlto* [25], SOAP3-dp [26], and BWA-MEM [27]. All of these aligners employ the *seed-and-extend* heuristic [28], in which a read is aligned by first identifying seeds, i.e., short matches, on the genome and then extending the alignment to the full read length using *dynamic programming* [28]. In order to reduce search space, we usually exert some constraints and filtration while refining and extending alignments. On the other hand, the seeding approach is also a key factor to the performance of an aligner.

In this chapter, we present the CUSHAW suite, a software package for parallel and efficient alignment of NGS reads to the reference genome. This suite consists of three individual software tools, namely CUSHAW [15], CUSHAW2 [20], and CUSHAW3 [22], all of which are designed based on the *seed-and-extend* heuristic. CUSHAW is the first distribution of our suite, which for the first time has introduced a complete alignment pipeline utilizing *Graphics Processing Unit* (GPU) computing for *paired-end* (PE) reads. This aligner employs a quality-aware bounded search approach based on the *Burrows-Wheeler Transform* (BWT) [29] and the FM-index [30] and harnesses *Compute Unified Device Architecture* (CUDA)-enabled GPUs to accelerate the alignment process. However, this aligner does not provide support for gapped alignments for *single-end* (SE) reads and merely enables gapped alignments while rescuing read mates for PE reads. CUSHAW2 is the second distribution of our suite and employs MEM seeds to find gapped alignments. This aligner has been further accelerated using CUDA, leading to the open-source CUSHAW2-GPU [21]. CUSHAW3 is the third distribution of our suite, which introduces a hybrid seeding approach in order to further improve the alignment quality for base-space reads. Meanwhile, it provides new support for the alignment of color-space reads.

The performance of our aligners have been evaluated by aligning both simulated and real reads to the human genome. In our evaluations, CUSHAW has been excluded since it merely supports ungapped alignments and has inferior alignment quality to both CUSHAW2 and CUSHAW3. In our assessment, the base-space alignment quality of our aligners is compared to that of *Novoalign* [31], BWA-MEM, *Bowtie2*, and GEM, and the color-space alignment quality, to that of SHRiM2 [32] and BFAST [12].

The rest of the chapter is organized as follows. In Sect. 10.2, we briefly introduce the essential techniques constituting the foundation of our algorithms. Subsequently, we present the algorithmic and implementation details of CUSHAW in Sect. 10.3, CUSHAW2 in Sect. 10.4, and CUSHAW3 in Sect. 10.5, respectively. In Sect. 10.6, we evaluate the performance of our algorithms and further compare them with existing state-of-the-art works. Finally, we conclude this chapter in Sect. 10.7.

10.2 Essential Techniques

Given a sequence S, define $|S|$ to denote the length of S, $S[i]$ to denote the character at position i, and $S[i,j]$ to denote the substring of S starting at i and ending at j for $0 \leq i < |S|$ and $0 \leq j < |S|$.

10.2.1 Seed-and-Extend Heuristic

The *seed-and-extend* heuristic [28] is based on the observation that significant alignments are likely to include homologous regions containing exact or inexact short matches between two sequences. It generally works in three steps.

- Firstly, the seed-and-extend heuristic generates seeds (represented as short matches indicating highly similar regions) between a query sequence, e.g., a NGS read, and a target sequence, e.g., a reference genome.
- Secondly, it extends and refines the seeds under certain constraints, such as maximal edit distance, minimal percentage identity, and extension length, to filter out noisy seeds.
- Finally, it employs more sophisticated dynamic programming algorithms to obtain the final alignments, typically the Needleman-Wunsch algorithm [33] or the *Smith-Waterman* (SW) algorithm [34].

Different seeding polices may be employed by different aligners. Several types of seeds have been proposed, which can be generally classified into three categories: fixed-length seeds, variable-length seeds, and hybrid seeds [22] (see Table 10.1). Fixed-length seeds are the most widely used seed type. The simplest instance of fixed-length seeds is an exact k-mer (a substring of k characters) match. Some improvements have been suggested by allowing mismatches or indels in the k-mers, including inexact k-mers (allowing only mismatches at any position of a k-mer) [15], gapped k-mers (allowing both mismatches and indels at any position of a k-mer) [14], spaced seeds (allowing only mismatches at predefined positions) [37], and q-gram filters (allowing mismatches and indels at any position given a maximal error rate $\epsilon > 0$) [38]. Unlike fixed-length seeds, variable-length seeds have varied seed lengths, just as indicated by the name. *Maximal Exact Matches* (MEMs) [20], long gapped seeds [17], and adaptive seeds [36] are example variable-length seeds. MEMs are exact matches that cannot be extended in either direction without allowing a mismatch. Long gapped seeds and adaptive seeds are very similar, and

Table 10.1 Seeding policies for next-generation sequencing read alignment

Seed type		Example aligners
Fixed-length seeds	Exact k-mers	GASSST [18]
	Inexact k-mers	CUSHAW [15], *Bowtie2* [19]
	Gapped k-mers	BWA [14]
	Spaced seeds	BFAST [12]
	q-Gram filters	SARUMAN [35]
Variable-length seeds	Adaptive seeds	LAST [36]
	Long gapped seeds	BWA-SW [17]
	Maximal exact match seeds	CUSHAW2 [20]
Hybrid seeds		CUSHAW3 [22]

both of them allow mismatches or indels but with some additional constraints such as minimum optimal alignment score and the number of seed occurrences in the target sequence. Hybrid seeds [22] work by combining multiple types of seeds, such as exact k-mer seeds and variable-length seeds, to realize more sophisticated seeding.

10.2.2 Burrows-Wheeler Transform and FM-Index

The BWT of T starts from the construction of a conceptual matrix M_T, whose rows are all cyclic rotations of a new sequence $T\$$ sorted in lexicographical order. $T\$$ is formed by appending to the end of T the special character \$ that is lexicographically smaller than any character in Σ. After getting M_T, the last column of the matrix is taken to form the transformed text B_T, i.e., the BWT of T. B_T is a permutation of T and thus occupies the same memory size $|T|\lceil \log_2(|\Sigma|)\rceil$ bits as T. M_T has a property called *Last-to-First Column Mapping*, which means that the i-th occurrence of a character in the last column corresponds to the i-th occurrence of the same character in the first column. Table 10.2 shows an example matrix M_T and B_T.

The FM-index consists of a vector $C(\cdot)$ and an occurrence array $Occ(\cdot)$, both of which are constructed from B_T. $C(\cdot)$ contains $|\Sigma|$ elements with each element $C(x)$ representing the number of characters in T that are lexicographically smaller than $x \in \Sigma$. $Occ(\cdot)$ is an array of size $|\Sigma| \times |T|$ with each element $Occ(x, i)$ representing the number of occurrences of x in $B_T[0, i]$. $C(\cdot)$ requires only $|\Sigma\lceil| \log_2(|T|)\rceil$ bits, while $Occ(\cdot)$ requires $|\Sigma||T|\lceil \log_2(|T|)\rceil$ bits. Thus, $Occ(\cdot)$ dominates the overall memory footprint of the FM-index. An approach to trade off speed and memory overhead is to use a reduced occurrence array (ROcc) [15], which only stores the elements whose indices in Occ are multiples of μ and then calculates the missing elements with the help of B_T at the runtime. Through the use of *ROcc*,

Table 10.2 M_T, B_T, and *SA* of an example sequence *cattattagga*

Index	Cyclic rotation	Suffix	Matrix M_T	Sorted suffix	B_T	*SA*
0	cattattagga$	cattattagga$	$cattattagga	$	a	11
1	attattagga$c	attattagga$	a$cattattagg	a$	g	10
2	ttattagga$ca	ttattagga$	agga$cattatt	agga$	t	7
3	tattagga$cat	tattagga$	attagga$catt	attagga$	t	4
4	attagga$catt	attagga$	attattagga$c	attattagga$	c	1
5	ttagga$catta	ttagga$	cattattagga$	cattattagga$	$	0
6	tagga$cattat	tagga$	ga$cattattag	ga$	g	9
7	agga$cattatt	agga$	gga$cattatta	gga$	a	8
8	gga$cattatta	gga$	tagga$cattat	tagga$	t	6
9	ga$cattattag	ga$	tattagga$cat	tattagga$	t	3
10	a$cattattagg	a$	ttagga$catta	ttagga$	a	5
11	$cattattagga	$	ttattagga$ca	ttattagga$	a	2

we are able to reduce the memory footprint of the FM-index to $|\Sigma| \log_2(|T|) + |T|(|\Sigma|\lceil \log_2(|T|)\rceil/\mu + \lceil \log_2(|\Sigma|)\rceil)$ bits.

10.2.3 Search for Exact Matches

The suffix array *SA* of *T* stores the starting positions of all suffixes of *T* in lexicographically ascending order (Table 10.2 shows an example *SA*). In other words, $SA[i] = j$ means that the *i*-th lexicographically smallest suffix (among all suffixes of *T*) starts at position *j* in *T*. The *SA* of *T* has an overall memory footprint of $|T|\lceil \log_2(|T|)\rceil$ bits.

Given a substring *S* of *T*, we can find all of its occurrences within an *SA* interval, which is an index range $[I_1, I_2]$ with I_1 and I_2 representing the indices in *SA* of the lexicographically smallest and largest suffixes of *T* with *S* as a prefix, respectively. Based on $C(\cdot)$ and $Occ(\cdot)$, the *SA* interval $[I_1, I_2]$ can be recursively calculated from the rightmost to the leftmost suffixes of *S*, by:

$$\begin{cases} I_1(i) = C(S[i]) + Occ(S[i], I_1(i+1) - 1) + 1, & 0 \le i < |S| \\ I_2(i) = C(S[i]) + Occ(S[i], I_2(i+1)), & 0 \le i < |S| \end{cases} \tag{10.1}$$

where $I_1(i)$ and $I_2(i)$ represent the starting and end indices of the *SA* interval for suffix $S[i, |S| - 1]$. $I_1(|S|)$ and $I_2(|S|)$ are initialized as 0 and $|T|$, respectively. The calculation stops if it encounters $I_1(i+1) > I_2(i+1)$. The condition $I_1(i) \le I_2(i)$ holds if and only if suffix $S[i, |S| - 1]$ is a substring of *T*. The total number of the occurrences is calculated as $I_1(0) - I_2(0) + 1$ if $I_1(0) \le I_2(0)$ and 0, otherwise. After getting the *SA* interval, the location of each occurrence can be determined by directly looking up *SA* with a constant time complexity. Hence, the time complexity for finding n occurrences of S is $O(|S|+n)$ when using Occ and $O(\mu \cdot |S|+n)$ when using $ROcc$.

10.2.4 Locating the Occurrences

After getting the SA interval, the position of each occurrence in T can be determined by directly looking up the SA. However, it will take $|T|\lceil \log_2(|T|)\rceil$ bits if the entire SA is loaded into memory. This large memory consumption is prohibitive for large genomes. Fortunately, we can reconstruct the entire SA from parts of it. Ferragina and Manzini [30] have shown that an unknown value $SA[i]$, can be reestablished from a known $SA[j]$ using Eq. (10.2):

$$\begin{cases} SA[i] = SA[j] + d \\ j = \beta^{(d)}(i) \end{cases} \tag{10.2}$$

where $\beta^{(d)}(i)$ means repeatedly applying the function $\beta(i)$ d times. The $\beta(i)$ function implements *Last-to-First Column Mapping* for the i-th row of M_T and is calculated by:

$$\beta(i) = C(T[i]) + Occ(T[i], i) \tag{10.3}$$

In Eq. (10.2), d is actually the distance between the two starting positions (i.e., $SA[i]$ and $SA[j]$) in T. Thus, for any unknown i-th element of SA, we can calculate its value $SA[i]$ in $\leq v$ iterations if we store the SA elements whose values are multiples of a constant number v (i.e., the starting positions in T are multiples of v). However, this will complicate the storage of SA by introducing additional data structures for occurrence locating. Instead, we construct a *Reduced Suffix Array* (RSA) [15] by simply storing $SA[i]$ whose index i is a multiple of v. This approach reduces the total memory size of a suffix array to $|T|\lceil \log_2(|T|)\rceil/v$ bits but cannot guarantee to complete each computation within v iterations. This is because a maximal distance v between i and j does not mean a maximal distance v for starting positions $SA[i]$ and $SA[j]$ in T. The selection of v is a trade-off between look-up time and memory space. For a suffix array index i that is not a multiple of v, we repeat d iterations using Eq. (10.3) until j is a multiple of v, where $SA[j]$ is equal to $RSA[j/v]$, and then calculate $SA[i]$ as $SA[j] + d$ following Eq. (10.2). In this case, we have an approximate time complexity of $O(v \cdot n)$ for locating n occurrences of S.

By combining $ROcc$ and RSA, we can arrive at a significantly smaller memory footprint of $|\Sigma|\lceil \log_2(|T|)\rceil + |T|\lceil \log_2(|\Sigma|)\rceil + |T|(|\Sigma|/\mu + 1/v)\lceil \log_2(|T|)\rceil$ bits. Additionally, the time complexity of $O(\mu \cdot |S| + v \cdot n)$ for finding n occurrences of S is still acceptable.

10.3 CUSHAW: Ungapped Alignment Using Inexact Matches

CUSHAW [15] is the first distribution of our CUSHAW software suite, and its source code is freely and publicly available at http://cushaw.sourceforge.net. This aligner targets ungapped alignments and employs CUDA-enabled GPUs to accelerate short-read alignment.

10.3.1 Search for Inexact Matches

To align a read, CUSHAW only considers substitutions allowed to occur at any position, excluding the support for indels. This constraint enables us to transform the search for inexact matches to the search for exact matches for all permutations of all possible bases at all positions of a read. All of the permutations can be represented by a complete 4-ary tree (see Fig. 10.1), where each permutation corresponds to a path routing from the root to a leaf (the root node is Ø, meaning an empty string). Each node in the path corresponds to a base in a read that has the same position. Hence, all inexact matches can be found by traversing all paths using either *Depth-First Search* (DFS) or *Breadth-First Search* (BFS) approaches. BFS requires a large amount of memory to store the results of all valid nodes at the same depth. Hence, this approach is infeasible for GPU computing, because we need to launch hundreds of thousands of threads to leverage the compute power of GPUs and thus can only assign a small amount of device memory to each thread. Alternatively, CUSHAW

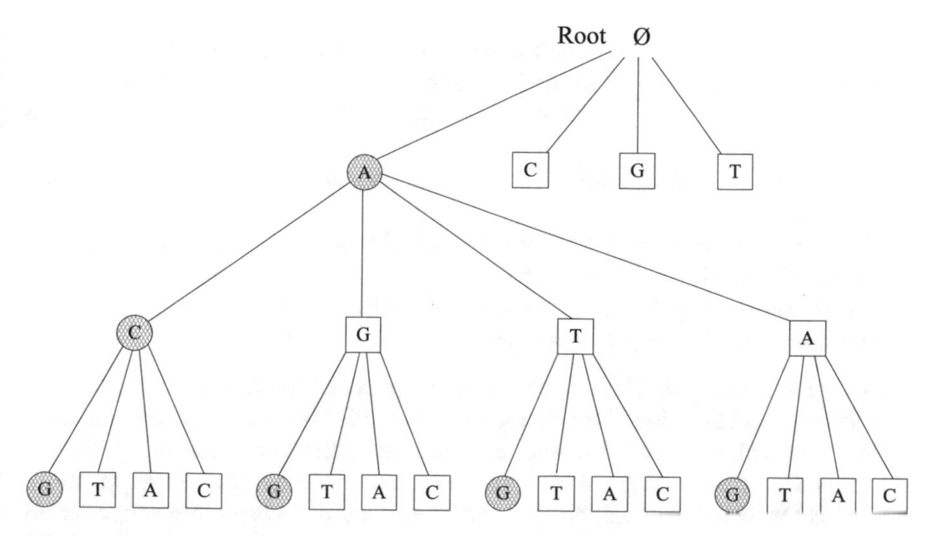

Fig. 10.1 An equivalent complete 4-ary tree for the search of all inexact matches of the sequence "ACG" using a reverse BWT: *circles* mean the original bases, *rectangles* mean the mutated bases and it only expands one subtree of the root

chooses to use the DFS approach, whose memory consumption is small and directly proportional to the depth of the tree (i.e., the full length of a read). In addition, we have used a stack framework to implement the DFS approach.

10.3.2 Quality-Aware Ungapped Alignment

CUSHAW employs a quality-aware bounded search approach to reduce the search space and guarantee alignment. It supports two types of alignments: seeded alignment and end-to-end alignment. The seeded alignment starts by finding inexact k-mer seeds at the high-quality end, and then extends the seeds to the full read length by only allowing mismatches on some conditions. The end-to-end alignment is considered as a special case of the seeded alignment which considers the full length of a read as the seed size. The reverse complement of a read is also incorporated and for clarity; the following discussions only refer to the forward strand. The quality-aware property exploits the base-calling quality scores in a read. A base quality score Q_b is computed as $-10\,\lg(p)$ following the PHRED [39] definition, where p is the probability of the base being miscalled. Lower quality scores indicate higher probabilities of base miscalling.

The bounded search is able to reduce the search space by exerting several constraints on sums of quality scores and maximal allowable number of mismatches. The use of these constraints can prune branches of the complete tree ahead of time and thus significantly reduce the number of backtracks. The constraints have been used in other aligners, e.g. *Bowtie*, BWA and SOAP2, in part or whole, and are detailed as

- *MMS*: maximal number of mismatches allowed in the seed (default = 2)
- *MMR*: maximal number of mismatches allowed in the full length of a read, which is calculated as $MMS + \lfloor err \times |S| \rfloor$, where *err* is the uniform base error rate (default = 4%)
- *QSS*: maximal sum of quality scores at all mismatched positions in the seed (default = 70)
- *QSR*: maximal number of quality scores at all mismatched positions in the full length of a read (default = 3 × *QSS*)
- *QSRB*: maximal *QSR* among the currently selected best alignments, which is updated as the aligning process goes on

Increasing *MMS* and *MMR* might enable the alignment of more reads but will result in a longer execution time. Decreasing *QSS* and *QSR* focuses the aligner more on mismatches with low-quality scores and can thus facilitate earlier pruning of some branches. However, smaller values also carry the risk of pruning real alignments.

CUSHAW outputs the alignments with the smallest quality score sum over all mismatched positions in the full-length read alignment. All possible alignments are compared and enumerated by exerting the above constraints. Our alignment approach is different from the ones used in *Bowtie* and BWA. *Bowtie* allows any number of mismatches in the non-seed region and outputs the "best" alignment after a specified maximum number of search attempts. BWA does not use base quality scores when performing alignments but assigns different penalties on mismatches and gaps. It then reports the alignment with the best score that is calculated from the number of mismatches and gaps in the alignment.

10.3.3 Progressive Constraint Approach

As mentioned above, CUSHAW attempts to find the best alignments in the full-length read alignment by enumerating and evaluating all possible alignments. Given a read of length l and a specific *MMR*, the total number of possible alignments is $\Sigma_{k=0}^{MMR} 3^k C_l^k$, which increases polynomially with l (postulating *MMR* is fixed) and increases exponentially with *MMR* (postulating l is fixed). Furthermore, to miss as few correct alignments as possible, we have to accordingly increase *MMR* as l becomes larger. Even though the search space can be significantly reduced by exerting the other four constraints, the number of possible alignments that needs to be evaluated grows significantly for larger l and *MMR*. In this context, we have introduced a new progressive constraint approach, with respect to *MMR*, in order to further reduce the search space for longer reads.

The progressive constraint approach works by exerting an additional constraint on the maximal allowable number of mismatches MMR' in the current alignment path of length l' ($1 \leq l' \leq l$), not only in the full-length read alignment, as the alignment goes. This approach is likely to lower the SE alignment quality, since it excludes the evaluation of those possibly correct alignments with mismatches clustering in small regions, but does improve the execution speed by further reducing the search space. For each l', its MMR' is independent of reads to be aligned and is pre-calculated before alignment using the following method.

We consider the number of base errors m in a read as a random variable and postulate that all bases have the same error probability p (2.5% in our case), for simplicity. Thus, the probability of having k base errors in a sequence of length l' is:

$$P(m = k) = C_{l'}^k p^k (1 - p)^{l' - k} \tag{10.4}$$

where m follows a binomial distribution. In our case, this probability can be approximated using a Poisson distribution with the mean $\lambda = l'p$. If l' is less than or equal to the seed size, MMR' is set to *MMS*, and otherwise, MMR' is calculated by:

$$MMR' = \max\{MMS + 1, \min\{k | P(m > k) < err\}\} \tag{10.5}$$

where err is the uniform base error rate specified for the input reads. By default, CUSHAW supports a maximal read length of 128 (can be configured up to 256).

The $MMR's$ for all l' ($1 \leq l' \leq 128$) are pre-calculated on the host prior to alignment and are then loaded into cached constant memory on the GPU devices.

10.3.4 Paired-End Mapping

For a read pair S_1 and S_2, CUSHAW supports paired-end mapping and completes it in the following three steps:

- Firstly, if both S_1 and S_2 have matches to the reference genome, we iterate each mapping position of S_1 and calculate the distance to each mapping position of S_2. If a distance satisfies the maximal insert size, the read pair is considered paired and the corresponding mapping positions are output, finishing the pairing of S_1 and S_2.
- Secondly, if S_1 has matches to the genome (no matter whether S_2 has or not), it iterates some mapping positions (at most 2 by default) of S_1 and estimates the region in the genome to the right of S_1, where S_2 is likely to have a match, using the maximal insert size. The SW algorithm is used to find the optimal local alignment for S_2 and the genome region. If we find an alignment satisfying the constraints specified by the user, such as the maximal number of unknown bases in the short read and the minimal number of bases in the optimal local alignment for both the short read and the genome region, the read pair is considered paired and otherwise, we will continue the pairing process.
- Finally, if S_2 has matches to the genome (no matter whether S_1 has or not), it uses the same pairing method as in the previous step, except that the estimated genome region is to the left of S_2. If we still fail to find an optimal local alignment satisfying the constraints specified by the user in this step, the read pair is considered unpaired.

The SW algorithm is most time consuming in the pairing process, having a quadratic time complexity with respect to sequence length. Thus, the more reads that are paired in the first step, the smaller the overall execution time. Moreover, the maximal insert size also has an impact on the number of reads that are paired and the overall execution time. This pairing process has been multi-threaded in order to benefit from multi-core CPUs.

10.4 CUSHAW2: Gapped Alignment Using Maximal Exact Matches

CUSHAW2 [20] is the second distribution of our CUSHAW suite, and its source code is freely and publicly available at http://cushaw2.sourceforge.net. This aligner employs MEM seeds to enable gapped alignments of reads to the reference genome.

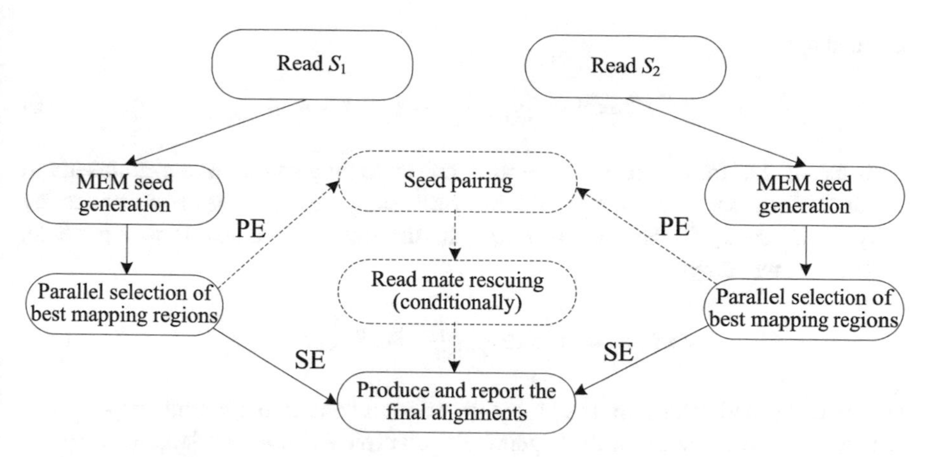

Fig. 10.2 CUSHAW2 pipeline: the *dashed lines* show the additional two stages for the PE alignment

For the SE alignment, CUSHAW2 works in three stages: (1) generate MEM seeds; (2) select the best mapping regions on the genome; and (3) produce and report the final alignments. For the PE alignment, we introduce two additional stages before producing the final alignments: one is the seed pairing stage and the other is the read rescuing stage. Figure 10.2 illustrates the pipelines of CUSHAW2 for both the SE and the PE alignments.

10.4.1 Generation of Maximal Exact Match Seeds

Herein, we will investigate how to generate MEMs based on the succinct BWT and FM-index data structures.

10.4.1.1 Estimation of the Minimal Seed Size

We are only interested in the MEM seeds whose lengths are not less than a minimal seed size *MSS*. Decreasing *MSS* generally increases the sensitivity by finding more hits in homologous regions but may have a higher probability of generating more noisy hits. Increasing *MSS* generally decreases the number of hits, thus improving speed, but may undergo the decrease in sensitivity. Many seed-based aligners therefore require users to carefully tune *MSS*. However, this tuning work is tedious. To address this issue, we propose an automatic estimation of *MSS* according to a given read length.

Our estimation of *MSS* is based on the q-gram lemma [38] and a simplified error model. The q-gram lemma states that two aligned sequences S_1 and S_2 with an edit distance of e (the number of errors) share at least N_q q-grams, where N_q is defined by:

$$N_q = \max(|S_1|, |S_2|) - q + 1 - q \cdot e \tag{10.6}$$

This means that for overlapping q-grams, one error may cause up to q-grams not to be shared by the two reads, and for non-overlapping q-grams, one error can destroy only one q-gram [35]. Hence, given the edit distance e of S aligned to the genome, MSS is estimated by:

$$MSS = \min\left\{\max\left\{\lfloor \frac{|S|}{e+1} \rfloor, MSS_{lo}\right\}, MSS_{hi}\right\} \tag{10.7}$$

where MSS_{lo} and MSS_{hi} are the global lower bound and upper bound, respectively. The estimation is based on the pigeonhole principle for non-overlapping q-grams, meaning that at least one q-gram of length MSS is shared by S and its aligned substring mate on the genome. By default, our aligner sets $MSS_{lo} = 13$ and $MSS_{hi} = 49$.

Since the error model for gapped alignments is quite complicated, we employ a simplified error model for ungapped alignments to estimate e. Supposing that the number of substitutions w in the full-length alignment of S is a random variable and each base in S has the same error probability p (default = 2%), the probability of having z substitutions is calculated by:

$$P(w = z) = C_{|S|}^{z} p^{z}(1 - p)^{|S|-z} \tag{10.8}$$

where w follows a binomial distribution. By specifying a missing probability m (default = 4%), e can be estimated by:

$$e = \min\{z | P(w > z) < m\} \tag{10.9}$$

Our simplified error model results in the following values: $MSS = 16$ for 100-bp reads, $MSS = 22$ for 200-bp reads and $MSS = 35$ for 500-bp reads. In addition, we also provide parameters to allow users to customize MSS.

10.4.1.2 Search for Maximal Exact Matches

We represent an exact match between two sequences S_1 and S_2 as a triplet (p, q, k), where k is the length of the exact match and the substring $S_1[p, p+k-1]$ is identical to the substring $S_2[q, q+k-1]$. An exact match is called right maximal if $p+1 = |S_1|$ or $q + 1 = |S_2|$ or $S_1[p + k] \neq S_2[q + k]$ and left maximal if $p = 0$ or $q = 0$ or $S_1[p - 1] \neq S_2[q - 1]$. An exact match is called MEM if it is both left maximal and right maximal.

To identify MEMs between S and T, we advance the starting position p in S from left to right to find the *Longest Exact Matches* (LEMs) using the BWT (the reverse one) and the FM-index. According to the above definitions, we know that the identified LEMs are right maximal. We know that the LEMs starting at the beginning of S are both left maximal and right maximal. This means that when advancing the starting positions from the beginning to the end of S, the identified LEMs are also left maximal if it is not part of any previously identified MEM. In this way, only unidirectional substring search is required. Since we are only concerned about MEMs of sufficient lengths, we discard the MEMs whose lengths are less than a minimum seed size threshold MSS.

For large genomes, it is possible to find a lot of occurrences of a MEM starting at a certain position of S. In this case, we only keep its first h ($h = 1024$ by default) occurrences and discard the others. However, it is also observed that we sometimes fail to find any MEM seeds for some reads using MSS. To improve sensitivity, we therefore attempt to rescue them by reconducting the MEM identification procedure using a new and smaller minimal seed size MSS_{new} by:

$$MSS_{new} = \frac{1}{2}(MSS + MSS_{lo}) \tag{10.10}$$

10.4.2 Determination and Selection of Mapping Regions

For local alignment with affine gap penalty, the positive score for a match is usually smaller than the penalty charged for a substitution or for an indel. Using such type of scoring schemes, the length of the optimal local alignment of S to the genome cannot be greater than $2|S|$ as a local alignment requires a positive alignment score. This conclusion forms the foundation of our genome mapping region determination approach for each identified MEM seed. In our aligner, we employ a commonly used scoring scheme (e.g., also used in BLAST [28] and BWA-SW) with a match score 1, a mismatch penalty 3, a gap opening penalty 5 and a gap extension penalty 2.

For a read, a MEM indicates a mapping region on the genome, which includes the seed and potentially contains the correct alignment of the full read. We can determine the range of the mapping region by extending the MEM in both directions by a certain number of bases. Since the optimal local alignment length of S cannot be greater than $2|S|$ in our aligner, it is safe to determine the mapping region range by extending the MEM by $2|S|$ bases in each direction. This extension does work but will result in lower speed due to the introduced redundancy. Hence, we attempt to compute a smaller mapping region with as little loss of sensitivity as possible. We define P_m to denote the starting position of a MEM in S, T_m to denote the mapping position of the MEM on the genome and L_m to denote the MEM length. Assuming that the MEM is included in the final alignment, our aligner estimates the mapping region range $[T_1, T_2]$ by:

$$\begin{cases} T_1 = T_m - 2(P_m + 1) \\ T_2 = T_m + L_m + 2(|S| - P_m - L_m) \end{cases} \quad (10.11)$$

Our aligner computes the optimal local alignment scores in all determined mapping regions of S using the SW algorithm and then builds a sorted list of all mapping regions in the descending order of score. Mapping regions whose scores are less than a minimal score threshold (default = 30) are removed from the sorted list. Subsequently, the sorted list of qualified mapping regions is used in the SE and the PE alignment for determining the final alignments and approximating the mapping quality scores.

10.4.3 Paired-End Mapping

The alignment of two paired reads generally has two constraints: alignment strand and mapping distance. For the alignment strand constraint, our aligner requires the two reads to be aligned to the genome from different strands. For the mapping distance constraint, our aligner requires that the mapping distance of the two reads cannot exceed a maximal mapping distance threshold defined by the insert size information of a library. Assuming that the mean insert size is $\overline{X}$ and the standard deviation of the insert size is σ, we calculate the maximal mapping distance threshold as $\overline{X} + 4\sigma$. For the PE mapping, our aligner has two stages: (1) pairing qualified mapping regions in order to find the correct alignments for both ends and (2) rescuing unaligned reads through their aligned read mates.

For any aligned read pair, we can first compare their mapping distance on the genome (calculated from the positions of the best alignments of the two reads) to the insert size constraint. If this comparison is within the mapping distance threshold, the corresponding alignment is output. Otherwise, we could calculate the mapping distance for each mapping position pair from all qualified mapping regions in the sorted list. However, the associated computational overhead cannot be tolerated since we need to obtain the alignment paths for all qualified mapping regions of a read pair. Hence, we introduce a seed-pairing approach to heuristically accelerate the read pairing.

The seed-pairing heuristic works by enumerating each seed pair of S_1 and S_2 in order to find all potential seed pairs. If the seed pair has different alignment strands and locates on the same genome fragment, it will be used to estimate the mapping distance of S_1 and S_2 and otherwise will be discarded. In our aligner, the mapping position T_p of S is estimated from one of its MEMs by:

$$T_p = T_m - \begin{cases} P_m, & \text{if the strand is forward} \\ |S| - P_m + 1, & \text{if the strand is reverse} \end{cases} \quad (10.12)$$

where we assume that S is aligned to the genome without gaps. To compensate for the difference between the estimated mapping distance and the correct one, we employ a larger maximal insert size threshold $\overline{X} + 4\sigma + 2e$ for the seed-pairing heuristic. If the estimated mapping distance does not exceed the maximal insert size threshold, this seed pair is considered qualified and will be saved for future use. After finding all qualified seed pairs, we enumerate each qualified seed pair to compute the real mapping distance of S_1 and S_2, which is compared to the maximal insert size threshold $\overline{X}+4\sigma$. If the insert size constraint is met, S_1 and S_2 are reported as paired. Otherwise, we will compute the best alignment for S_1 (or S_2) to rescue its mate by employing the insert size information to determine the potential mapping region of its mate. This rescuing procedure is also applied when only one read of S_1 and S_2 is aligned.

10.4.4 Approximation of Mapping Quality Scores

Since the introduction of mapping quality scores in MAQ [11] to indicate the probability of the correctness of alignments, the concept of mapping quality scores has been frequently used in many NGS read aligners. Generally, a higher mapping quality score indicates a higher confidence in the correctness of an alignment.

As stated in BWA-SW, if an aligner guarantees to find all local alignments of a read, the mapping quality score M_q is determined by these local alignments only. Although our aligner does not find all local alignments of the read, the sorted list of qualified mapping regions still provides sufficient information to approximate M_q. In our aligner, we employ two equations to approximate M_q for the SE and the PE alignment. For the SE alignment, M_q is approximated by:

$$M_q = 250 \times \frac{b_1 - b_2}{b_1} \times r \tag{10.13}$$

Equation (10.13) is similar to the mapping quality approximation in BWA-SW, where b_1 is the best local alignment score, b_2 is the second best local alignment score, and r is calculated by dividing the number of bases of the read in the final alignment by the read length. For the PE alignment, the calculation of M_q depends on two conditions. If the two reads are correctly paired through the seed-pairing heuristic, the mapping quality score for each read is equal to its SE M_q. Otherwise, if one read is rescued by its mate, the mapping quality score of the read is approximated as $r \times M_{qmate}$, where M_{qmate} is the SE M_q of its mate.

10.4.5 Multi-Threaded and CUDA-Based Parallelization

In CUSHAW2, the most time-consuming part of SE alignment is the selection of the best mapping region using the SW algorithm. To accelerate its execution, on the CPU side, we have adopted the *Streaming SIMD Extensions* 2 (SSE2)-based parallel implementation of the SW algorithm in SWIPE [40]. Moreover, a multi-threaded design based on *Pthreads* has been introduced to parallelize the alignment process. We use a dynamic scheduling policy to assign reads to threads, which allows one thread to immediately start a new alignment without waiting for the completion of the other threads. For the SE alignment, a thread aligns a single read at a time and then reads a new read from the input file immediately after finishing the current alignment. For the PE alignment, we follow the same scheduling policy with the difference that one read pair is assigned at a time. Locks are appropriately used to ensure mutually exclusive accesses to both the input and output files.

We have further accelerated CUSHAW2 using the CUDA-enabled GPU. In this parallelization, we have investigated an inter-task hybrid CPU-GPU parallelism mode to efficiently harness the compute power of both the CPU and the GPU, which conducts concurrent alignments of different reads on the two types of processing units. Moreover, a tile-based SW alignment backtracking algorithm using CUDA [41] is introduced to facilitate fast alignments on the GPU. More implementation details can be obtained from our open-source CUSHAW2-GPU [21] software.

10.5 CUSHAW3: Gapped Alignment Using Hybrid Seeds

CUSHAW3 [22] is the third distribution of our CUSHAW suite, and its source code is freely and publicly available at http://cushaw3.sourceforge.net. This aligner provides new support for color-space read alignment in addition to base-space alignment and has introduced a hybrid seeding approach in order to improve alignment quality. This hybrid seeding approach incorporates three different seed types: MEM seeds, exact k-mer seeds, and variable-length seeds derived from local alignments, into the alignment pipeline. Furthermore, three techniques, namely weighted seed-pairing heuristic, PE alignment pair ranking, and sophisticated read mate rescuing, have been proposed to facilitate accurate PE alignment.

10.5.1 *Hybrid Seeding*

Our hybrid seeding approach incorporates MEM seeds, exact-match k-mer seeds, and variable-length seeds at different phases of the alignment pipeline. For a single read, the alignment pipeline generally works as follows (see Fig. 10.3):

- Firstly, we produce the MEM seeds for both strands of the read based on BWT and FM-index.

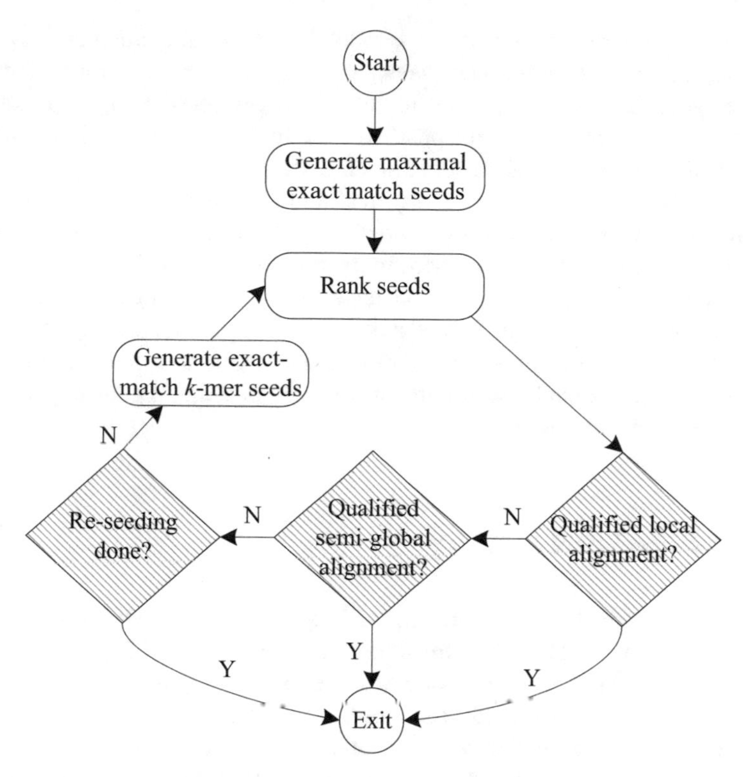

Fig. 10.3 Program workflow of the single-end alignment using hybrid seeding

- Secondly, from each seed, we determine on the genome a potential mapping region for the read and then perform the SW algorithm to gain the optimal local alignment score between the read and the mapping region. All seeds are subsequently ranked in terms of optimal local alignment score, where greater scores mean higher ranks.
- Thirdly, dynamic programming is employed to identify the optimal local alignment of the read to the genome from the highest-ranking seeds. If satisfying the local alignment constraints, including minimal percentage identity (default = 90%) and aligned base proportion per read (default = 80%), the optimal local alignment will be considered as qualified. Otherwise, we will attempt to rescue the read using a semi-global alignment approach. As an optimal local alignment usually indicates the most similar region on the genome, our semi-global alignment approach takes the optimal local alignment as a variable-length seed, re-computes a new mapping region on the genome and then performs semi-global alignment between the read and the new mapping region to obtain an optimal semi-global alignment. If the optimal semi-global alignment satisfies the global alignment constraints, including minimal percentage identity (default = 65%) and aligned base proportion per read (default = 80%), this

alignment will be deemed to be qualified. This double-alignment approach enables us to capture the alignments with more continuous mismatches and longer gaps. This is because we might fail to get good enough optimal local alignments in such cases, as the positive score for a match is usually smaller than the penalty charged for mismatches and indels.

- Finally, when we still fail to get any qualified alignment, this likely means that the true alignment is implied by none of the evaluated MEM seeds. In this case, we attempt to rescue the alignment by reseeding the read using exact-match k-mer seeds. To improve speed, we search all non-overlapping k-mers of the read against the genome to identify seed matches. Subsequently, we employ the k-mer seeds to repeat the aforementioned alignment process to rescue the read. If we still fail to gain a qualified alignment, we will stop the alignment process and then report this read as unaligned.

10.5.2 Paired-End Mapping

In comparison with the SE alignment, the long-range positional information contained in PE reads usually allow for more accurate short-read alignment by either disambiguating alignments when one of the two ends aligns to repetitive regions or rescuing one end from its aligned mate. In addition, for aligners based on the seed-and-extend heuristic, the PE information, such as alignment orientations and insert size of both ends, can aid to significantly reduce the number of noisy seeds prior to the time-consuming alignment extensions. This filtration can be realized through a seed-pairing heuristic proposed in CUSHAW2, as a seed determines the alignment orientation of a read and the mapping distance constraint on seed pairs can be inferred from the insert size of read pairs.

For a read pair S_1 and S_2, our PE alignment pipeline generally works as follows (see Fig. 10.4):

- Firstly, we generate and rank the MEM seeds, following the same procedure as in SE alignment.
- Secondly, a weighted seed-pairing heuristic is introduced to pair seeds, where only high-quality seeds, whose scores are not less than a minimal score threshold (default = 30), will be taken into account. This heuristic enumerates each high-quality seed pair of S_1 and S_2 to identify all qualified seed pairs that meet the alignment orientation and insert size requirements. To distinguish all qualified seed pairs in terms of quality, we have calculated a weight for each qualified seed pair and further ranked all of them by a max-heap data structure. This quality-aware feature allows for us to visit all qualified seed pairs in the decreasing order of quality.
- Thirdly, if failed to find any qualified seed pair, we will resort to the reseeding based on exact-match k-mers by sequentially checking both ends to see if either of them has not yet been reseeded. If so, the k-mer seeds will be produced for

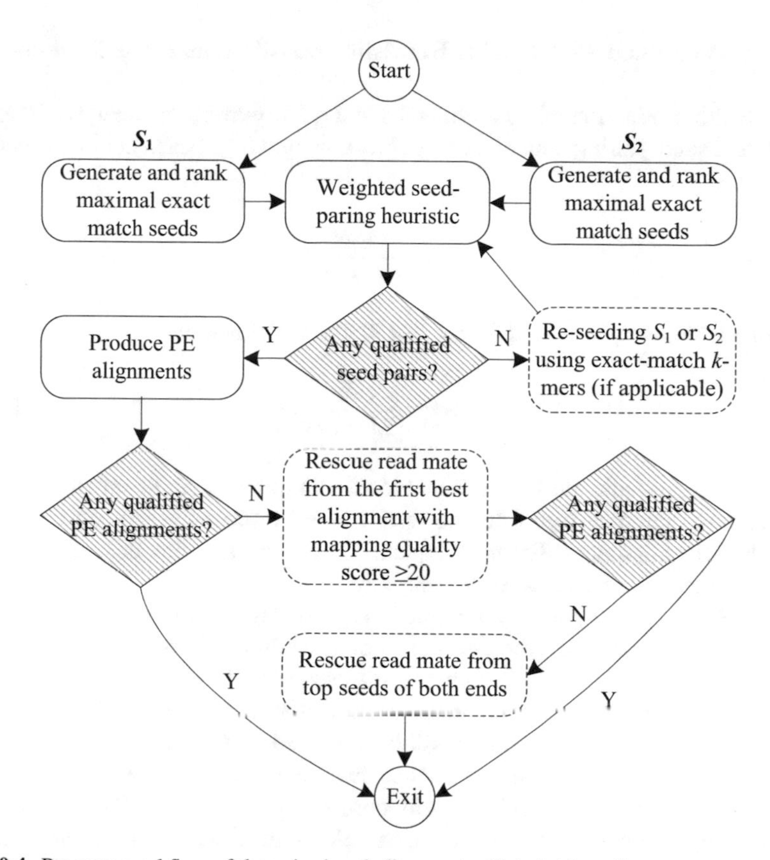

Fig. 10.4 Program workflow of the paired-end alignment with hybrid seeding

that end and all new seeds will be ranked in the same way as for MEM seeds. Subsequently, we merge all high-quality k-mer seeds with the high-quality MEM seeds and then re-rank all seeds. The seed merge is used because some significant alignments, which are not covered by MEM seeds, may be reflected by k-mer seeds, and vice versa. After getting the new list of seeds, we repeat the weighted seed-pairing heuristic to gain qualified seed pairs. The seed-pairing and reseeding process will be repetitively continued until either both ends have been reseeded or any qualified seed pair has been identified.

- Fourthly, we compute the real alignments of both ends from the qualified seed pairs. An alignment pair will be considered qualified if their mapping position distance satisfies the insert size constraint. Similar to the weighted seed-pairing approach, we have also ranked all qualified alignment pairs by means of a max-heap data structure. In this manner, we would expect better alignment pairs to come out earlier in the output.
- Finally, we attempt to rescue read mates from the best alignments of each end when failed to pair reads in previous steps.

10.5.2.1 Weighted Seed-Pairing Heuristic and Alignment Pair Ranking

To guide the production of PE alignments in a quality-aware manner, we introduce a weighted seed-pairing heuristic computing a weight ω for each qualified seed pair by:

$$\omega = \frac{2\omega_1\omega_2}{\omega_1 + \omega_2} \tag{10.14}$$

where ω_i is the weight of read S_i ($i = 1$ or 2) and is defined by:

$$\omega_i = \frac{AS_i}{|S_i| \times MS} \tag{10.15}$$

In Eq. (10.15), AS_i denotes the optimal local alignment score between read S_i and the mapping region derived from the seed, and MS represents the positive score for a match in the alignment. To rank all qualified seed pairs, a max-heap data structure has been used with ω as the key of each entry.

In addition to seed pairs, all qualified alignment pairs have been further ranked in terms of weight and edit distance. For an alignment pair, we calculate its weight following Eq. (10.14) with the difference that AS_i is not definitely the optimal local alignment score but might be the optimal semi-global alignment score. This is because an alignment is possibly produced from a semi-global alignment as mentioned above. Furthermore, when two qualified alignment pairs hold the same weights, we further rank them by comparing the sums of the edit distances of each alignment pair. In this case, smaller edit distance sums mean higher ranks.

10.5.2.2 Sophisticated Read Mate Rescuing

For unpaired reads, we have employed a sophisticated read mate rescuing procedure, which attempts to rescue one read from the top hits of its aligned mate by using the paired-end long-range distance information. In general, our rescuing procedure works as follows:

- Firstly, the best alignments of the two reads are computed (if available). The read, whose best alignment has a *MAPping Quality* (MAPQ) exceeding a minimum threshold (default = 20), will be used to rescue its mate. If an optimal alignment satisfying the aforementioned constraints has been gained for the mate, the two reads are considered as paired. Otherwise, we will continue the rescuing process using the alignments with smaller MAPQs.
- Secondly, if the two reads have not yet been properly paired, we will attempt to pair them from more top hits of both reads. The rescuing process will not stop until the two reads have been properly paired or have reached the maximum number (default = 100) of top seeds for each read.

- Finally, for unpaired reads, we will report their best alignments (if available) in an SE alignment mode.

This read mate rescuing is usually time consuming mainly due to two factors. One is the dynamic-programming-based alignment with quadratic time complexity. The other is the maximal insert size of a read pair, which basically determines the mapping region size of the mate on the genome. In sum, the more reads are paired by seed-pairing heuristic, the less time is taken by the read mate rescuing procedure.

10.5.3 Color-Space Alignment

Most existing color-space aligners encode a nucleotide-based genome as a color sequence and then identify potential short-read alignment hits in color space. However, different approaches may be used to produce the final base-space alignments. For a color-space read, one approach is to identify a final color-space alignment and then convert the color sequence to nucleotides under the guidance of the alignment using dynamic programming [14]. An alternative is to directly perform color-aware dynamic-programming-based alignment by simultaneously aligning all four possible translations [12, 32].

In CUSHAW3, we convert a nucleotide-based genome to a color sequence and perform short-read alignment in color space basically following the same workflow as the base-space alignment (mentioned above). For a color-space read, after obtaining a qualified color-space alignment, we must convert the color sequence into a nucleotide sequence. This conversion is accomplished by adopting the dynamic programming approach proposed by Li and Durbin [14]. Subsequently, the translated nucleotide sequence will be re-aligned to the nucleotide-based genome using either local or semi-global alignment depending on how its parent alignment has been produced.

10.6 Results and Discussion

To measure alignment quality, we have used the sensitivity metric for both simulated and real data. Sensitivity is calculated by dividing the number of aligned reads by the total number of reads. Since the ground truth is known beforehand for simulated data, we have further used the recall metric. Recall is calculated by dividing the number of correctly aligned reads by the total number of reads. For simulated reads, an alignment is deemed to be correct if its mapping position has a distance of ≤ 10 to the true position on the genome.

Considering that for a read, GEM reports all of the detected alignments and BWA-MEM might produce multiple primary alignments, we define that a read is deemed to be correctly aligned if any of its reported alignments is correct. For fair comparison, we have configured CUSHAW3, CUSHAW2, and *Bowtie2* to report at most 10 alignments per read and *Novoalign* to report all repetitive alignments. All of our tests have been conducted on a workstation with dual Indel Xeon X5650 hex-core 2.67 GHz CPUs and 96 GB RAM running *Linux* (*Ubuntu* 12.04 LTS). In addition, we have highlighted in bold all of the best values in the following tables.

10.6.1 Evaluation on Base-Space Reads

As mentioned above, CUSHAW has been excluded from the following evaluations, since it has inferior alignment quality to both CUSHAW2 and CUSHAW3. We have evaluated the performance of CUSHAW2 (v2.1.10) and CUSHAW3 (v3.0.2) by aligning both simulated and real reads to the human genome (hg19) and have further compared the performance to that of four other leading aligners: *Novoalign* (v3.00.04), BWA-MEM (v0.7.3a), *Bowtie2* (v2.1.0), and GEM (v1.376).

10.6.1.1 On Simulated Data

We have simulated three *Illumina*-like PE datasets simulated from the human genome (hg19) using the `wgsim` program in SAMtools (v0.1.18) [42]. All of the three datasets have the same read lengths of 100 but with different mean base error rates: 2%, 4%, and 6%. Each dataset consists of one million read pairs with insert sizes drawn from a normal distribution $N(500, 50)$.

Firstly, we have compared the alignment quality of all evaluated aligners by considering all reported alignments (see Table 10.3) and setting the minimum MAPQ to 0. For the SE alignment, *Novoalign* achieves the best sensitivity and recall for each dataset. CUSHAW3 holds the equally best sensitivity for the dataset with 2% error rate and the second best sensitivity for the rest. As the error rate increases, each aligner has underwent some performance drop in terms of both sensitivity and recall. *Novoalign* has the smallest sensitivity (recall) decrease by 0.02% (2.95%), whereas *Bowtie2* undergoes the most significant sensitivity (recall) decrease by 18.10% (21.66%). CUSHAW3 gives the second smallest performance drop with a sensitivity (recall) decrease by 0.74% (3.76%). With PE information, each aligner is able to improve alignment quality over the SE alignment for each case. CUSHAW3, *Novoalign* and BWA-MEM are consistently the top three aligners for all datasets, while *Bowtie2* is the worst in terms of sensitivity. As for recall, CUSHAW3 is superior to the other aligners on the dataset with 6% error rate, while *Novoalign* is best for the remaining datasets. CUSHAW3 outperforms CUSHAW2, BWA-MEM, *Bowtie2*, and GEM for each dataset. Similar to the SE alignment, error rate also has a significant impact on the alignment quality of each aligner. As the

Table 10.3 Alignment quality on simulated reads (in %)

Aligner	Error rate 2%		Error rate 4%		Error rate 6%	
	Sensitivity	Recall	Sensitivity	Recall	Sensitivity	Recall
SE						
CUSHAW3	**100.0**	99.04	99.92	97.85	99.26	95.28
CUSHAW2	99.95	99.00	99.33	97.61	95.45	92.84
Novoalign	**100.0**	**99.59**	**99.97**	**98.81**	**99.98**	**96.65**
BWA-MEM	99.99	95.95	99.59	94.33	97.38	89.86
Bowtie2	99.30	95.69	93.64	87.59	81.20	74.03
GEM	99.76	99.02	97.08	92.28	90.46	77.64
PE						
CUSHAW3	**100.0**	99.54	**100.0**	99.14	99.96	**98.06**
CUSHAW2	99.73	99.43	99.36	98.71	96.47	95.07
Novoalign	**100.0**	**99.87**	**100.0**	**99.23**	**100.0**	97.13
BWA-MEM	**100.0**	97.59	**100.0**	97.11	99.88	95.55
Bowtie2	99.45	98.53	93.54	91.52	80.29	77.37
GEM	**100.0**	99.20	99.79	98.06	97.99	93.24

error rate grows higher, *Novoalign* has the least significant performance drop and CUSHAW3, the second least in terms of sensitivity. In terms of recall, nonetheless, CUSHAW3 has the smallest performance decrease.

Secondly, we have generated the *Receiver Operating Characteristic* (ROC) curves by plotting the *True Positive Rate* (TPR) against the *False Positive Rate* (FPR), where the minimum MAPQ threshold is set to be greater than 0. For each dataset, we first sort all of the alignments in descending order of MAPQ. At each mapping quality score $q > 0$, TPR is computed by dividing the number of correctly aligned reads of MAPQs $\leq q$ by the total number of reads, and FPR by dividing the number of incorrectly aligned reads of MAPQs $\leq q$ by the number of aligned reads of MAPQs $\leq q$. As GEM does not compute MAPQs, it has been excluded. For *Bowtie2*, we have disabled the option `-k` to enable meaningful MAPQ and have used the default setting to report ≤ 1 alignment per read. CUSHAW2 and CUSHAW3 have both been configured to report ≤ 1 alignment per read for the SE and PE alignments. For *Novoalign*, we have used the `-r Random` parameter to report ≤ 1 alignment for a single read. Figure 10.5 shows the ROC curves using simulated data, where *Novoalign* yields the most significant MAPQs for each case.

10.6.1.2 On Real Data

Thirdly, we have assessed all aligners using three real PE datasets produced from *Illumina* sequencers (see Table 10.4). All datasets are publicly available and named after their accession numbers in the NCBI sequence read archive. In this evaluation,

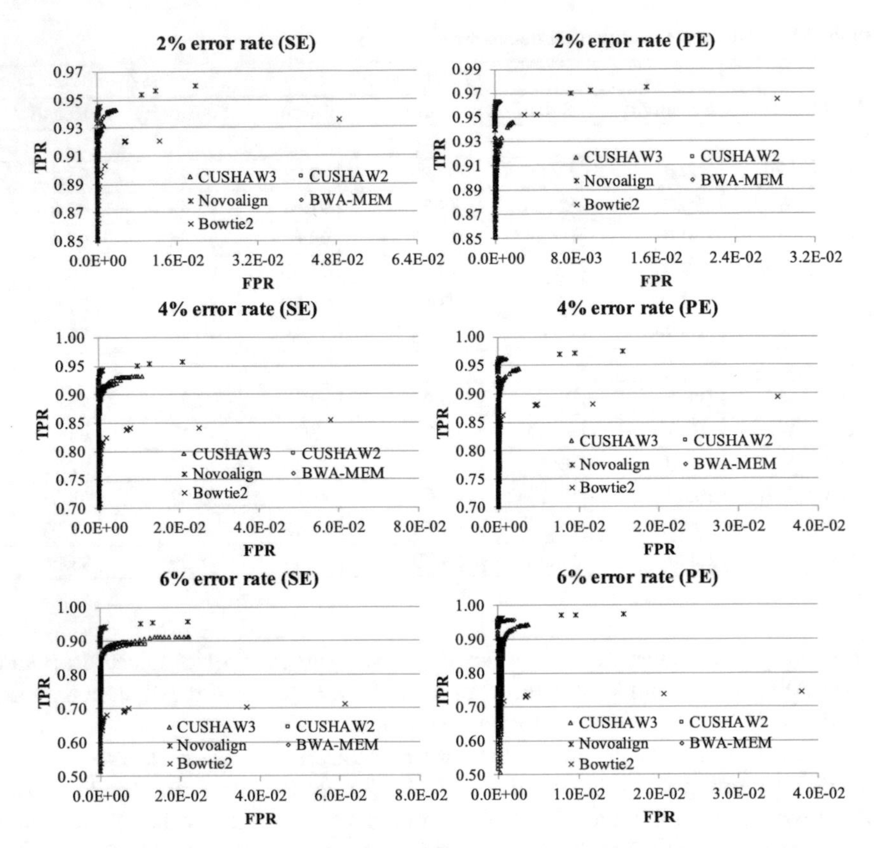

Fig. 10.5 ROC curves of all evaluated aligners on the simulated data with the minimum MAPQ > 0

Table 10.4 Real dataset information

Name	Type	Length	Number of reads	Mean insert size
SRR034939	PE	100	36,201,642	525
SRR211279	PE	100	50,937,050	302
ERR024139	PE	100	53,653,010	313

the sensitivity is computed by setting the minimum MAPQ to 0 (see Table 10.5). For the SE alignment, CUSHAW3 aligned the most reads for each dataset and GEM is worst. As for the PE alignment, BWA-MEM gives the best sensitivity and CUSHAW3, the second best for all datasets.

10.6.1.3 On GCAT Benchmarks

Finally, we have assessed the performance of all evaluated aligners using the public benchmarks at GCAT [43], which is a free collaborative platform for comparing

Table 10.5 Alignment quality on real reads (in %)

Type	Aligner	SRR034939	SRR211279	ERR024139
SE	CUSHAW3	**98.48**	**99.25**	**99.12**
	CUSHAW2	93.86	96.76	96.74
	Novoalign	96.80	98.44	98.49
	BWA-MEM	98.30	99.17	99.07
	Bowtie2	95.56	97.13	97.20
	GEM	93.69	95.10	94.82
PE	CUSHAW3	98.92	99.46	99.33
	CUSHAW2	94.38	96.94	96.92
	Novoalign	98.00	99.25	99.13
	BWA-MEM	**99.06**	**99.49**	**99.36**
	Bowtie2	96.23	97.31	97.39
	GEM	95.52	96.16	96.15

Table 10.6 Alignment results on the GCAT benchmark

Type	Dataset	Measure	CUSHAW3	CUSHAW2	*Novoalign*	BWA-MEM
SE	Small indels	Sensitivity	**100.0**	99.86	97.56	99.99
		Recall	**97.52**	**97.52**	97.47	**97.52**
	Large indels	Sensitivity	**100.0**	99.50	97.56	99.99
		Recall	97.37	97.04	97.35	**97.40**
PE	Small indels	Sensitivity	**100.0**	99.99	98.85	**100.0**
		Recall	99.06	99.05	98.83	**99.22**
	Large indels	Sensitivity	**100.0**	99.71	98.84	**100.0**
		Recall	98.91	98.62	98.69	**99.08**

multiple genome analysis tools across a standard set of metrics. In this evaluation, we have compared CUSHAW3, CUSHAW2, BWA-MEM, and *Novoalign* with respect to alignment quality and variant calling. In addition, the evaluation results can also be obtained from the project homepage of CUSHAW3 (http://cushaw3.sourceforge.net) for each aligner.

To assess alignment quality, we have used two *Illumina*-like SE datasets and two *Illumina*-like PE datasets. For the two datasets of each alignment type, one has small indels in reads (the small-indel dataset) and the other contains large indels (the large-indel dataset). All of the four datasets are simulated from the human genome and have read length 100, where there are 11,945,249 reads in each SE dataset and 11,945,250 reads in each PE dataset. To be consistent with the GCAT standard evaluations, both CUSHAW2 and CUSHAW3 are configured to report ≤ 1 alignment per read for both the SE and PE alignments. Table 10.6 shows the alignment results of all evaluated aligners. In terms of SE alignment, CUSHAW3 yields the best sensitivity for both datasets. The best recall is achieved by CUSHAW3, CUSHAW2, and BWA-MEM on the small-indel dataset and by BWA-MEM on the large-indel dataset. CUSHAW3 performs better than *Novoalign* for each case.

Table 10.7 Variant calling results on the GCAT benchmark

Aligner	Sensitivity	Specificity	T_i/T_v	Correct SNP	Correct Indel
CUSHAW3	83.74	99.9930	2.285	115,709	5974
CUSHAW2	83.51	99.9930	**2.323**	112,727	5841
Novoalign	84.10	**99.9951**	2.289	121,992	**9416**
BWA-MEM	**85.30**	99.9926	2.285	**124,459**	9232

Sensitivity=$TP/(TP + FN)$, specificity=$TN/(TN + FP)$ and T_i/T_v is the ratio of transitions to transversions in SNPs. *TP* is true positive, *TN* is true negative, *FP* is false positive, and *FN* is false negative

CUSHAW2 outperforms *Novoalign* on the small-indel dataset in terms of both sensitivity and recall while yielding smaller recall on the large-indel dataset. In terms of PE alignment, BWA-MEM performs best for each case and CUSHAW3 is the second best. On the small-indel dataset, CUSHAW2 outperforms *Novoalign* in terms of both sensitivity and recall. On the large-indel dataset, CUSHAW2 yields better sensitivity than *Novoalign*, while *Novoalign* gives better recall.

To assess how an aligner affects the performance of variant calling, we have used a real exome sequencing dataset. This dataset is comprised of *Illumina* 100-bp PE reads and has 30× the coverage of the human exome. In this test, SAMtools is used as the variant caller. Table 10.7 shows the variant calling results, where the novel *Single Nucleotide Polymorphisms* (SNPs) in dbSNP [44] are not taken into account. BWA-MEM yields the best sensitivity and *Novoalign*, the second best. In terms of specificity, *Novoalign* achieves the best performance, while CUSHAW2 and CUSHAW3 tie for the second place. As for T_i/T_v ratio, CUSHAW2 produces the largest value of 2.323 and *Novoalign* gives the second best value of 2.289. CUSHAW3 and BWA-MEM are joint third. BWA-MEM identifies the most correct SNPs, while *Novoalign* yields the most correct indels. Compared to CUSHAW2, CUSHAW3 holds a smaller T_i/T_v ratio but has an improved sensitivity as well as identifies more correct SNPs and indels.

10.6.1.4 Speed and Memory Comparison

Besides alignment quality, the speed of each aligner has been evaluated using the aforementioned simulated and real data. We have run each aligner with 12 threads on the aforementioned workstation. For fair comparison, GEM has counted in the SAM format conversion time (sometimes takes >50% of the overall runtime), as every other aligner reports alignments in SAM format. In addition, all runtimes are measured in wall clock time.

Table 10.8 shows the runtime (in minutes) of all evaluated aligners on both simulated and real data. For the simulated data, *Novoalign* is the slowest for nearly all cases, with an exception that CUSHAW3 performs worst in terms of PE alignment for the dataset with 4% error rate. For the SE alignment, BWA-MEM runs fastest on the datasets with 2% and 4% error rates, while *Bowtie2* performs best for

Table 10.8 Runtimes (in minutes) on simulated and real base-space reads

	Error rate 2%		*Error rate 4%*		*Error rate 6%*	
Simulated	*SE*	*PE*	*SE*	*PE*	*SE*	*PE*
CUSHAW3	3.4	6.2	3.7	8.1	3.9	10.7
CUSHAW2	2.5	2.5	2.8	2.9	2.9	3.1
Novoalign	6.7	6.6	38.1	7.0	131.7	12.6
BWA-MEM	**1.4**	**2.3**	**1.9**	**1.9**	2.0	**2.1**
Bowtie2	2.1	3.6	2.0	2.7	**1.7**	2.2
GEM	5.7	2.4	5.9	**1.9**	5.4	2.0
	SRR034939		*SRR211279*		*ERR024139*	
Real	*SE*	*PE*	*SE*	*PE*	*SE*	*PE*
CUSHAW3	62.0	292.4	78.6	317.9	85.1	264.1
CUSHAW2	38.0	38.5	47.2	49.0	51.4	50.5
Novoalign	862.1	497.6	2024.0	1243.8	754.2	460.3
BWA-MEM	**25.2**	**25.9**	**24.6**	**26.1**	**27.7**	**30.9**
Bowtie2	50.4	55.9	79.1	69.5	78.0	72.7
GEM	53.0	34.4	72.2	44.7	68.3	51.0

the dataset with 6% error rate. For the PE alignment, BWA-MEM is superior to all other aligners for each dataset, with an exception that GEM has a tie with BWA-MEM for the dataset with 4% error rate. In addition, the runtimes of both *Novoalign* and CUSHAW3 are more sensitive to the error rates compared to other aligners. For the real data, BWA-MEM is consistently the fastest for each case and *Novoalign* is the worst.

As for memory consumption, the peak resident memory of each aligner has been calculated by performing PE alignment on the dataset with 2% error rate using a single CPU thread. *Bowtie2* takes the least memory of 3.2 GB and *Novoalign* consumes the most memory of 7.9 GB. CUSHAW3 and CUSHAW2 have a memory footprint of 3.3 GB and 3.5 GB, respectively. For BWA-MEM and GEM, the peak resident memory is 5.2 GB and 4.1 GB, respectively.

10.6.2 Evaluation on Color-Space Reads

In addition to base-space alignment, we have evaluated the performance of CUSHAW3 for color-space alignment on the aforementioned workstation and have further compared our aligner to SHRiMP2 (v2.2.3) and BFAST (v0.7.0a). In this evaluation, we have simulated two mate-paired datasets (read lengths are 50 and 75) from the human genome using the ART (v1.0.1) simulator [45]. Each dataset has 10% coverage of the human genome (resulting in 6,274,322 reads in the 50-bp dataset and 4,182,886 reads in the 75-bp dataset) and has an insert size 200 ± 20.

Table 10.9 Alignment quality and runtimes (in minutes) on color-space reads

Dataset	Measure	CUSHAW3	SHRiMP2	BFAST
50-bp	Sensitivity	**92.13**	91.55	88.94
	Recall	86.28	**88.58**	81.01
	Time	**41**	227	160
75-bp	Sensitivity	92.27	92.33	**93.44**
	Recall	91.16	**91.24**	86.14
	Time	**20**	263	389

Both CUSHAW3 and SHRiMP2 are configured to report up to 10 alignments per read and BFAST to report all alignments with the best score for each read. Each aligner conducts mate-paired alignments and runs with 12 threads on the aforementioned workstation. Table 10.9 shows the alignment quality and the runtimes of the three aligners. In terms of sensitivity, CUSHAW3 outperforms both SHRiMP2 and BFAST for the 50-bp dataset, while BFAST is the best for the 75-bp dataset. When considering all reported alignments, SHRiMP2 produces the best recall and CUSHAW3 performs second best for every dataset. When only considering the first alignment occurrence per read, CUSHAW3 is superior to both SHRiMP2 and BFAST for each dataset. In terms of speed, CUSHAW3 is the fastest for each case. On average, CUSHAW3 achieves a speedup of 9.5 (and 11.9) over SHRiMP2 (and BFAST). In particular, for the 75-bp dataset, our aligner runs 13.5× and 19.9× faster than SHRiMP2 and BFAST, respectively. In addition, for each aligner, the recall gets improved as read length increases.

10.7 Conclusion

In this chapter, we have presented CUSHAW software suite for parallel, sensitive, and accurate NGS read alignment to large genomes such as the human genome. This suite is open-source and comprises three individual aligners: `CUSHAW`, `CUSHAW2`, and `CUSHAW3`. All of three aligners are designed based on the well-known seed-and-extend heuristic but have different seeding policies. CUSHAW employs inexact k-mer seeds, CUSHAW2 adopts MEM seeds, and CUSHAW3 introduces hybrid seeds incorporating three different seed types, i.e., MEM seeds, exact-match k-mer seeds, and variable-length seeds derived from local alignments. Our aligners accept NGS reads represented in FASTA and FASTQ formats, which can be uncompressed or zlib-compressed, and provide an easy-to-use and well-structured interface, as well as a more detailed documentation about the installation and usage. Furthermore, our aligners intend to generate PHRED-compliant MAPQs for the produced alignments and then report them in SAM format, enabling seamless integration with established downstream analysis tools like SAMtools and GATK [46]. In addition, support for color-space read alignment has been introduced in CUSHAW3.

We have evaluated the performance of CUSHAW2 and CUSHAW3 by aligning both simulated and real-case sequence data to the human genome. For base-space alignment, both of our aligners have been compared to other top-performing aligners, including *Novoalign*, BWA-MEM, *Bowtie2*, and GEM. On simulated data, CUSHAW3 achieves consistently better alignment quality (by considering all reported alignments) than CUSHAW2, BWA-MEM, *Bowtie2*, and GEM in terms of both SE and PE alignment. Compared to *Novoalign*, CUSHAW3 has comparable PE alignment performance for the reads with low error rates but performs better for the reads with high error rates. On real data, CUSHAW3 achieves the highest SE sensitivity and BWA-MEM yields the best PE sensitivity for each dataset. As for speed, CUSHAW3 does not have any advantage over CUSHAW2, BWA-MEM, *Bowtie2*, and GEM but shows to be almost always faster than *Novoalign*. In terms of color-space alignment, CUSHAW3 is compared to the leading SHRiMP2 and BFAST by aligning simulated mate-paired color-space reads to the human genome. The results show that CUSHAW3 is consistently one of the best color-space aligners in terms of alignment quality. Moreover, on average, CUSHAW3 is one order of magnitude faster than both SHRiMP2 and BFAST on the same hardware configurations. Finally, as shown in our evaluations, the hybrid seeding approach does improve accuracy but at the expense of speed. To significantly reduce the runtime, one promising solution is the use of accelerators such as GPUs and *Xeon Phis*, as existing work, e.g., [47–49] on GPUs and [50–52] on *Xeon Phis*, has shown that sequence alignment can be significantly accelerated using special hardware beyond general-purpose CPUs.

Acknowledgements We thank the *Novocraft* Technologies Company for granting a trial license of *Novoalign*.

References

1. Zerbino, D.R., Birney, E.: Velvet: algorithms for de novo short read assembly using de Bruijn graphs. Genome Res. **18**, 821–829 (2008)
2. Simpson, J.T., Wong, K., Jackman, S.D., et al.: ABySS: a parallel assembler for short read sequence data. Genome Res. **19**, 1117–1123 (2009)
3. Liu, Y., Schmidt, B., Maskell, D.L.: Parallelized short read assembly of large genomes using de Bruijn graphs. BMC Bioinformatics **12**, 354 (2011)
4. Luo, R., Liu, B., Xie, Y., et al.: SOAPdenovo2: an empirically improved memory-efficient short-read de novo assembler. Gigascience **1**, 18 (2012)
5. Li, H., Homer, N.: A survey of sequence alignment algorithms for next-generation sequencing. Brief. Bioinform. **11**, 473–83 (2010)
6. Yang, X., Chockalingam, S.P., Aluru, S.: A survey of error-correction methods for next-generation sequencing. Brief. Bioinform. **14**, 56–66 (2013)
7. Peng, Y., Leung, H.C., Yiu, S.M., et al.: Meta-IDBA: a de novo assembler for metagenomic data. Bioinformatics **27**, 194–1101 (2011)
8. Yang, X., Zola, J., Aluru, S.: Parallel metagenomic sequence clustering via sketching and quasi-clique enumeration on map-reduce clouds. In: 25th International Parallel and Distributed Processing Symposium, pp. 1223–1233 (2011)

9. Nguyen, T.D., Schmidt, B., Kwoh, C.K.: Fast Dendrogram-based OTU Clustering using Sequence Embedding. In: 5th ACM Conference on Bioinformatics, Computational Biology, and Health Informatics, pp. 63–72 (2014)
10. Smith, A.D., Xuan, Z., Zhang, M.Q.: Using quality scores and longer reads improves accuracy of Solexa read mapping. BMC Bioinformatics **9**, 128 (2008)
11. Li, H., Ruan, J., Durbin, R.: Mapping short DNA sequencing reads and calling variants using mapping quality scores. Genome Res. **18**, 1851–1858 (2008)
12. Homer, N., Merriman, B., Nelson, S.F.: BFAST: an alignment tool for large scale genome resequencing. PLoS ONE **4**, e7767 (2009)
13. Langmead, B., Trapnell, C., Pop, M., et al.: Ultrafast and memory-efficient alignment of short DNA sequences to the human genome. Genome Biol. **10**, R25 (2009)
14. Li, H., Durbin, R.: Fast and accurate short read alignment with Burrows-Wheeler transform. Bioinformatics **25**, 1755–1760 (2009)
15. Liu, Y., Schmidt, B., Maskell, D.L.: CUSHAW: a CUDA compatible short read aligner to large genomes based on the Burrows-Wheeler transform. Bioinformatics **28**, 1830–1837 (2012)
16. Li, R., Yu, C., Li, Y., et al.: SOAP2: an improved ultrafast tool for short read alignment. Bioinformatics **25**, 1966–1967 (2009)
17. Li, H., Durbin, R.: Fast and accurate long-read alignment with Burrows-Wheeler transform. Bioinformatics **26**, 589–595 (2010)
18. Rizk, G., Lavenier, D.: GASSST: global alignment short sequence search tool. Bioinformatics **26**, 2534–2540 (2010)
19. Langmead, B., Salzberg, S.: Fast gapped-read alignment with Bowtie 2. Nat. Methods **9**, 357–359 (2012)
20. Liu, Y., Schmidt, B.: Long read alignment based on maximal exact match seeds. Bioinformatics **28**, i318–i324 (2012)
21. Liu, Y., Schmidt, B.: CUSHAW2-GPU: empowering faster gapped short-read alignment using GPU computing. IEEE Des. Test **31**, 31–39 (2014)
22. Liu, Y., Popp, B., Schmidt, B.: CUSHAW3: sensitive and accurate base-space and color-space short-read alignment with hybrid seeding. PLoS ONE **9**, e86869 (2014)
23. González-Domínguez, J., Liu, Y., Schmidt, B.: Parallel and scalable short-read alignment on multi-core clusters Using UPC++. PLoS ONE **11**, e0145490 (2016)
24. Marco-Sola, S., Sammeth, M., Guigó, R., et al.: The GEM mapper: fast, accurate and versatile alignment by filtration. Nat. Methods **9**, 1885–1888 (2012)
25. Mu, J.C., Jiang, H., Kiani, A., et al.: Fast and accurate read alignment for resequencing. Bioinformatics **28**, 2366–2373 (2012)
26. Luo, R., Wong, T., Zhu, J., et al.: SOAP3-dp: fast, accurate and sensitive GPU-based short-read aligner. PLoS ONE **8**, e65632 (2013)
27. Li, H.: Aligning sequence reads, clone sequences and assembly contigs with BWA-MEM (2013). arXiv:1303.3997 [q-bio.GN]
28. Altschul, S.F., Gish, W., Miller, W., et al.: Basic local alignment search tool. J. Mol. Biol. **215**, 403–410 (1990)
29. Burrows, M., Wheeler, D.J.: A block sorting lossless data compression algorithm. Technical Report 124, Digital Equipment Corporation, Palo Alto, CA (1994)
30. Ferragina, P., Manzini, G.: Indexing compressed text. J. ACM **52**, 4 (2005)
31. Novoalign, http://www.novocraft.com/products/novoalign
32. David, M., Dzamba, M., Lister, D., et al.: SHRiMP2: sensitive yet practical short read mapping. Bioinformatics **27**, 1011–1012 (2011)
33. Needleman, S.B., Wunsch, C.D.: A general method applicable to the search for similarities in the amino acid sequence of two proteins. J. Mol. Biol. **48**, 443–453 (1970)
34. Smith, T.F., Waterman, M.S.: Identification of common molecular subsequences. J. Mol. Biol. **147**, 195–197 (1981)
35. Blom, J., Jakobi, T., Doppmeier, D., et al.: Exact and complete short read alignment to microbial genomes using Graphics Processing Unit programming. Bioinformatics **27**, 1351–1358 (2011)

36. Kiełbasa, S.M., Wan, R., Sato, K., et al.: Adaptive seeds tame genomic sequence comparison. Genome Res. **21**, 487–493 (2011)
37. Ma, B., Tromp, J., Li, M.: PatternHunter: faster and more sensitive homology search. Bioinformatics **18**, 440–445 (2002)
38. Rasmussen, K.R., Stoye, J., Myers, E.W.: Efficient q-gram filters for finding all epsilon-matches over a given length. J. Comput. Biol. **13**, 296–308 (2006)
39. Ewing, B., Green, P.: Base-calling of automated sequencer traces using phred. II. Error probabilities. Genome Res. **8**, 186–194 (1998)
40. Rognes, T.: Faster Smith-Waterman database searches with inter-sequence SIMD parallelisation. BMC Bioinformatics **12**, 221 (2011)
41. Liu, Y., Schmidt, B.: GSWABE: faster GPU-accelerated sequence alignment with optimal alignment retrieval for short DNA sequences. Concurr. Comput. Pract. Exp. **27**, 958–972 (2014). doi:10.1002/cpe.3371
42. Li, H., Handsaker, B., Wysoker, A., et al.: The sequence alignment/map format and SAMtools. Bioinformatics **25**, 2078–2079 (2009)
43. Highnam, G., Wang, J.J., Kusler, D., et al.: An analytical framework for optimizing variant discovery from personal genomes. Nat. Commun. **6**, 6275 (2015)
44. Sherry, S.T., Ward, M.H., Kholodov, M., et al.: dbSNP: the NCBI database of genetic variation. Nucleic Acids Res. **29**, 308–311 (2001)
45. Huang, W., Li, L., Myers, J.R., et al.: ART: a next-generation sequencing read simulator. Bioinformatics **28**, 593–594 (2012)
46. McKenna, A., Hanna, M., Banks, E., et al.: The Genome Analysis Toolkit: a MapReduce framework for analyzing next generation DNA sequencing data. Genome Res. **20**, 1297–1303 (2010)
47. Liu, Y., Schmidt, B., Maskell, D.L.: MSA-CUDA: multiple sequence alignment on graphics processing units with CUDA. In: 20th IEEE International Conference on Application-specific Systems, Architectures and Processors, pp. 121–128 (2009)
48. Liu, Y., Maskelml, D.L., Schmidt, B.: CUDASW++: optimizing Smith-Waterman sequence database searches for CUDA-enabled graphics processing units. BMC Res. Notes **2**, 73 (2009)
49. Alachiotis, N., Berger, S.A., Stamatakis, A.: Coupling SIMD and SIMT architectures to boost performance of a phylogeny-aware alignment kernel. BMC Bioinformatics **13**, 196 (2012)
50. Liu, Y., Schmidt, B.: SWAPHI: Smith-Waterman protein database search on Xeon Phi coprocessors. In: 25th IEEE International Conference on Application-specific Systems, Architectures and Processors, pp. 184–185 (2014)
51. Liu, Y., Tran, T.T., Lauenroth, F., et al.: SWAPHI-LS: Smith-Waterman algorithm on Xeon Phi coprocessors for long DNA sequences. In: 2014 IEEE International Conference on Cluster Computing, pp. 257–265 (2014)
52. Wang, L., Chan, Y., Duan, X., et al.: XSW: accelerating biological database search on Xeon Phi. In: 28th IEEE International Symposium on Parallel and Distributed Processing Workshops and Phd Forum, pp. 950–957 (2014)

Chapter 11
String-Matching and Alignment Algorithms for Finding Motifs in NGS Data

Giulia Fiscon and Emanuel Weitschek

11.1 Introduction

The development of *Next Generation Sequencing* (NGS) technologies allows the extraction at low cost of an extremely large amount of biological sequences in the form of *reads*, i.e., short fragments of an organism's genome. The length of such reads is very small when compared to the length of a whole genome: it may range from 40 to 1000 *base pairs* (bp), i.e., characters, while the length of a simple genome (e.g., bacteria) is in the order of *Megabase pairs* (Mbp). Four main NGS technologies are currently used [13]: *Roche 454*, *Illumina*, *Pacific Bioscience*, and *Ion Torrent*. At present, *Illumina* technology performances are 600 *Gigabase pairs* (Gbp) per day at a low cost per bp [3] with reads an average length of 150 bp; *Roche 454* performances are 1 Gbp per day at a higher cost with reads at an average length of 700 bp [1]; and *Ion Torrent* machines [4] produce reads of 200 bp with a throughput of 130 Gbp per day at a low cost per bp [31]. Recently, *Pacific Bioscience* [5] introduced sequencers able to produce longer reads (in the order of 10,000 nucleotides). Typically, the number of reads produced by NGS experiments reaches several millions or more, depending on the depth of the sequencing coverage. The use of NGS machines results in much larger sets of reads to be analyzed, posing new problems for computer scientists and bioinformaticians, whose task is to design algorithms that align and merge the reads for an effective reconstruction of the

G. Fiscon (✉)
Institute of Systems Analysis and Computer Science, National Research Council, Via dei Taurini 19, 00185 Rome, Italy
e-mail: giulia.fiscon@iasi.cnr.it

E. Weitschek
Department of Engineering, Uninettuno International University, Corso Vittorio Emanuele II 39, 00186 Rome, Italy
e-mail: emanuel@iasi.cnr.it

M. Elloumi (ed.), *Algorithms for Next-Generation Sequencing Data*,
DOI 10.1007/978-3-319-59826-0_11

genome (or large portions of it) with sufficient precision and speed [33]. Therefore, it is extremely important to develop methods that can quickly establish whether two reads or NGS sequences are similar or not.

In this chapter, we focus on string-matching and alignment algorithms to address this problem and to analyze biological sequences extracted from NGS machines. Additionally, we describe the most widespread string-matching, alignment-based, and alignment-free techniques. Section 11.2 presents an overview of the string-matching algorithms, both exact and approximate ones. Sections 11.3 and 11.4 deal with the state-of-the-art and up-to-date alignment algorithms, which are able to process NGS reads and sequences. In Sect. 11.5 we introduce the techniques for NGS read mapping. Moreover, in Sect. 11.6, we describe a practical approach for NGS reads and sequences comparisons, alignment-free algorithms, which are very effective due to their speed and scalability. Finally, in Sect. 11.7 we draw the conclusions.

The aim of this chapter is to provide the reader with a wide overview of the main string-matching and alignment methods for NGS sequence analysis.

11.2 String Matching

An array of characters is commonly called a *string*. A string can be used in the biological framework to describe a *sequence* of nucleotides. There are several problems that deal with the manipulation of strings (e.g., the *distance* among the strings or the string-matching problem), which have many applications in NGS for the DNA or RNA sequence analysis [16, 36, 40, 46].

The aim of *string matching* is to find where one or more strings (also known as *pattern*) are placed within a longer string, or a *text*; it tries to find the occurrences of a *pattern* or a *motif* in a text. Therefore, a pattern or a motif can be defined as one or a set of shorter strings that has to be searched in the original text.

Addressing the problem of looking for a string in a text, i.e., string matching, is not limited to word processing, but it is widely used in computational biology to search for molecular sequences within longer sequences, to look for specific characteristics in the DNA sequences, or to determine how different two sequences are. In NGS, pattern matching is widely used in DNA assembly, whose aim is to reconstruct a sequence of DNA from small fragments called reads obtained from a NGS experiment.

11.2.1 Regular Expressions

A specified pattern that is the synthetic representation of a set of strings is called a *regular expression* (*regexp*) [22]. In order to define these sets, rules are often shorter than lists of set members. These expressions consist of constants and operator symbols, which denote sets of strings and operations over these sets, respectively [24]. Formally, if there exists at least one *regexp* that matches a particular set, then there exists an infinite number of such expressions.

One of the key points of regular expressions is the ability to search for a *character class* (or one to be chosen from a provided list of characters), instead of searching for a specific character. Indeed, character classes allow to try an infinite number of different combinations. Furthermore, another essential feature of *regexp* is the repetition of characters. In fact, for each expression "a," i.e., a simple character or character class, we can specify how many times the character must be present in the text. Thus, for example the *regexp* $ab * (c \mid \epsilon)$ denotes the set of strings starting with "a," then zero or more "b," and lastly, optionally a "c," i.e., {a, ac, ab, abc, abb, abbc, ... }. Finally, several advanced features exist (e.g., modifiers, statements, groups) that can extend their simple basic syntax.

It is worth noting that regular expressions are only for describing a pattern but do not influence the complexity of pattern searching algorithms. Therefore, by using the regular expressions, we can flexibly manipulate the character strings since we are able to generally describe several comparisons or substitutions within other strings through a unique pattern, i.e., a construction scheme of a string.

11.2.2 Exact Pattern Matching

Given a pattern $P[1 \cdots m]$ and a text $T[1 \cdots n]$, where $m < n$, the pattern matching problem is equal to find all occurrences of P in T. The elements of P and T are characters drawn from a finite alphabet set. Therefore, the character arrays of P and T are also referred to as strings of characters [12]. In greater detail, the pattern P is said to occur with shift s in text T.

If $0 \leq s \leq n - m$ and $T[s + 1 \cdots s + m] = P[1 \cdots m]$ or $T[s + j] = P[j]$ for $1 \leq j \leq m$, such a shift s is called a *valid shift*.

The string-matching problem is hence the problem of finding all valid shifts with which a given pattern P occurs in a given text T. It should be noticed that the occurrences of P in T could be overlapped. In order to solve this problem, very different algorithms have been developed, which are explained in the following subsections.

11.2.2.1 Brute-Force Algorithm

There is always a naive solution that sooner or later finds the correct solution to the string-matching problem. Clearly, such a solution is extremely inefficient, and this inefficiency increases with the length of the considered text T [12]. Thus, the first solution for the string-matching problem is based on a trivial algorithm which aligns the first character of the pattern P to the first one of the text T. The algorithm finds all valid shifts or possible values of s (i.e., $n - m + 1$) so that $P[1 \cdots m] = T[s + 1 \cdots s + m]$. The algorithm steps are the following:

1. Align the first character of P with the first character of T.
2. From left to right, compare the corresponding characters of P and T until a mismatch has been found or the end of P has been reached.
3. If the end of P has been reached, return the character position of T corresponding to the first character of P.
4. P moves one place to the right.
5. If the last character of P is beyond the end of T, stop; otherwise, repeat.

In the worst case, this algorithm takes $\Theta((n-m+1)m)$, and hence, its running time increases with the length of the text T. Moreover, whenever a character mismatch occurs after the matching of several characters, the comparison begins by going back in T from the character which follows the last beginning one. The algorithm would perform better if it did not go back to T. A *mismatch* position k is a position in which the text T and the pattern P have different symbols. In conclusion, the Brute-Force Algorithm is the least efficient one among other algorithms. However, it should be noticed that it is a competitive solution when the first occurrence of a pattern is found at the beginning of the text or when we look for a pattern of three to four characters (unlikely for a pattern of DNA). In such a case, the additional cost of other algorithms presented hereinafter (such as KMP and BM) could not repay the advantage of a fast scan of the text as it stops in the initial positions.

11.2.2.2 Knuth–Morris–Pratt Algorithm

The algorithm of *Knuth*, *Morris*, and *Pratt* (KMP) [23] proposes the answer on how to increase the shifting without losing occurrences yields by using a preprocessing phase in order to "learn" the internal structure of the pattern P and of the text T. Mostly, this step decreases the running time of the algorithm. Indeed, as opposed to the Brute-Force Algorithm, it is not always necessary to move the P of one location.

Let us define the prefix and the suffix of a string S:

prefix $S[1 \cdots i]$ of a string S is a substring that begins at position 1 of S and ends at position i.

suffix $S[i \cdots |S|]$ of a string S is a substring that begins at position i and ends at the final position of S.

Suppose we have the setting drawn in Fig. 11.1 with P in $s + 1$. It should be noticed that within the matching of length $q = 5$, there is a substring $P[3 \cdots 5] =$ aga, which coincides with the prefix $P[1 \cdots 3]$.

Intuitively, it is clear that we can move P in $s' = s+1+(q-k)+1$, making sure that there exists a first matching of length $k = 3$ for the prefix $P[1 \cdots 3]$ as shown in Fig. 11.2. Since the prefix $P[1 \cdots q]$ coincides with the substring $T[s + 1 \cdots s + q]$, we want to know what is the minimum deviation $s' > s$ such that $P[1 \cdots k] = T[s' + 1 \cdots s' + k]$. We find $s' = s + q - k$, showing that the comparison among the first k characters of P are not necessary.

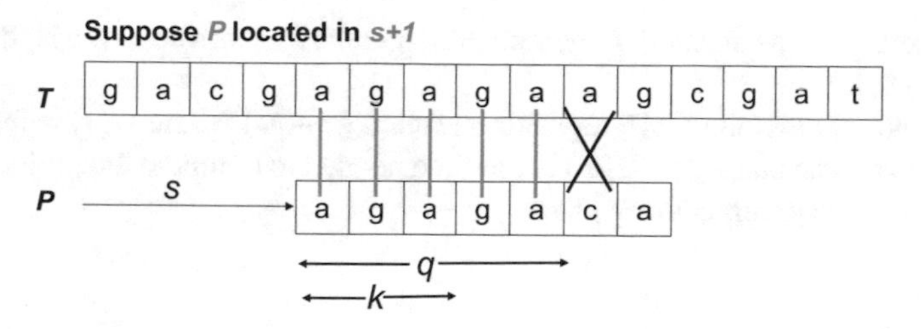

Fig. 11.1 An example of string matching (I)

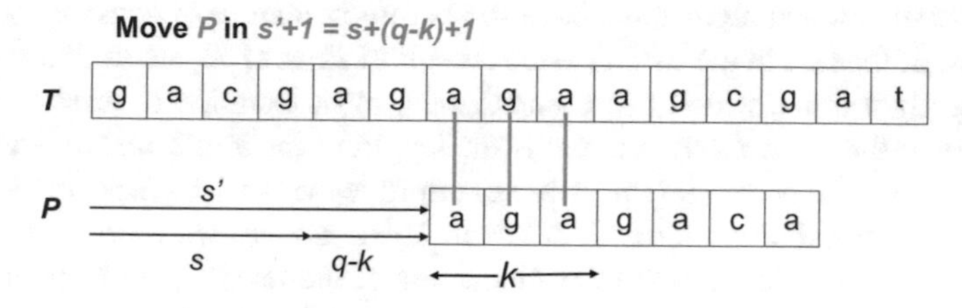

Fig. 11.2 An example of string matching (II)

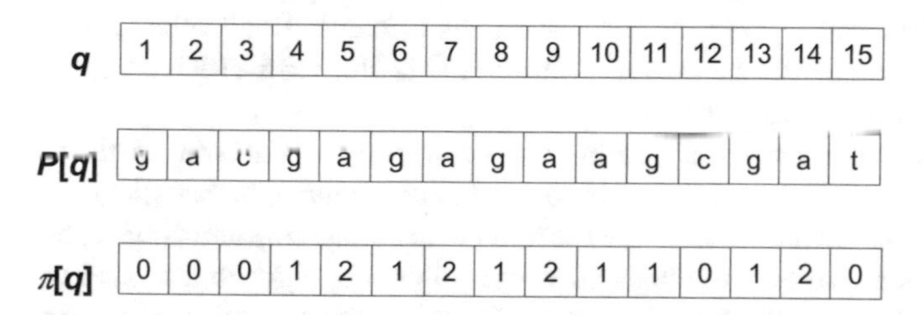

Fig. 11.3 Example of the computation of the prefix function $\pi[q]$ starting from a pattern P with q matched characters

Thus, after a character of P (such as q) matches with T and then a mismatch occurs, the matched q characters allow us to determine immediately that certain shifts are invalid, and hence, we can directly go to the shift which is potentially valid. In fact, the matched characters in T are prefixes of P, so just from P, it is sufficient to determine whether a shift is invalid or not.

Formally, given a pattern $P[1, \ldots, m]$, its prefix function has been computed as follows:

$$\pi : \{1, 2, \ldots, m\} \rightarrow \{0, 1, \ldots, m-1\} \tag{11.1}$$

$\pi[q] = \max\{k : k < q$ and $P[1 \cdots k]$ is a suffix and must be found such that $P[1 \cdots q]\}$ where $\pi[q]$ is the length of the longest prefix of P that is also a suffix for $P[1 \cdots q]$. In Fig. 11.3, we show an example of the prefix function computation. We can see that the KMP algorithm is divided in two phases: the *preprocessing*, where the prefix function is computed, and the *search*, where the effective search for the matches is performed. In general, the pattern P is moved to the right by more than one position.

In the worst case, the KMP algorithm takes $O(n + m)$ where $O(n)$ is the running time of the search phase and $O(m)$ is the time needed to compute the prefix function, according to the preprocessing phase.

11.2.2.3 Boyer–Moore Algorithm

The above mentioned algorithms have the common feature to consider at least all characters of the text in the worst case. A result of *Rivest* [38] states that the pattern matching algorithms, which use comparisons, must examine at least $n - m + 1$ characters in the worst case, where n is the length of the text T and m is the length of the pattern P. However, it is possible to improve the performances in the average case. Indeed, simultaneously with KMP in 1991, *Boyer and Moore* proposed an algorithm that for each portion of m characters of the text T, compares on average only $O(\log n)$ [18].

The KMP algorithm is quite fast with short patterns since it increases the probability that a prefix of the pattern has a match. In all other cases (in practice the most frequent), the BM algorithm [11] is the most efficient, specifically with large texts and large patterns.

In particular, the BM algorithm makes the comparison between the text T and the pattern P from right to left and incorporates two heuristics that allow avoiding much of the work that the previously described string-matching algorithms perform. These heuristics are so effective that they often allow the algorithm to skip altogether the examination of many text characters. These heuristics, known as the *bad-character heuristics* and the *good-suffix heuristics* can be viewed as operating independently in parallel; when a mismatch occurs, each heuristics proposes an amount by which s can safely be increased in order to perform a valid shift. Then, the BM algorithm chooses the largest amount and increases s by the latter.

Bad-Character Heuristics

When a mismatch occurs, the bad-character heuristics uses the information about where the bad text character $T[s + j]$ occurs in the pattern in order to propose a new shift. In the best case, the mismatch occurs on the first comparison ($P[m] \neq T[s+m]$) and the bad character $T[s + m]$ does not occur in the pattern at all. Let us consider searching for a pattern P of length m in a text string T of length n. In this case, we can increase the shift s by m since any shift smaller than $s + m$ will align some pattern characters against the bad character, causing a mismatch. If the best case occurs repeatedly, the BM algorithm examines only a fraction $\frac{1}{m}$ of the text characters since each examined text character yields a mismatch, thus causing s to increase by m. This best-case behavior illustrates the power of matching right to left instead of left to right.

Formally, given a pattern P, *the function of the bad character* is defined as follows:

$$\Lambda : s_i, s_2, \ldots, s_{|S|} \rightarrow 1, 2, \ldots, m \tag{11.2}$$

$$\Lambda : [s_i] = \max\{k : 1 \leq k \leq m, P[k] = s_i\} \tag{11.3}$$

where s_i is the i-th symbol of the alphabet Σ.

Three different cases can be presented:

1. The bad character is not in P (see Fig. 11.4); the shift is performed in order to align the first character of P with the ones of T, and the bad character follows as shown in Fig. 11.5:
2. The rightmost occurrence of the bad character of P is in a position k before the index j, which corresponds to the character of P aligned with the bad character (see Fig. 11.6); in this case, the shift is performed in order to align $P[k]$ with the bad character of T, as shown in Fig. 11.7:
3. The rightmost occurrence of the bad character of P is in a position k after the index j which corresponds to the character of P aligned with the bad character (see Fig. 11.8); the only feasible shift is performed in order to move P one place to the right, as shown in Fig. 11.9.

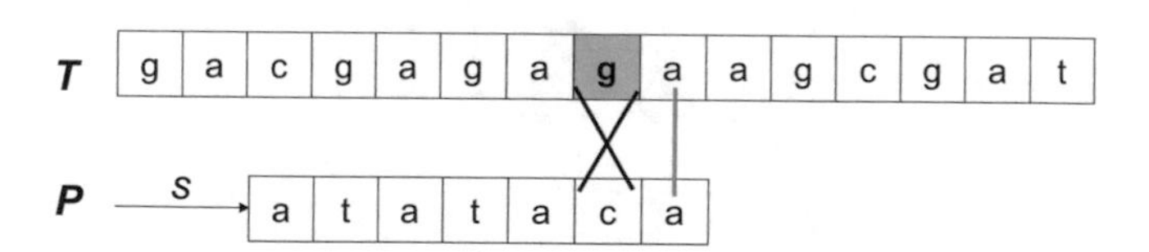

Fig. 11.4 Example of the bad-character heuristics: case 1

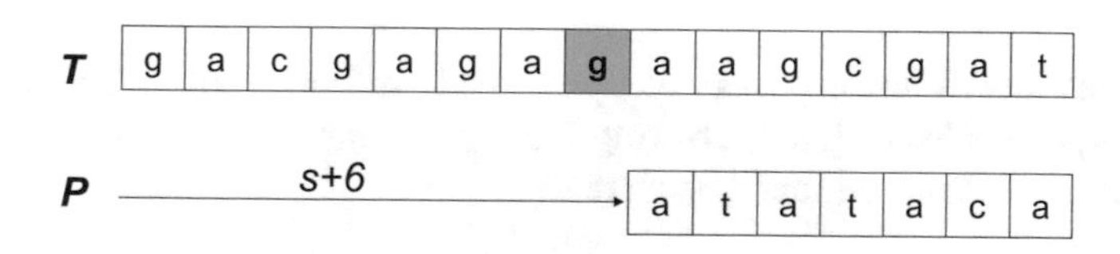

Fig. 11.5 Example of the bad-character heuristics: case 1—character shift

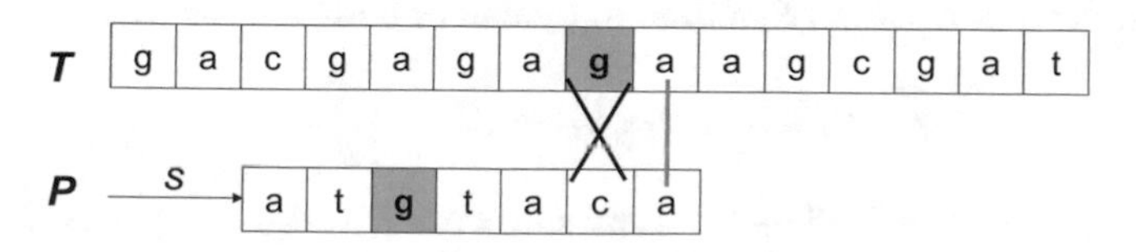

Fig. 11.6 Example of the bad-character heuristics: case 2

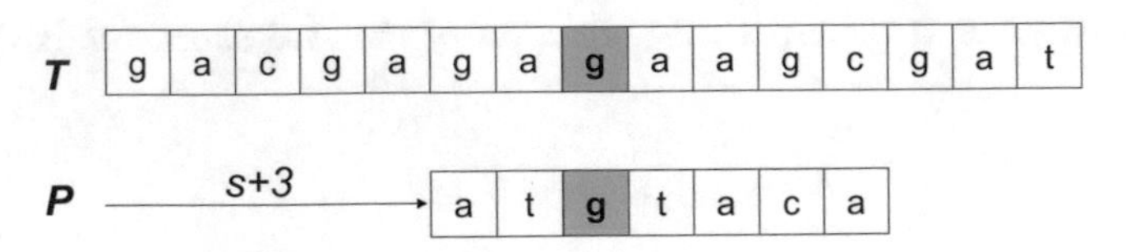

Fig. 11.7 Example of the bad-character heuristics: case 2—character shift

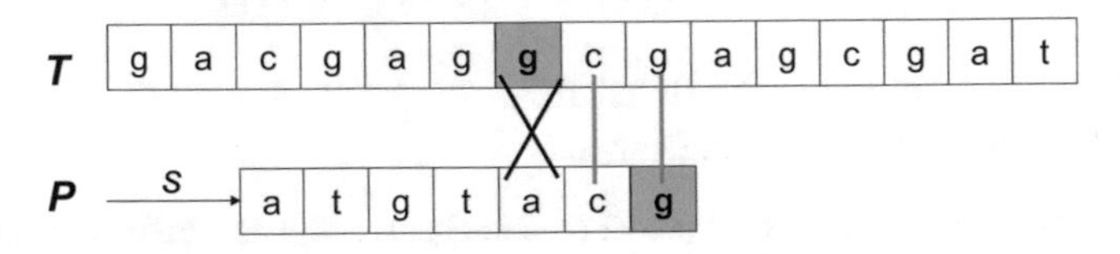

Fig. 11.8 Example of the bad-character heuristics: case 3

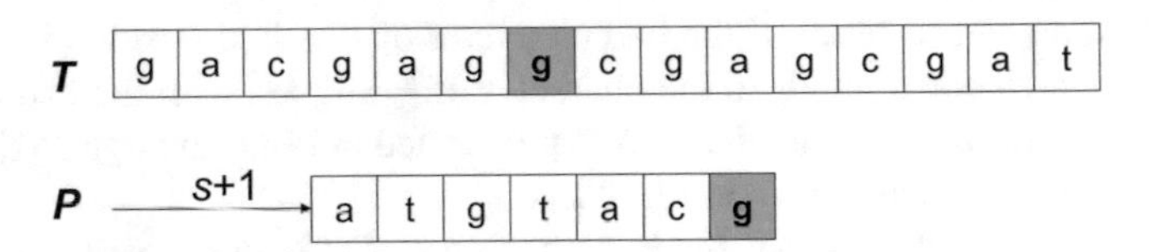

Fig. 11.9 Example of the bad-character heuristics: case 3—character shift

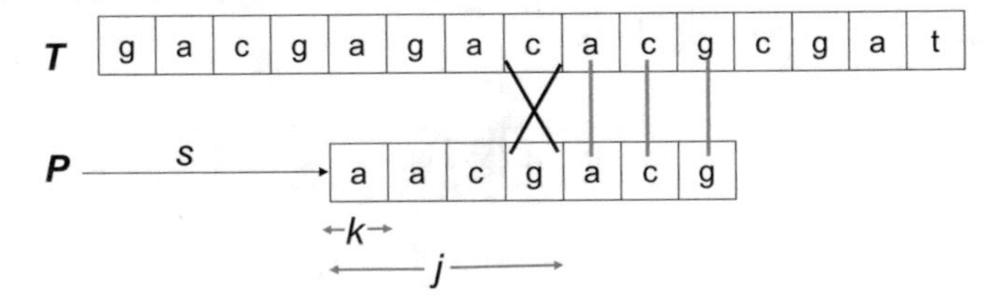

Fig. 11.10 Example of a good-suffix heuristics

Good-Suffix Heuristics

We analyze the following example presented in Fig. 11.10. Since the suffix $P[j+1, m]$ coincides with the substring $T[s+j+1, s+m]$, the rightmost position $k < j$ (if it exists) must be found, such that:

$$P[k] \neq P[j] \text{ and } P[k+1, k+m-j] = T[s+j+1, s+m] \tag{11.4}$$

and then shift P in $s' + 1$ such that: $s' + k = s + j$.

Formally, given a pattern P, a function γ can be computed as follows:

$$\gamma : \{0, 1, \ldots, m-1\} \rightarrow \{0, 1, \ldots, m-1\} \tag{11.5}$$

$\gamma[j] = \max\{k : k < j+1, P[j+1, \ldots, m]$ suffix of $P[1 \cdots k+m-j]$ and $P[k] \neq P[j]\}$.

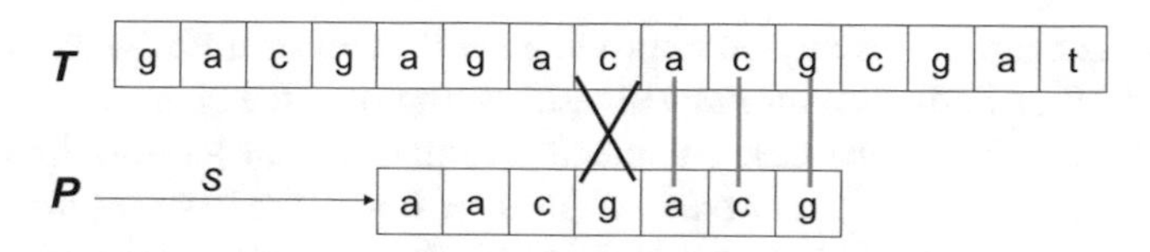

Fig. 11.11 Example of a good-suffix heuristics: case 1—the position k exists

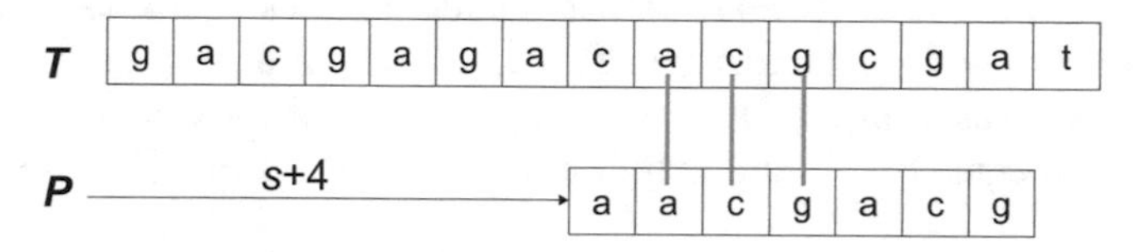

Fig. 11.12 Example of a good-suffix heuristics: case 2—the position k does not exist

The $\gamma[j]$ is the lowest amount by which we can advance s and not cause any characters in the "good suffix" $T[s + j + 1 \cdots s + m]$ to be mismatched against the new alignment of the pattern. The function γ is defined for all j and is called *good-suffix function* for the pattern P.

Two different cases can be presented:

1. The position k does not exist: in this case, P moves up to match its prefix with a suffix of $T[s + j + 1 \ldots s + m]$ or m steps if it is a prefix of P that is suffix of $T[s + j + 1 \ldots s + m]$ does not exist.
2. The position k exists (see Fig. 11.11) and P moves the lowest number of steps in order to match its prefix with a suffix of the occurrence of P in T, or if this one does not exist, P moves m steps, as shown in Fig. 11.12.

In conclusion, even the BM algorithm can be divided in two phases: the *preprocessing*, where the two heuristics functions are computed, and the *search*, where the effective search for the matches is performed. In general, the pattern P is moved by more than one position to the right.

The BM algorithm takes $O(nm)$ in the worst case, where $O(n-m+1) \cdot m = O(nm)$ is the running time of the search phase and $O(|\Sigma| + m) + O(m) = O(|\Sigma| + m)$ is the time needed to compute the prefix function, according to the preprocessing phase.

To date, the BM algorithm is the most frequently used for the detection of exact matches in DNA and NGS sequences, due to its performances.

11.2.2.4 Rabin–Karp Algorithm

The *Rabin–Karp* (RK) algorithm [21] may be considered a numerical algorithm rather than a comparison-based algorithm. In spite of the fact that other algorithms have a better worst-case complexity, the RK algorithm is still interesting because it performs well in practice and also because it easily generalizes to other problems, such as the two-dimensional pattern matching and the many-patterns problem.

Whereas most algorithms rely on improving the number of shifts by examining the structure of P or T, the *Rabin–Karp* algorithm improves the character-to-character comparison step. The worst-case complexity is still $O((n-m+1)\cdot m)$, like the naive algorithm, and even $O(n^2)$ for certain choices of P and T. However, the algorithm has a good average-case behavior, which lowers time complexity to $O(n+m)$.

Let's assume the alphabet as follows: $\Sigma = \{0,1,2,\ldots,9\}$, a string of k consecutive characters can be considered as representing a length-k decimal number. Let p denote the decimal number for $P[1\cdots m]$, and let t_s denote the decimal value of the length-m substring $T[s+1\cdots s+m]$ of $T[1\cdots n]$ for $s = 0,1,\ldots,n-m$. The value t_s is equal to p if and only if $T[s+1\cdots s+m] = P[1\cdots m]$ and s is a valid shift.

The number p and t_s can be computed in $O(m)$ time as follows:

$$p = P[m] + P[m-1]10^1 + \cdots + P[1]10^{m-1} \tag{11.6}$$

$$t_s = T[s+m] + T[s+m-1]10^1 + \cdots + T[s+1]10^{m-1} \tag{11.7}$$

According to the theorem that claims if $t_s = p$, pattern P occurs in text T at position s, and the numerical equality between t_s and p is equivalent to find an occurrence of P in T.

Regarding the running time needed to compute p and $t_0, t_1, \ldots, t_{n-m}$, all occurrences of the pattern $P[1\cdots m]$ in the text $T[1\cdots n]$ can be found in time $O(n+m)$.

However, p and t_s may be too large to work, and another easier solution using a suitable modulus q has been found.

In general, with a d-ary alphabet $\{0,1,\ldots,d-1\}$, q is chosen such that $d\times q$ fits within a computer word. The recurrence equation can be written as:

$$t_{s+1} = (d(t_s - T[s+1]h) + T[s+m+1]) \bmod q \tag{11.8}$$

where $h = d^{m-1} (\bmod q)$ is the value of the digit "1" in the high order position of an m-digit text window.

Since two numbers are equal, modulo q does not imply they are actually equal, so false matches between P and T can be performed (i.e., $t_s = \bmod q$ but $t_s \neq p$); the latter forced the algorithm to explicitly perform the comparison $T[s+1\ \cdots\ n] = P[1\ \cdots\ m]$ when $t_s = p \bmod q$

In conclusion, the running time of the algorithm is the sum of two addends: the first to compute p and $t_0, \ldots, t_{n-m}$ and the second one to perform $n-m+1$ comparisons. The overall time complexity is hence the following: $O(n+m) + O(n-m) = O(n+m)$.

11.2.2.5 String-Matching Automata

In order to scan the text string T for all occurrences of the pattern P, a lot of string-matching algorithms build a finite automaton. These string-matching automata are very efficient since they examine each text character exactly once, taking constant time per text character.

In particular, a *finite automaton* M is a 5-tuple $(Q, q_0, A, \Sigma, \delta)$, where:

1. Q is a finite set of states.
2. $q_0 \in Q$ is the start state.
3. A (subset of Q) is a set of accepting states.
4. Σ is a finite input alphabet.
5. δ is a function from $Q \times \Sigma$ into Q, called the *transition function* of M.

The idea is to use a finite state automaton that recognizes the words that have our pattern P as a suffix and during the input scanning whenever the automaton makes a transition that leads to the final state; it is recognized that up to this point, one occurrence of the pattern has been accessed and the starting position of the occurrence about the shift s is returned.

Let P be a pattern such as $|P| = m$, and let P_q be defined as $P[1 \cdots q]$ and representing the prefix of the string P of length q.

Thus, given a pattern P, let's define an auxiliary function σ, called *suffix function*, corresponding to P. The function is a mapping from Σ to $\{0, 1, \ldots, m\}$ such that $\sigma(x)$ is the length of the longest prefix of P that is a suffix of x.

The string-matching automaton M $(Q, q_0, A, \Sigma, \delta)$ to be used that corresponds to a given pattern P is built as follows:

1. The state set Q is $0, 1, \ldots, m$, where $q_0 = 0$ and m is the only accepted state.
2. $A = m$.
3. The transition function δ is defined for any state q and character a as follows: $\delta(q, a) = \sigma(P_q a)$.

11.2.3 Multiple Pattern Matching

The multiple pattern matching is defined as the search of more than one pattern in a sequence. Several approaches exist to address this issue. The most widespread ones are explained in this section.

11.2.3.1 Naive Algorithm

Likewise, in one single pattern matching, the easiest solution is a naive approach, or rather, a repeating a single algorithm for pattern matching (e.g., the algorithm "brute force") r times, one for each pattern. Since there is not an optimization based on any affinity between the searched patterns, the complexity of the algorithm is equal to its own multiplied by the number of r patterns. Such an approach is very simple to implement; however, it has a prohibitive complexity equal to $O(|S|mn)$, where S is the set of pattern to look for, and hence, it is not able to deal with very long sequences.

11.2.3.2 Tree-Data Structure

Another approach to address a multiple pattern matching is based on structuring the information in a tree. In particular, we introduce the *prefix tree* [14] and *suffix tree* [32, 48, 51], which allow an optimization of the search for patterns, exploiting the similarity among the subsequences of the patterns.

Prefix Tree

A multiple prefix tree, also named *trie*, is a tree whose edges are labeled with a character that will be different if one vertex has two different outgoing edges. The use of the prefix tree can be divided in two parts: (1) the tree *construction* and (2) the *search* for matching [14]. The *generation* of a trie is a preprocessing phase (Fig. 11.13), where the pattern P belonging to the set S to look for in the text T should form a prefix tree. The leading idea is that the more the branches among the patterns that share a common prefix will be away from the root, the more the common prefix will be long. In greater detail, the construction of the prefix tree is structured as follows: firstly, the root vertex (not labeled) is created, and then, each pattern P_i ($\in S$) is added to the tree. Starting from the root node, the path labeled with the character of the pattern P_i is followed. Thus, if the path ends before P_i, a new edge and a new vertex for each remaining character of P_i are added, labeling the new edge with the sum of characters. Lastly, the P_i identifier i in the last node of the path is stored.

The *search* phase (Fig. 11.14) is performed by scrolling the text along the tree, becoming aware whether one or more patterns belong to the latter. Along the prefix tree, we have some vertices that represent the end of a specific pattern, labeled with the pattern identifier. Once one of those vertices have been reached during the scrolling, it is sufficient to find an occurrence in the text of the pattern to which such a vertex belongs. Finally, the computation of the first phase of tree generation requires a complexity equal to $O(\mid S \mid)$, while the second phase can be performed with a simple algorithm of complexity equal to $O(\mid S \mid \cdot m)$, where m is the length of the pattern P. This data structure is particularly suited for NGS to align efficiently the same substring to a reference because this operation is performed once (identical substrings are represented by a single path in the trie) [29].

A widespread NGS analysis technique based on prefix trees is the *Ferragina* and *Manzini* index (FM-index) [29], which allows a constant time search phase for locating exact matches. The index is memory efficient and used for reference genome indexing.

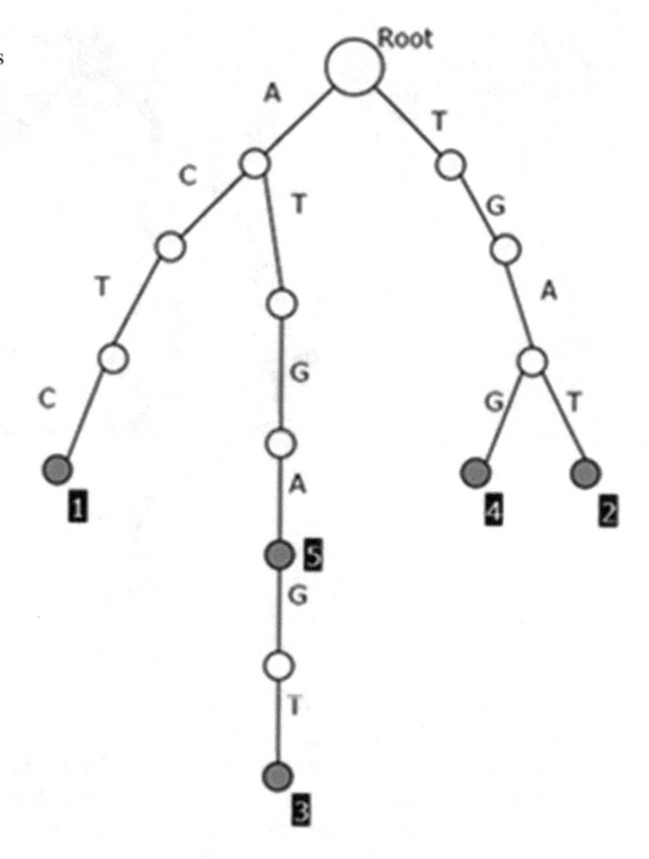

Fig. 11.13 The Prefix tree: Generation of the trie. In this example the set $S = \{P_1, P_2, P_3, P_4, P_5\} = \{ACTC, TGAT, ATGAGT, TGAG, ATGA\}$

Suffix Tree

Instead of the prefix tree, a *suffix tree* is a data structure that does not stem from the set of patterns S but from the text T; the tree generation is based on all the suffixes of T. This solution, unlike the previous one can be used not only in order to look for multiple patterns in a text T but even to find all occurrences of the single pattern P in T [51]. A suffix tree has leaves numbered from 1 to n, where n is the size of the text T. This leaf-associated value corresponds to the position in the text from which the suffix starts; the suffix is obtained by concatenating all the labels from the root to such a leaf. As a prefix tree, the use of suffix trees requires two phases: the tree *construction* and the *search* for matches. As mentioned, the *construction* of a suffix tree (Fig. 11.15) takes place starting from all the suffixes of a text T. This phase requires a time complexity equal to $O(n)$. Once the tree is generated, the *search* for patterns (Fig. 11.16) has to be performed as follows. The tree is scrolled for each pattern P, and either when a leaf has been reached before the end of P or when a path not belonging to P occurs, there are no occurrences of P in T [32]. Otherwise, all the sub-tree leaves of the path with P labels are the points of T, in

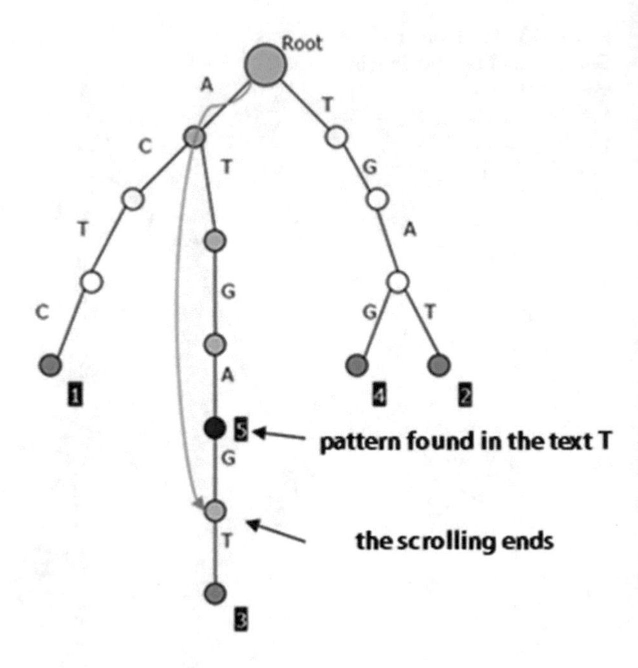

Fig. 11.14 The Prefix tree: Search for the patterns matching. In this example, the text $T = ATGAGCATGA$

which there is an occurrence of P in T. This phase requires a time complexity equal to $O(m)$, where m is the length of patterns [48].

In NGS analysis, suffix trees are often used for detecting and editing errors that occur in the reads because they are able to extract all their substring frequencies and build an index with them. This indexing allows the detection and correction of those erroneous reads that cannot be aligned through the identification of the under-represented branches (suffixes) [39].

11.2.4 Approximate Pattern Matching

Especially when searching for patterns in NGS sequences, an exact pattern matching is very hard to find and often not so useful to be investigated due to its complex structure. For this reason, an approximate solution for the pattern matching problem has been proposed.

In a pattern matching approximation, the issue is to seek whether a text subsequence is similar to the pattern searched for and, if the similarity overcomes a certain threshold, the search result is considered. Given a set of s sequences on the alphabet $\Sigma = \{A, C, G, T\}$, the approximate pattern matching aims to find all (M, k) patterns, i.e., patterns of length M that occur with at most k mismatches in at least q sequences of the set ($q \leq s$). In particular, a *mismatch* between the text T and the pattern P occurs for a given displacement of P, if one or more characters of P are different from the corresponding characters of T. The first simple

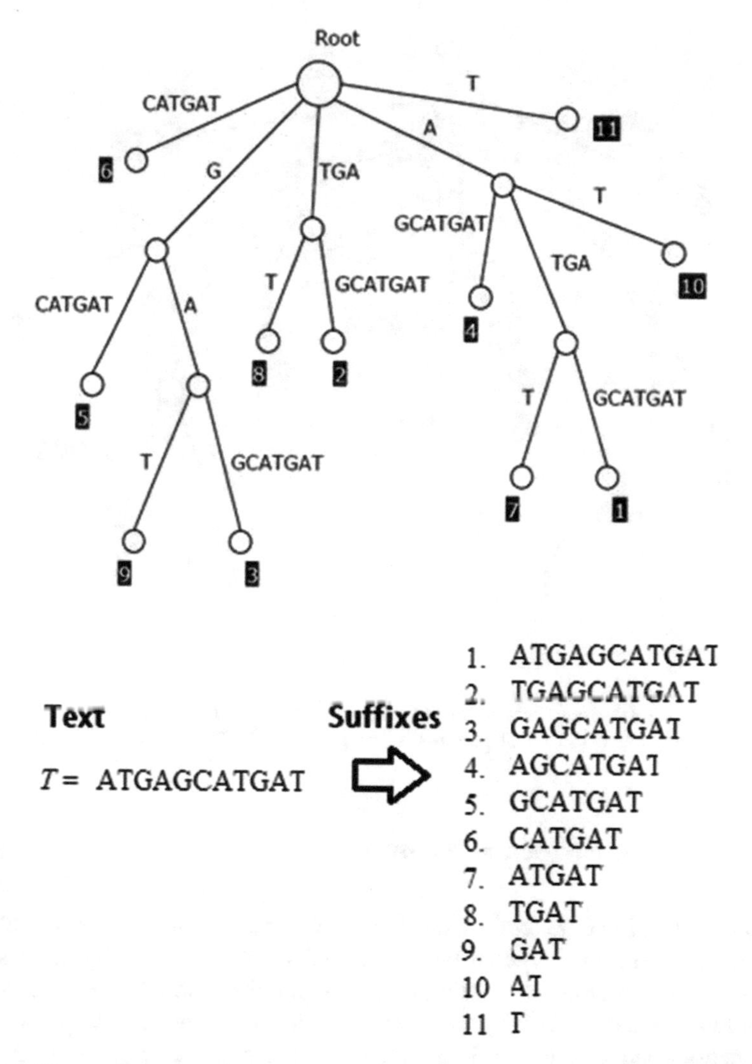

Fig. 11.15 The Suffix tree: Generation of the tree. In this example, the text $T = \{ATGAGCATGAT\}$

solution is a naive algorithm, which tries to find the pattern P in the text T with less than k errors, where k is equal to the highest number of allowed mismatches. The algorithm requires as input a pattern $P[1 \ldots m]$, a text $T[1 \ldots n]$, and the maximum number of mismatches k. The output of the algorithm is given by all positions $1 < i < (m-n+1)$, such that $T[i \cdots i+m]$ and $P[1 \cdots m]$ have at most k mismatches. Thus, the algorithm gives as output the set of valid shifts of the pattern P in the text T. This algorithm has a time complexity equal to $O(nm)$, where n and m are the T and P lengths, respectively.

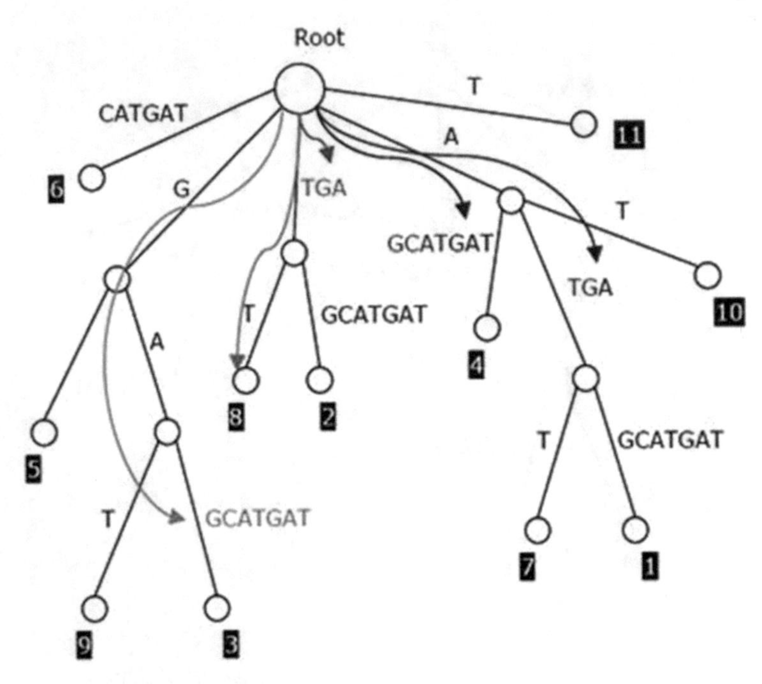

Fig. 11.16 The Suffix tree: Search for the patterns matching. In this example, the text $T = \{ATGAGCATGAT\}$ and the set of patterns $S = \{ATG, AGCATA, TGA, GAGCATGAT, TGATA\}$

11.3 Pairwise Sequence Alignment

The similarity of two genomic sequences of different organisms that are obtained by the NGS assembly process can be explained by the fact that they share a common ancestral DNA. According to this assumption, mutations occurred during the evolution, causing differences among families of contemporary species. Major modifications are due to local mutations, which specifically modify the DNA sequence. Such local changes between nucleotide sequences (or more generally between strings over an arbitrary alphabet) can be the following ones: (1) *insertion*, i.e., to insert one or more bases (letters) into the sequence; (2) *deletion*, i.e., to remove a base (or more) from the sequence; and (3) *substitution*, i.e., to replace a sequence base by another. Therefore, the analysis of DNA sequences can be approached in two different ways: to compute either a measure of distance or a measure of similarity. Following the concept of mutation events and, hence, of assigning weights to each mutation, the *distance* between two sequences is defined as the minimal sum of weights for a set of mutations that transform one into the other one. Following the theory of one ancestral ancient DNA and, hence, of assigning weights corresponding to the similarity, the *similarity* between two sequences is defined as the maximal sum of such weights [43].

```
tct---gtagaaggctcctaacgatg sequence #1
MMMIIIMMMMRMMMMMMMMDDDMMMM edit transcript
tctcctgtaggaggctcct---gatg sequence #2
```

Fig. 11.17 An example of Edit distance computation between two sequences

In the following, we provide some definitions needed to understand the alignment process.

The *edit distance* between two sequences is defined as the minimal number of *edit operations* needed to transform one sequence into the other one. The allowed edit operations can be I (insertion), D (deletion or removal), R (replacement or substitution), and M (match or correspondence or equality). An example is shown in Fig. 11.17. A string over an alphabet $\{I, D, R, M\}$ that describes a transformation from one string to another one is called *edit transcript* of the two strings. The *edit distance problem* is to compute the edit distance between two given strings such that the edit transcript is minimal.

The term *indel* means insertion or deletion which are usually indistinguishable. These two events come from evolution and often concerns more than one nucleotide, but it is still one event. Insertions and deletions are needed to accurately align very similar sequences.

The pairwise alignment is defined as the problem of comparing two sequences by allowing some matches between them [9, 19]. An alignment of two sequences (S and T) is obtained firstly by inserting the chosen spaces either into, at the ends of, or before S and T and then by placing the two resulting sequences one above the other, so that every character (or space (–)) in the sequence is opposite to a unique character (or a unique space) of the other one. A *match* is composed of two opposing identical characters, whereas a *mismatch* is composed of two opposing non-identical characters. A score (weight) is given to each two-character alignment and to – space character alignment.

The most widespread algorithms for pairwise sequence alignment in NGS (e.g., RMAP, SeqMap, MAQ, SHRiMP2, BWA, SOAP, SOAP2, Bowtie, SSAHA2) are evaluated and compared in [29, 41].

11.3.1 Global Alignment

A *global alignment problem* is defined as finding the best end-to-end alignment between two sequences. In greater detail, given two sequences $S = S_1, S_2, \ldots, S_m$ and $T = T_1, T_2, \ldots, T_n$ on the alphabet Σ, a *global alignment* of S and T consists of a pair of sequences $S' = S'_1, S'_2, \ldots, S'_k$ and $T = T'_1, T'_2, \ldots, T'_k$ on the alphabet $\sum \cup \{-\}$ ($-$ space character), which satisfy the following conditions:

1. $| S' |=| T' |= k$ where $\max(m, n) \leq k \leq (m + n)$.
2. S is obtained by removing all the spaces from S'.
3. T is obtained by removing all the spaces from T'.
4. If $S'_i = -$, then $T'_i \neq -$, and viceversa.

Therefore, a global alignment requires as input two sequences (S and T) of roughly the same length, and it investigates what is the maximum similarity between them, in order to find the best alignment. Unlike an edit distance, it aims to maximize the similarity instead of minimizing the distance. Moreover, it can be defined as a *weighted alignment*: a *score matrix* assigns to each pair of characters (a, b) a score d, representing the cost (or benefit) of replacing a with b. The aim of the alignment is, hence, finding an alignment (S', T') between S and T, whose score $A = \sum_{i=1}^{l} d(S'(i), T'(i))$ is *maximum*. In the following subsection, we will describe the widespread Needleman–Wunsch algorithm that is based on dynamic programming [35].

11.3.1.1 Needleman–Wunsch Algorithm

The *Needleman–Wunsch algorithm* [35] performs a global alignment on two sequences. Let's consider two sequences S and T of length m and n ($S = S_1, S_2, \ldots, S_m$ and $T = T_1, T_2, \ldots, T_n$) and a matrix $A(i,j)$ equal to the cost of the alignment between the prefix $S_1, S_2, \ldots, S_i$ and $T_1, T_2, \ldots, T_j$. Then, there exist three possibilities:

1. The character S_i is aligned with the character T_j, and hence, $A(i,j) = A(i-1, j-1) + d(S_i, T_j)$.
2. The character S_i is aligned with a space, and hence, $A(i,j) = A(i-1,j) + d(S_i, -)$.
3. The character $T_j$is aligned with a space, and hence, $A(i,j) = A(i,j-1) + d(-, T_j)$.

Searching for the maximum value, the recurrence that establishes a link between the generic subproblem $A[i,j]$ and subproblems $A[i-1, j-1], A[i-1, j]$ and $A[i, j-1]$, is the following one (*similarity score*):

$$A[i,j] = \max \begin{cases} A[i-1,j] + d(S_i, -), & \text{deletion} \\ A[i,j-1] + d(-, T_j), & \text{insertion} \\ A[i-1,j-1] + d[S_i, T_j], & \text{match or mismatch} \end{cases} \tag{11.9}$$

The initialization is the following:

$$\begin{cases} A[0,0] = 0 \\ A[i,0] = \sum_{k=1}^{i} d(s_k, -), & 0 < i < n \\ A[0,j] = \sum_{k=1}^{i} d(-, t_k), & 0 < j < m \end{cases} \tag{11.10}$$

The cost of the alignment is $A[m, n]$ and the complexity of such an algorithm is $O(mn)$, where m and n are the S and T lengths, respectively.

11.3.2 Local Alignment

Instead of the global alignment where the whole sequence must be aligned, a *local alignment* consists of aligning only parts of the analyzed sequence. A local alignment requires as input two sequences S and T, and it investigates what is the maximum similarity between a subsequence of S and a subsequence of T, in order to find the most similar subsequences. For example, this approach can be useful either to investigate unknown parts of the DNA or to find structural shared subunits among different proteins. To compute a local alignment, there are different techniques, such as the Smith–Waterman algorithm [42], the dotplot analysis (Dotter)[45], or heuristic approaches, like FASTA (based on Smith–Waterman algorithm), BLAST [7], and Bowtie [26, 27]. The latter is widely used in NGS analysis, particularly for read mapping and the assembly process.

11.3.2.1 Smith–Waterman Algorithm

The *Smith–Waterman algorithm* [42] performs a local pairwise sequences alignment. Given two sequences S and T (of length m and n), a matrix A of size $(m + 1) \cdot (n + 1)$ is built up, where $A[i, j]$ is the cost of an alignment between the suffix (possibly empty) $S[1 \cdots i]$ and the suffix $T[1 \cdots j]$. Then, it initializes the first row (row 0) and the first column (column 0) of A to zero. Finally, the recurrence that stems from the global alignment with the addition of 0 as the minimum value is the following:

$$A[i,j] = \max \begin{cases} 0 & \\ A[i-1,j] + d(S_i, -), & \text{deletion} \\ A[i,j-1] + d(-, T_j), & \text{insertion} \\ A[i-1,j-1] + d(S_i, T_j), & \text{match or substitution} \end{cases} \tag{11.11}$$

with $1 \leq i \leq m$, $1 \leq j \leq n$, and where:

$$d(S_i, T_j) = \begin{cases} \text{match value}, & \text{if } S_i = T_j \\ \text{mismatch value}, & \text{if } S_i \neq T_j \end{cases} \tag{11.12}$$

The Smith Waterman algorithm stops when a cell with zero value is reached and the final result is the one with the highest value within the entire matrix A [10].

11.3.2.2 Dotter

Dotter [45] is a "dotplot" program for a graphic comparison of two sequences. In particular, each base of a sequence is compared with respect to the bases of any other sequence. The value of a given base is a local value and also depends on bases around it. Moreover, a window flows along the two sequences to be aligned. The program computes a matrix whose points represent the value of homology for the pair bases with those coordinates. The homologous regions can be recognized by diagonal lines (when comparing a sequence against itself, the main diagonal alignment will have a 100%). *Dotter* is useful to graphically investigate the homology between two sequences to identify repeated regions in a single sequence. However, it is not able to align nucleotides against amino acids or vice versa. In conclusion, on one hand, *Dotter* is available for different architectures, it is visual and interactive, it allows observing the whole alignment, and it looks for repeated sequences. On the other hand, it is not automatized, it requires computing local resources, and it allows investigating few sequences at one time.

11.3.2.3 BLAST

Basic Local Alignment Search Tool (BLAST) [7] finds regions of similarity among sequences, comparing nucleotide or protein sequences to sequence databases and calculates the statistical significance of matches.

There exists also a version of BLAST called Mega-BLAST that uses a greedy algorithm for nucleotide sequence alignment search. This program is optimized for aligning sequences that differ slightly as a result of sequencing or other similar "errors." When a larger word size is used, it is up to ten times faster than more common sequence similarity programs. Mega-BLAST is also able to efficiently handle much longer DNA sequences than the traditional BLAST algorithm.

On one hand, BLAST is available online and does not require computing local resources and it is easy to read and interpret from a command line. Moreover, it is easily automatized. On the other hand, it is not interactive, it shows only the best homologies, and it is not a suitable tool to look for repeated sequences.

Furthermore, BLAST is used to search for homologies with proteins or known DNA sequences in huge databases, whereas Dotter is used to search for repeated regions or to search for homologies between few sequences. In conclusion, BLAST can be used in NGS for mapping a set of short query sequences against a long reference genome [29] or to assess the similarity among reads.

11.3.2.4 Bowtie

Bowtie [26, 27] is an ultra-fast and memory-efficient software, able to align short NGS DNA sequences (reads) to genomes. In particular, given a reference and a set of reads, it provides at least one good local alignment for each read if exists. It is

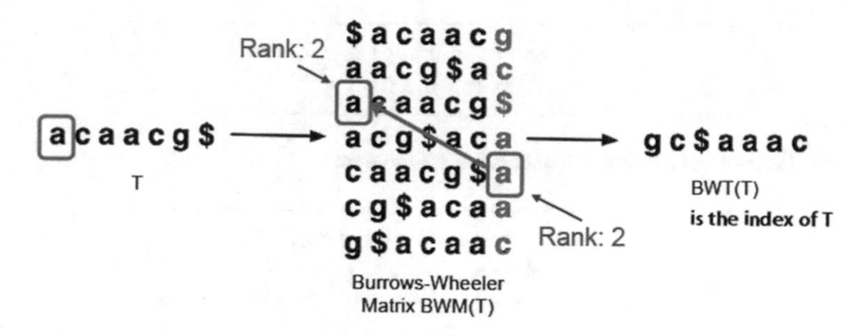

Fig. 11.18 An example of the BWT for the string T =acaacg. Starting from the suffix array or BWM, the BWT of the string T is the last column of the matrix, i.e., BWT(T)=*gc$aac*

worth noting that a "good alignment" implies an alignment with fewer mismatches, as well as the concept of "failing to align a low-quality base is better than failing to align an high-quality base."

Bowtie is based on *Burrows–Wheeler Transform* (BWT) [28] and makes use of suffix trees (described in section "Suffix Tree") to speed up the algorithm. BWT only finds exact matches; to map reads with differences, we must generate many alternate reads with different characters or indels at each position. Genomes and reads are too large for direct approaches like dynamic programming, and hence, Bowtie implements as indexing a reversible permutation of the characters in a text, known as BWT [28]. The BWT is based on the suffix array, also known as *Burrows–Wheeler Matrix* (BWM), i.e., an alphabetically ordered list of the suffixes of a string [43]. In particular, starting from an extended suffix array, the BWT represents the last column (see Fig. 11.18). It has the propriety that the k-th character of the given string in the first and last columns is the same character, which makes it easy to find the next character in a string by going backwards. The main advantages of BWT are that it requires small memory (1–3 GB) and that the same indexing can be used if the size of the reads changes. However, BWT only finds exact matches, and hence, in order to map reads with differences, it needs to generate many alternate reads with different characters or indels at each position [43].

11.3.3 Semi-global Alignment

Given as input two sequences S and T (possibly of different lengths), a *semi-global alignment* model aims to find a best alignment between the subsequences of S and T, when at least one of these subsequences is a prefix of the original sequence and one (not necessarily the other) is a suffix.

S = c a c t g t a c
T = g a c a c t t g

Fig. 11.19 Example of two input sequences for alignment

S = c a c - - t - g t a c
T = g a c a c t t g - - -

Fig. 11.20 Example of a global alignment for the input sequences S and T

S = - - c a c - t g t a c
T = g a c a c t t g - - -

Fig. 11.21 Example of a semi-global alignment for the input sequences S and T

In this variant of sequence alignment, any number of indel operations, i.e., insertions and deletions (either at the end or at the beginning of the alignment) contributes with a weight equal to zero. This drops the requirement that sequences have to start and end at the same place and allows us to align sequences that overlap or include another one. For example, let's consider the sequences shown in Fig. 11.19.

Assigning values of 2 for a match and -1 for an indel/substitution, the best *global alignment* will have a value of 1 and will be as shown in Fig. 11.20, while the semi-global alignment will have a value of 9 and will be as shown in Fig. 11.21. On one hand, it should be noticed that two subsequences coming from different parts of the original sequence will have a low *global alignment* score, as well as a low *semi-global alignment* score. On the other hand, two *overlapping subsequences* do not have the same starting position (nor the same end position) along the original sequence. Therefore, the *global alignment* score is lower than the *semi-global alignment* score because the two subsequences have an overlapping section.

11.3.4 Alignment with Gap Penalty

First of all, we define a *gap* as the maximal contiguous run of spaces in a single sequence within a given alignment. The *length* of a gap is the number of inserted or deleted subsequences. A *gap penalty function* is a function that measures the cost of a gap as a (non-linear) function of its length. Given as input two sequences S and T, it aims to find a best alignment between them by using the gap penalty function.

Gaps aid to build alignments that more closely fit biological models. The idea is to deal with a gap as a whole, rather than to give the same weight to each space. The concept of gap is relevant in many biological applications. In fact, insertions or deletions of an entire subsequence often occur as a single mutational event, which can create gaps of varying sizes. Therefore, when we try to align two sequences of DNA, a gap is needed to avoid assigning high cost to those mutations.

11.4 Multiple Sequence Alignment

The *Multiple Sequence Alignment* (MSA) problem is the process of taking three or more input sequences and forcing them to have the same length by inserting a universal gap symbol, in order to maximize their similarity as measured by a score function [50].

An optimal MSA can be useful to provide information on the evolutionary history of the species. Moreover, a local one can reveal conserved regions, which can be key functional regions, as well as priority targets for drug development.

A MSA of strings $S_1, S_2, \ldots, S_k$ is a series of strings with spaces $S = S'_1, S'_2, \ldots, S'_n$, such that: $| S'_1 |=| S'_2 |= \cdots =| S'_n |$ and S'_j is an extension of S_j, obtained by insertions of spaces. For example, a MSA for the three following sequences *ACBCDB*, *CADBD*, *ACABCD* is shown in Fig. 11.22. There are many available software packages to perform a MSA, e.g., CLUSTAL W [47], based on the Feng–Doolittle algorithm, which aligns sequences according to the order fixed by the phylogenetic tree, and *DIAlign* [34], which identifies the diagonals of the sequences, i.e., the subsequences without spaces, and builds the alignment starting from them. However, all the problems of multiple alignment are NP-hard [8, 20, 50], and hence, we need to find algorithms that provide good alignment with time-efficiency. The evaluation of a multiple alignment is based on a *scoring function*; the aim is to reach one of the alignments of maximum score. The rating of an alignment S' of S is given by the sum of the scores associated with all the columns of S'. The score is computed as explained in Sect. 11.3, extending the computation from pairwise to multiple alignment.

In NGS analysis, MSA is used to align and map the reads to the reference genome. However, sequencing errors, variations, and polymorphisms are present in the reference genome, and hence, a right choice of the MSA algorithm is crucial to reach effective results.

```
AC--BCDB
-CADB-D-
ACA-BCD-
```

Fig. 11.22 Example of a MSA with three input sequences (*ACBCDB*, *CADBD*, *ACABCD*)

11.5 NGS Read Mapping

The problem of read mapping is defined in the following way [43]. Given 100 million reads (or more) from a NGS experiment, for each find the genomic coordinates, chromosome, and first base, where it has the best match in a reference genome, either with the forward or reverse strand. The best match means zero or a small number of differences with the reference genome, which include mismatches and indels. Finally, determine if it has multiple matches or none at all.

An example of read mapping is drawn in Fig. 11.23a [43]. Structural variants are any rearrangements of the genome relative to a reference. They include:

1. Insertions/deletions
2. Inversions
3. Translocations
4. Tandem repeat variations

Examples of read mapping with insertions and deletions are sketched in Fig. 11.23b, c. Several mapping algorithms are available, i.e., *Bowtie*, BWA, SOAP2, SSAHA2, *Mosaik*, BFAST, *Zoom*, *SliderII*, and [41]. These algorithms can perform an indexing of the genome by using short words (seeds), which have a bounded expectation for failed mapping, or by using longer words (*seeds*), which guarantee a mapping with bounded number of errors. The methods for similarity detection involve some form of scanning the input sequences, usually with a window of fixed size (also called *seed*). Information about the contents of the window is stored, which is called *indexing*. The index is a list of all possible window contents together with a list, for each content, of where it occurs.

Alternatively, these algorithms can implement the BWT [28] of the genome, in order to obtain an exact matching or generating many alternate reads with different characters or indels at each position to map reads with differences.

11.6 Alignment-Free Algorithms

In this section, we focus on alignment-free techniques that have been proven to be effective in NGS sequence analysis [37, 49]. These techniques can be classified into two main groups: methods based on sequence compression and methods that rely on subsequence (oligomers) frequencies [49]. The aim of the methods belonging to the first group is to find the shortest possible description of the sequence. They compute the similarity of the sequences by analyzing their compressed representations. Current methods are based on the Kolmogorov complexity [30] and on *Universal Sequence Maps* [6]. An extensive review can be found in [15].

The methods based on oligomer frequencies rely on the computation of the substring frequencies of a given length k in the original sequences, called k-mers. Such k-mers represent the motifs of the NGS sequences or reads. Here, the similarity

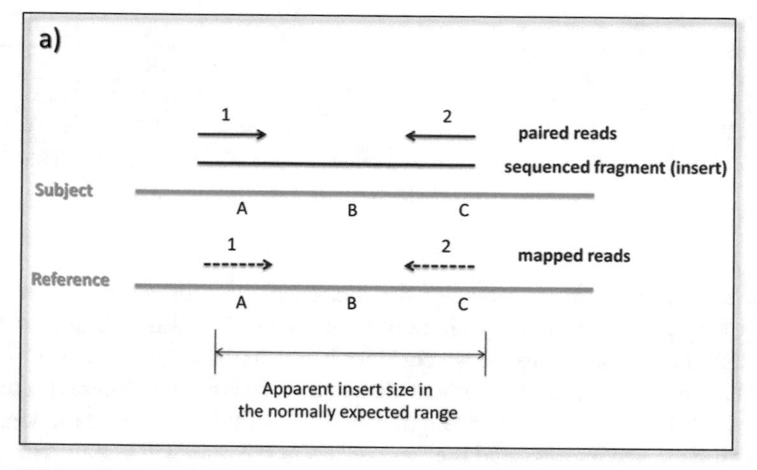

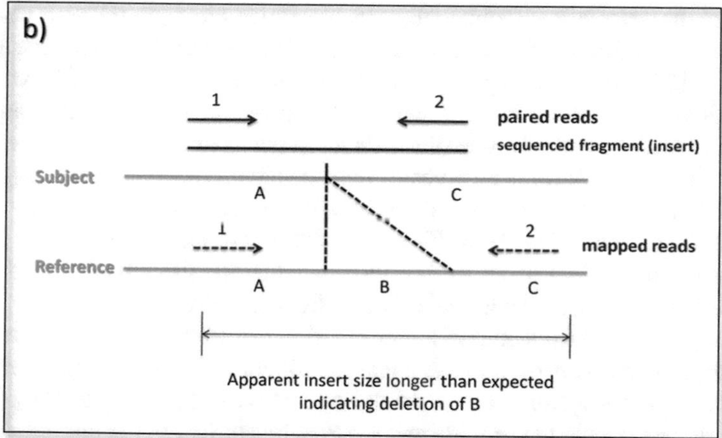

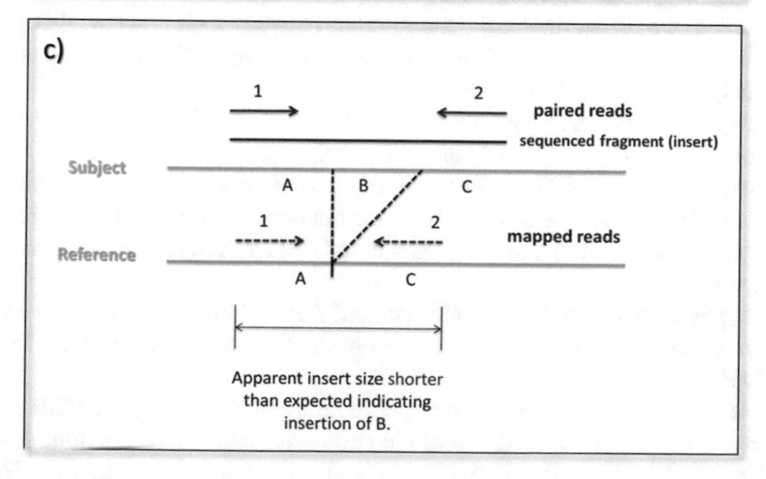

Fig. 11.23 Read mapping: (**a**) normal case, (**b**) case with deletion, (**c**) case with insertion [43]

Table 11.1 This table reports the Feature Vector (FeV) and Frequency Vector (FrV) of the sequence *AACGTAAC*

FeV		FrV	
AAC	2	AAC	0.333
ACG	1	ACG	0.166
CGT	1	CGT	0.166
GTA	1	GTA	0.166
TAA	1	TAA	0.166

of two sequences is based only on the dictionary of subsequences that appear in the strings, irrespective of their relative position [15]. An example of motifs computation is the following: if we consider $k = 3$ and the sequence *AACGTAAC*, the feature and frequency vector are depicted in Table 11.1. These vectors are obtained by sliding a window of length k (e.g., 3) on the considered sequence and by counting the occurrences and the frequencies of the obtained k-mers. A numeric coding of the sequences is obtained, which enables the analysis with statistical, mathematical, and computer science techniques. We cite the alignment-free distance adopted in [54], which is inspired to the k-mer frequency analysis [25], where the frequencies of the k-mers are represented in a real vector, and hence, they are easily tractable in a mathematical space: the distance between two reads is obtained by the distance between their frequency vector representations. A simple and easy way to compute a distance measure is the Euclidean distance, although others may be used (e.g., the $d2$ distance of [17]). In [54], the reliability of an alignment-free distance for read pairs similarity is evaluated and compared with respect to other read-to-read distances that are based on a global or local alignment of the two reads.

A straightforward implementation of the alignment-free distance **AF**, based on the Euclidean distance of the frequency distribution of k-mers, i.e., substrings composed of k consecutive bases, in the two reads is available in [54]. Such a distance is very simple to compute and requires linear time in the dimension of the reads. Given two reads, the Euclidean distance between their associated frequency vectors is an inverse measure of the similarity of the two reads and is defined as the AF distance between the two reads.

In [54], we described and tested a method to compare NGS DNA reads based on an alignment-free distance. The alignment-free distance outperforms alignment-based methods both in terms of computational time and of prediction performance and can be used effectively for read pairs comparison to derive more efficient sequence similarity assessment methods for DNA reads obtained from NGS sequencing.

Read pair comparison based on alignment-free distances may be conveniently used for DNA assembly [2] given its considerable speed, as well as for read classification [44], e.g., in metagenomics.

Another promising way to perform and detect common motifs in NGS reads is to combine their feature vector representation with supervised machine learning methods for classifying them [52, 53]. We cite the LAF technique [37, 52, 53], which uses feature vector representation and rule-based classification methods able

Table 11.2 An example of if-then rules (motifs) to classify bacterial genomes

A. baumannii	$f(\text{GTAC}) \geq 229.10 \wedge f(\text{TGCA}) \geq 515.63$
B. cereus	$384.04 \leq f(\text{CTCA}) < 490.11 \wedge 819.04 \leq f(\text{TCCA}) < 875.80$
B. animalis	$762.28 \leq f(\text{TCCA}) < 819.04 \wedge 469.35 \leq f(\text{TGCA}) < 515.63$
B. longum	$f(\text{GTAC}) \geq 229.10 \wedge 330.52 \leq f(\text{TGCA}) < 376.80$
B. aphidicola	$57.77 \leq f(\text{AGGC}) < 182.81$
C. jejuni	$490.11 \leq f(\text{CTCA}) < 596.17 \wedge 353.97 \leq f(\text{CTGA}) < 451.85$
C. trachomatis	$305.55 \leq f(\text{GGAC}) < 393.10 \wedge 875.80 \leq f(\text{TCCA}) < 932.56$
C. botulinum	$371.77 \leq f(\text{ACTC}) < 434.37 \wedge 112.00 \leq f(\text{GCAC}) < 261.71$
C. diphtheriae	$819.04 \leq f(\text{TCCA}) < 875.80 \wedge 423.07 \leq f(\text{TGCA}) < 469.35$
C. pseudotuberculosis	$875.80 \leq f(\text{TCCA}) < 932.56 \wedge 423.07 \leq f(\text{TGCA}) < 469.35$
E. coli	$710.86 \leq f(\text{GCAC}) < 860.58 \wedge 415.84 \leq f(\text{GCTA}) < 525.98$
F. tularensis	$592.00 \leq f(\text{TCCA}) < 648.76 \wedge 330.52 \leq f(\text{TGCA}) < 376.80$
H. influenzae	$549.73 \leq f(\text{CTGA}) < 647.60 \wedge 130.47 \leq f(\text{GGAC}) < 218.01$
H. pylori	$5.56 \leq f(\text{GTAC}) < 42.82$
L. monocytogenes	$411.43 \leq f(\text{GCAC}) < 561.15 \wedge 305.55 \leq f(\text{GGAC}) < 393.10$
M. tuberculosis	$649.71 \leq f(\text{ATCA}) < 772.78$
N. meningitidis	$590.29 \leq f(\text{GATA}) < 754.27 \wedge 376.80 \leq f(\text{TGCA}) < 423.07$
S. enterica	$525.98 \leq f(\text{GCTA}) < 636.11 \wedge 393.10 \leq f(\text{GGAC}) < 480.64$
S. aureus	$1082.23 \leq f(\text{GATA}) < 1246.22 \wedge f(\text{GTAC}) \geq 229.10$
S. pneumoniae	$393.10 \leq f(\text{GGAC}) < 480.64 \wedge 154.58 \leq f(\text{GTAC}) < 191.84$
S. pyogenes	$596.06 \leq f(\text{AGTA}) < 733.86 \wedge 1082.23 \leq f(\text{GATA}) < 1246.22$
S. suis	$918.25 \leq f(\text{GATA}) < 1082.23 \wedge 330.52 \leq f(\text{TGCA}) < 376.80$
S. islandicus	$218.01 \leq f(\text{GGAC}) < 305.55 \wedge 284.24 \leq f(\text{TGCA}) < 330.52$
Y. pestis	$596.17 \leq f(\text{CTCA}) < 702.24 \wedge f(\text{CTGA}) \geq 941.24$

$f(W)$ represents the relative frequency of substring W in a set of reads, multiplied by 10^5 for readability [53]

to extract a human readable model which contains the identified motifs of the NGS sequences. An example is drawn in Table 11.2, where if-then rules (motifs) to classify bacterial genomes starting from their reads are depicted.

11.7 Conclusions

Thanks to the new advances in NGS technologies, a large quantity of genomic sequences are produced daily by sequencing machines. In this chapter, efficient methods and algorithms for performing NGS sequence analysis have been illustrated and described. In particular, we focused on three types of analysis: *string-matching*, *sequence alignment*, and *alignment-free* techniques for NGS reads comparison. These methods and algorithms are effectively used for detecting patterns and motifs in NGS reads, which are the output of sequencing machines, as well as for whole genomes analysis. It is worth noting that many NGS analyses take advantage of sequence alignment, e.g., read mapping, genome assembly, and reads comparison. Most of the alignment algorithms must deal with short-read alignment against a reference sequence, i.e., genome. Furthermore, new sequencers are allowing to increase read length; thus, the design of novel efficient alignment algorithms for long reads is expected in the near future.

References

1. 454 - roche. http://www.454.com/
2. Dazzler assembler for pacbio reads. http://www.homolog.us/blogs/blog/2014/02/14/dazzle-assembler-pacbio-reads-gene-myers/
3. Illumina. http://www.illumina.com/
4. Ion torrent. https://www.thermofisher.com/it/en/home/brands/ion-torrent.html
5. Pacific bioscience. http://www.pacb.com/
6. Almeida, J.S., Vinga, S.: Universal sequence map (USM) of arbitrary discrete sequences. BMC Bioinf. **3**, 6 (2002)
7. Altschul, S.F., Gish, W., Miller, W., Myers, E.W., Lipman, D.J.: Basic local alignment search tool. J. Mol. Biol. **215**(3), 403–410 (1990)
8. Arora, S., Karger, D., Karpinski, M.: Polynomial time approximation schemes for dense instances of NP-hard problems. In: Proceedings of the Twenty-Seventh Annual ACM Symposium on Theory of Computing, pp. 284–293. ACM, New York (1995)
9. Bergeron, B.P.: Bioinformatics Computing. Prentice Hall Professional, Englewood Cliffs (2003)
10. Blazewicz, J., Frohmberg, W., Kierzynka, M., Pesch, E., Wojciechowski, P.: Protein alignment algorithms with an efficient backtracking routine on multiple GPUs. BMC Bioinf. **12**(1), 181 (2011)
11. Boyer, R.S., Moore, J.S.: A fast string searching algorithm. Commun. ACM **20**(10), 762–772 (1977)
12. Cormen, T.H., Leiserson, C.E., Rivest, R.L., Stein, C., et al.: Introduction to Algorithms, vol. 2. MIT Press, Cambridge (2001)
13. Eisenstein, M.: The battle for sequencing supremacy. Nat. Biotechnol. **30**(11), 1023–1026 (2012)
14. Fredkin, E.: Trie memory. Commun. ACM **3**(9), 490–499 (1960)
15. Giancarlo, R., Scaturro, D., Utro, F.: Textual data compression in computational biology: synopsis. Bioinformatics **25**(13), 1575–1586 (2009)
16. Gusfield, D.: Algorithms on Strings, Trees and Sequences: Computer Science and Computational Biology. Cambridge University Press, Cambridge (1997)
17. Hide, W., Burke, J., da Vison, D.B.: Biological evaluation of d2, an algorithm for high-performance sequence comparison. J. Comput. Biol. **1**(3), 199–215 (1994)
18. Hume, A., Sunday, D.: Fast string searching. Softw.: Pract. Exp. **21**(11), 1221–1248 (1991)
19. Jones, N.C., Pevzner, P.: An Introduction to Bioinformatics Algorithms. MIT Press, Cambridge (2004)
20. Just, W.: Computational complexity of multiple sequence alignment with SP-score. J. Comput. Biol. **8**(6), 615–623 (2001)
21. Karp, R.M., Rabin, M.O.: Efficient randomized pattern-matching algorithms. IBM J. Res. Dev. **31**(2), 249–260 (1987)
22. Kleene, S.C.: Representation of events in nerve nets and finite automata. Tech. rep., DTIC Document (1951)
23. Knuth, D.E., Morris, J.H. Jr., Pratt, V.R.: Fast pattern matching in strings. SIAM J. Comput. **6**(2), 323–350 (1977)

24. Kozen, D.: A completeness theorem for kleene algebras and the algebra of regular events. Inf. Comput. **110**(2), 366–390 (1994)
25. Kuksa, P., Pavlovic, V.: Efficient alignment-free DNA barcode analytics. BMC Bioinf. **10**(Suppl. 14), S9 (2009). doi:10.1186/1471-2105-10-S14-S9. http://dx.doi.org/10.1186/1471-2105-10-S14-S9
26. Langmead, B., Salzberg, S.L.: Fast gapped-read alignment with bowtie 2. Nat. Methods **9**(4), 357–359 (2012)
27. Langmead, B., Trapnell, C., Pop, M., Salzberg, S.: Ultrafast and memory-efficient alignment of short DNA sequences to the human genome. Genome Biol. **10**(3), R25 (2009)
28. Li, H., Durbin, R.: Fast and accurate short read alignment with Burrows–Wheeler transform. Bioinformatics **25**(14), 1754–1760 (2009)
29. Li, H., Homer, N.: A survey of sequence alignment algorithms for next-generation sequencing. Brief. Bioinform. **11**(5), 473–483 (2010)
30. Li, M., Vitnyi, P.M.: An Introduction to Kolmogorov Complexity and Its Applications, 3rd edn. Springer Publishing Company, New York (2008)
31. Liu, L., Li, Y., Li, S., Hu, N., He, Y., Pong, R., Lin, D., Lu, L., Law, M.: Comparison of next-generation sequencing systems. J. Biomed. Biotechnol. **2012**, 251364 (2012). doi:10.1155/2012/251364
32. McCreight, E.M.: A space-economical suffix tree construction algorithm. J. ACM **23**(2), 262–272 (1976)
33. Metzker, M.L.: Sequencing technologies - the next generation. Nat. Rev. Genet. **11**(1), 31–46 (2010). doi:10.1038/nrg2626. http://dx.doi.org/10.1038/nrg2626
34. Morgenstern, B., Frech, K., Dress, A., Werner, T.: Dialign: finding local similarities by multiple sequence alignment. Bioinformatics **14**(3), 290–294 (1998)
35. Needleman, S.B., Wunsch, C.D.: A general method applicable to the search for similarities in the amino acid sequence of two proteins. J. Mol. Biol. **48**(3), 443–453 (1970)
36. Pevzner, P.: Computational Molecular Biology: An Algorithmic Approach. MIT Press, Cambridge (2000)
37. Polychronopoulos, D., Weitschek, E., Dimitrieva, S., Bucher, P., Felici, G., Almirantis, Y.: Classification of selectively constrained DNA elements using feature vectors and rule-based classifiers. Genomics **104**(2), 79–86 (2014)
38. Rivest, R.L.: Partial-match retrieval algorithms. SIAM J. Comput. **5**(1), 19–50 (1976)
39. Savel, D.M., LaFramboise, T., Grama, A., Koyutürk, M.: Suffix-tree based error correction of NGS reads using multiple manifestations of an error. In: Proceedings of the International Conference on Bioinformatics, Computational Biology and Biomedical Informatics, BCB'13, pp. 351:351–351:358. ACM, New York (2013). doi:10.1145/2506583.2506644. http://doi.acm.org/10.1145/2506583.2506644
40. Setubal, J.C., Meidanis, J.: Introduction to Computational Molecular Biology. PWS Publishing Company, Boston (1997)
41. Shang, J., Zhu, F., Vongsangnak, W., Tang, Y., Zhang, W., Shen, B.: Evaluation and comparison of multiple aligners for next-generation sequencing data analysis. Biomed. Res. Int. **2014**(309650) (2014)
42. Smith, T.F., Waterman, M.S.: Identification of common molecular subsequences. J. Mol. Biol. **147**(1), 195–197 (1981)
43. Sokol, D., Benson, G., Tojeira, J.: Tandem repeats over the edit distance. Bioinformatics **23**(2), e30–e35 (2007)
44. Song, K., Ren, J., Zhai, Z., Liu, X., Deng, M., Sun, F.: Alignment-free sequence comparison based on next generation sequencing reads. J. Comput. Biol. **20**(2), 64–79 (2013)
45. Sonnhammer, E.L., Durbin, R.: A dot-matrix program with dynamic threshold control suited for genomic DNA and protein sequence analysis. Gene **167**(1), GC1–GC10 (1995)
46. Stephen, G.A.: String Searching Algorithms. World Scientific Publishing Company, Singapore (1994)

47. Thompson, J.D., Higgins, D.G., Gibson, T.J.: CLUSTAL W: improving the sensitivity of progressive multiple sequence alignment through sequence weighting, position-specific gap penalties and weight matrix choice. Nucleic Acids Res. **22**(22), 4673–4680 (1994)
48. Ukkonen, E.: On-line construction of suffix trees. Algorithmica **14**(3), 249–260 (1995)
49. Vinga, S., Almeida, J.: Alignment-free sequence comparison—a review. Bioinformatics **19**(4), 513–523 (2003)
50. Wang, L., Jiang, T.: On the complexity of multiple sequence alignment. J. Comput. Biol. **1**(4), 337–348 (1994)
51. Weiner, P.: Linear pattern matching algorithms. In: IEEE Conference Record of 14th Annual Symposium on Switching and Automata Theory 1973, SWAT'08, pp. 1–11. IEEE, New York (1973)
52. Weitschek, E., Cunial, F., Felici, G.: Classifying bacterial genomes on k-mer frequencies with compact logic formulas. In: Database and Expert Systems Applications (DEXA)- 25th International Workshop on Biological Knowledge Discovery, pp. 69–73. IEEE Computer Society, Washington (2014)
53. Weitschek, E., Cunial, F., Felici, G.: LAF: logic alignment free and its application to bacterial genomes classification. BioData Min. **8**(1), 1 (2015)
54. Weitschek, E., Santoni, D., Fiscon, G., De Cola, M.C., Bertolazzi, P., Felici, G.: Next generation sequencing reads comparison with an alignment-free distance. BMC. Res. Notes **7**(1), 869 (2014)

Part IV
Assembly of NGS Data

Chapter 12
The Contig Assembly Problem and Its Algorithmic Solutions

Géraldine Jean, Andreea Radulescu, and Irena Rusu

12.1 Introduction

DNA sequencing, assuming no prior knowledge on the target DNA fragment, may be roughly described as the succession of two steps. The first of them uses some sequencing technology to output, for a given DNA fragment (not necessarily a whole genome), a collection of possibly overlapping sequences (called *reads*) representing small parts of the initial DNA fragment. The second one aims at recovering the sequence of the entire DNA fragment by assembling the reads. The quality of the resulting DNA sequence (or *assembled sequence*) depends on the quality of the reads, on the specific information at hand, and on the assembly method. We briefly discuss each of them below. (Note that we do not consider here the assembly of DNA fragments from RNAseq or metagenomic experiments, which have an additional dimension due to the different gene sequences and genomes present in the sample.)

The quality of the reads is expressed in terms of read length, the rate and type of sequencing errors, and the average number of reads covering an arbitrary position in the target DNA fragment (the latter one is termed *coverage*). These parameters directly depend on the sequencing technology used to obtain the reads. The classical Sanger technology produces relatively long reads of sizes between 500 and 1000 bp, with a low error rate. The research for lower-cost sequencing methods led to the emergence in 2005 of the next-generation sequencing (NGS for short) platforms (see [1] for a description), producing billions of reads with a very low cost. However, cost and time reduction came with a modification of read length and quality. The length of the reads ranges from 35 to 800 bp, depending on the technology, thus allowing us to speak about *very short reads* (up to 50 bp), *short reads* (between 50

G. Jean (✉) • A. Radulescu • I. Rusu
LS2N, University of Nantes, Nantes, France
e-mail: Geraldine.Jean@univ-nantes.fr; Andreea.Radulescu@univ-nantes.fr; Irena.Rusu@univ-nantes.fr

M. Elloumi (ed.), *Algorithms for Next-Generation Sequencing Data*,
DOI 10.1007/978-3-319-59826-0_12

and 250 bp), and *longer reads* (more than 250 bp). In order to counterbalance the loss of information due to smaller fragments, a higher coverage of the sequenced DNA fragment is needed. Note that new high-throughput platforms appeared more recently, allowing to speak about third-generation sequencing. In this chapter, we assume NGS data concerns very short or short reads, whatever the platform which generated them.

In addition to the collection of reads, the assembly step may benefit from supplementary information specific to the DNA fragment to be assembled. Most NGS platforms propose sequencing protocols providing pairs of reads (called *paired reads*) for which an approximate distance (called *insert size*) separating them on the DNA fragment is known. Another important piece of information, when it is available, is a reference genome that is used as a template for the assembly, allowing reads to be mapped against it [48]. This type of assembly is termed *comparative assembly*. At the opposite, when the assembly of the reads uses no reference genome, we are in a de novo context.

Finally, the assembly method uses the information at hand and reads, as well as reference genomes when available, to construct an assembled sequence aiming at being as close as possible to the real one. Assembly is a very complex task, dealing with a variety of problems and requiring a wide range of skills. Consequently, an assembler is usually a multistage pipeline including as separate stages the *pre-assembly error correction* in reads, the *contig assembly* in which overlapping reads are grouped into larger sequences called contigs, and the *contig scaffolding* resulting into the assembled sequence, where contigs are ordered and separated by gaps representing the uncovered regions. Another important task, but not necessarily identified as a separate stage, is *repeat solving*, which constitutes one of the bottlenecks of assembly and requires to correctly locate on the assembled sequence the regions which have identical sequence but different locations. The performances of an assembler are therefore the result of an important number of methods the assembler implements.

Three main approaches are known for dealing with (very) short data, namely greedy, *Overlap-Layout-Consensus* (OLC), and de Bruijn methods. Whereas the former two of them, and especially the OLC methods, are widely (since successfully) used for computing assemblies with Sanger data, only the latter of them is able to reach good performances on very short or short data, and particularly when the size of the genomes is large. This is due to the very large amount of data to handle in the case of very short or short reads.

In this chapter, we propose a state-of-the-art of the algorithmic aspects presented by de novo whole-genome assembly using NGS data, focused on the crucial step of contig construction. Our aim here is threefold: firstly, to propose a tutorial allowing the reader, with some help from us, to make himself a methodology-based comparison of the assembly algorithms that is complementary to the numerous experimental comparisons of the assemblers [5, 16, 49, 61]; secondly, to point out the omnipresent heuristics, whose legitimacy is based on many true arguments (like erroneous data, huge data, NP-complete problems, not enough criteria to discriminate predictions between right or wrong, too many criteria to evaluate and so on) but which are still very limited from an algorithmic point of view; and thirdly, to support the idea that, since assembly is a collection of problems, better assemblers should result from a fine assembly of best solutions to each problem. It is worth noticing here that this study is focused on algorithmic issues and their performances, rather than on assemblers and their performances, meaning that we chose to present the principles of the assemblers and not to compare the assemblers. As a consequence, assemblers are mentioned in order to illustrate the algorithmic issues we present here but are certainly not limited to these issues (many of them are far too complex to be deeply described in a few pages).

The organization of the chapter is as follows. In Sect. 12.2 we give the main definitions, formulate the problems to solve, and fix the context of our tutorial. In Sects. 12.3–12.5, we present respectively the general features of the greedy, OLC, and de Bruijn approaches, whose different algorithmic developments are given. Section 12.6 contains the perspectives and the conclusion.

12.2 Setting the Context

In this section, we present the general assembly problem, discuss the role of contig assembly in solving it, and fix the context of our chapter.

Given a sequence σ over the alphabet $\Sigma = \{A, C, G, T\}$, where A, C, G, T respectively stand for the nucleotides adenine, cytosine, guanine, and thymine, its *length* h is the number of its characters. We write $|\sigma| = h$ and we also say that σ is a *h-mer*. Its i-th character is denoted $\sigma[i]$, whereas the subsequence of characters between the positions i and j is denoted $\sigma[i \ldots j]$. The *reverse complement* $\overline{\sigma}$ of σ is obtained by first reversing the order of its characters and then replacing every A by a T, every G by a C, every T by an A, and every C by a G. Furthermore, given $k \leq h$, we say that a k-mer x *occurs* in σ if there exist positions i and j such that $x = \sigma[i \ldots j]$. Equivalently, we also say that σ *contains* x. The *k-spectrum* of σ is the set $S_k(\sigma)$ of k-mers occurring into σ. A *read* is possibly any sequence of the alphabet $\Sigma = \{A, C, G, T\}$.

Then, the assembly problem, in its most general form, requires to reconstruct a DNA fragment given a set R of reads from this fragment, in the most accurate way possible. Depending on the modeling of the problem, a sequence S is considered an accurate reconstruction from R if it solves an optimization problem (see for instance, [36, 40]), that is, if it maximizes or minimizes a given *reconstruction score* $rec(S, R)$. In other words, the precise problem to solve depends on the modeling, but as a generality, the assembly problem may be stated as follows:

GENERAL ASSEMBLY PROBLEM

Input: A non-empty set of reads R over Σ, of length l each, resulting after sequencing a DNA fragment D; a reconstruction score *rec*

Requires: Find an *assembled sequence* S over Σ containing each read in R or the respective complement of each read such that $rec(S, R)$ is optimal.

Notice that the GENERAL ASSEMBLY PROBLEM allows any read or its complement (and even both) to be part of the assembled sequence since the DNA fragment has two complementary strands and the original strand of a read (i.e., its *orientation*) is not known. Moreover, the read is not required to have an exact copy in S (assemblers correct reads before including them in the assembled sequence). Also note that the assumption that all reads have the same length is not always true in practice but is not a real restriction and allows us to greatly simplify the presentation.

Assembly algorithms strongly rely on the overlaps between reads. Formally, overlaps are defined as follows. Let $s(x, y)$ be a similarity measure between reads (usually the Smith–Waterman similarity measure [56]), and call a *prefix* (respectively a *suffix*) *of length* j, with $j > 0$, of a read x the subsequence of x made of the first (respectively last) j characters of x. Denote, respectively, $Pref(x, j)$ and $Suff(x, j)$ the prefix and suffix of x of length j.

Definition 1 (Overlap) Let r_1 and r_2 be two reads, and t be a positive real value. Then r_1 *overlaps* r_2 (equivalently, there is an *overlap* between r_1 and r_2) with overlap weight t if there is a suffix sr_1 of r_1 and a prefix pr_2 of r_2 such that $s(sr_1, pr_2) = t$. When $sr_1 = pr_2$, the overlap is called *exact*. Otherwise, it is called *approximate*. We note $w(r_1, r_2) = \max\{t \mid r_1 \text{ overlaps } r_2 \text{ with weight } t\}$, the maximum overlap weight between r_1 and r_2, and $len(r_1, r_2) = \min\{|sr_1|, |pr_2|\}$, their overlap length.

The most natural approach in solving the GENERAL ASSEMBLY PROBLEM is to compare reads so as to identify overlaps between them and to infer that overlapping reads should be glued together in the assembled sequence.

Definition 2 (Extension) Let r_1 and r_2 be two reads of length l with an exact overlap of length p. Then, the *extension* of r_1 with r_2 is the sequence obtained by concatenating to r_1 the suffix of r_2 of length $l - p$.

When the overlap is approximate, an extension may also be defined, but it must be *context-dependent*, that is, it must take into account the other reads which overlap with the two initial reads. Due to many difficulties (that are discussed in the remainder of the chapter), and particularly repeats in the target DNA fragment, the sequences resulting from a series of extensions (called *contigs*) do not cover the entire genome. Therefore, the problem to solve is slightly different from the GENERAL ASSEMBLY PROBLEM formulated above. Here, a *length score* is a function evaluating the combined length of a set of sequences over Σ.

CONTIG ASSEMBLY PROBLEM

Input: A non-empty set of reads R over Σ of length l each, resulting after sequencing a DNA fragment D; a reconstruction score rec; and a length score ls

Requires: Find a set C of sequences over Σ, named *contigs*, such that $ls(C)$ is maximum and, for each sequence C_i in C, there exists a set R_i of reads from R such that $rec(C_i, R_i)$ is optimal.

Equivalently, each contig C_i from C is required to be an accurate reconstruction of a region from D, and the total length of the contigs is required to be maximum.

Contigs are then parts of the target sequence D, for which the order, the orientation, and the distances between them along the assembled sequence are computed during the scaffolding step, which finishes the construction of the assembled sequence.

The main source of ambiguity in finding the contigs comes from repeats, i.e., sequences having several identical copies in the target genome. In the set of reads, one loses the information about the location of each copy on the target genome and, moreover, each copy appears in several reads, either as a whole or only as fragments. Resolving repeats thus means correctly assembling the reads containing the same copy (or fragments of it), implying to correctly choose them among those containing other copies. This is a very hard task, especially when complicated repeats (i.e., included in each other, or much longer than the reads etc.) exist. Therefore, the treatment of repeats is essential in every assembler and has a major effect on the quality of the resulting assembled sequence.

The particular context we deal with in this chapter is as follows. We consider the de novo assembly problem in the context of whole-genome sequencing, i.e., when the DNA fragment to sequence is an entire genome and there is no reference genome. We focus on the CONTIG ASSEMBLY PROBLEM (or contig assembly, for short), but we present the existing algorithms within the general context of the assemblers implementing them. Therefore, the pre-assembly error correction procedure is detailed when such a correction exists since it is prior to contig assembly and thus has an effect on it, whereas the scaffolding stage is only mentioned in our algorithms.

As most assemblers do, we assume a priori that we have no information concerning paired reads at the contig assembly level. Such information, when available, is usually exploited at the scaffolding level, although some assemblers [2, 8, 37, 39] are able to use it at the contig assembly level. As the orientation of the reads is not known, the set R is assumed to contain both the reads obtained after sequencing and their reverse complements. As a consequence, the assembly is always performed symmetrically on both strands.

We distinguish three different approaches for contig assembly of NGS data (see Fig. 12.1). The greedy approach adopts a local contig extension procedure avoiding all-to-all comparisons between reads but whose performances are limited by the local choices. The OLC approach benefits from global information obtained by all-to-all comparisons between reads but needs important computer resources. The de Bruijn approach splits the reads into overlapping k-mers from which a so-called de Bruijn graph is built that captures the overlaps between k-mers but which is also very sensitive to errors in reads. Figure 12.1 also shows some hybrid approaches (combining different features, usually greedy and OLC) that we mention without giving details. While reading the following sections, the reader should be aware that whereas the contig assembly problem is the core of the assembly, an assembler is an implementation of the algorithm which adds many features impossible to describe here, many of them related much more to engineering than to algorithmics.

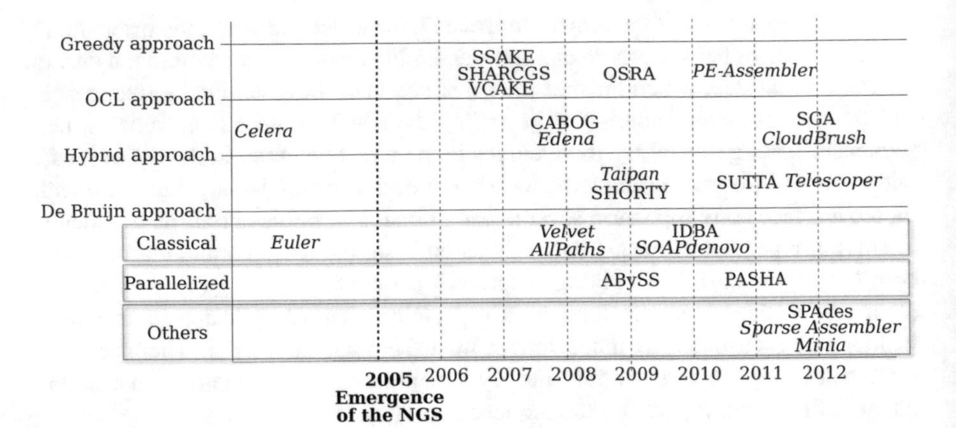

Fig. 12.1 Overview of the approaches and several examples of assemblers

12.3 The Greedy Approach

The greedy approach is the most intuitive method for building contigs using the overlaps between reads. Based on the idea that the larger the overlap between two reads, the stronger the confirmation that the two reads are overlapping in the target genome, the greedy approach successively makes local choices of the read which best extends a current contig (itself made of overlapping reads). See Fig. 12.2a, b.

12.3.1 General Pattern

As indicated in Sect. 12.2, we assume we are given a set R of l-length reads over the alphabet $\Sigma = \{A, C, G, T\}$, containing all the reads obtained after sequencing, and their reverse complements. An overlap between a contig c and a read e is defined similarly to an overlap between reads (see Definition 1) since a contig is made of overlapping reads. When such an overlap has been identified, the *extension* of c with e, according to this overlap, is the contig obtained by concatenating to c the suffix of e not overlapping c.

Algorithm 1 presents the general lines of a greedy algorithm, which always follows a *seed-and-extend* principle. It first computes a set S of *seeds*, which are reads or even contigs identified as interesting for extension. Then, the seeds are successively considered for extension, each of them defining an initial contig c. Notice that in Algorithm 1, once a seed c is used, both it and every read in it (as well as its reverse complement) are removed from all the sets they belonged to (here, S and/or R). This is done by the function $Remove(c, \overline{c}, R, S)$. Then, the greedy algorithm first considers the possible extensions of c to the right using the reads in R and successively extends c either by one read or, in some versions, by one character

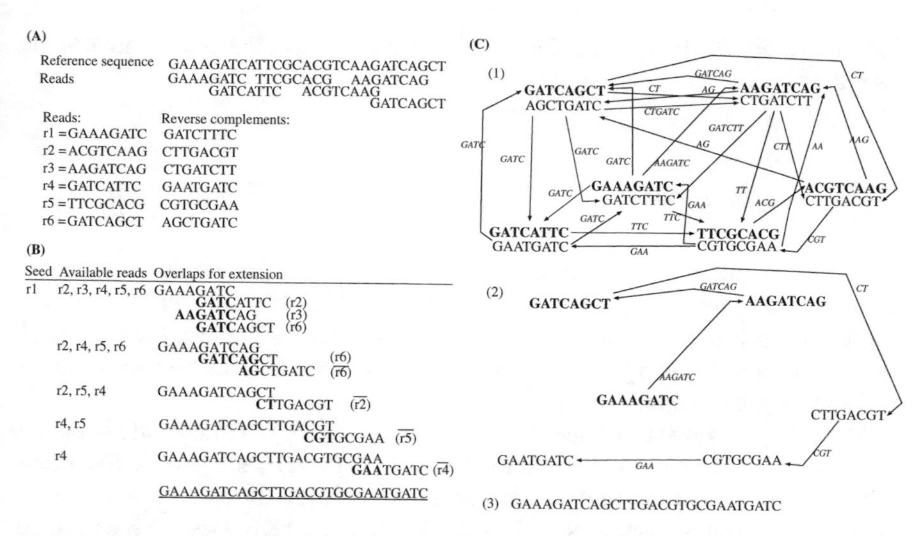

Fig. 12.2 Illustration of the greedy and OLC approaches. (**a**) The reads (here, assumed to occur on the same strand of the genome) and their complements, of length $l = 8$. (**b**) The greedy method, where a read is available if and only if its reverse complement is available. (**c**) The OLC method, where the reads only (and not their complements) are in *bold*. Here, the overlaps are exact, and—for each pair of reads—only the longest overlap of length at least 2 is represented. (**c.1**) The overlap graph built in the overlap step. (**c.2**) A path found in the layout step. (**c.3**) The consensus sequence corresponding to the path, computed in the consensus step, which is also—in this case—the assembled sequence

---

Algorithm 1 Generic greedy

Input: Set R of reads of length $l > 0$
Output: An assembled sequence

Step 1: Data preparation
  1.1: Frequency based sorting of reads in R
  1.2: Compute the seed set S and update R
Step 2: Build contigs by left-right seed extensions

  $C \leftarrow 0$;
  while $R \neq \emptyset$ **do**
    2.1: choose a seed c from S; $Remove(c, \overline{c}, R, S)$
    2.2: repeatedly extend c, then $\overline{c}$, to the right
    $C \leftarrow C \cup \{c\}$
  end while
Step 3: (optional) Scaffolding
  Use paired reads to solve repeats and connect contigs

---

at a time. Second, the algorithm searches in the same way the extensions to the left of c by taking the reverse complement $\overline{c}$ of c (not of R since it contains both a read and its complement). Once the two extension steps are finished, c is added to the set C of contigs. Scaffolding is the last step, meant to orient and order the contigs to deduce an assembled sequence. This step is often missing in the first versions of the assemblers.

12.3.2 Refinement

Hereafter, we present—as a representative example—the first de novo algorithm based on the greedy approach, called SSAKE [57]. We then briefly discuss improvements brought by other assemblers. In all these assemblers, the storage of the reads is done using a *prefix tree*, whose branches spell all the prefixes of reads in R up to a given length.

SSAKE [57] uses the set of seeds $S = R$, but we chose to redundantly keep the notation S both in order to preserve the framework given by the generic algorithm and because later versions of the published algorithm allow the use of a different set of seeds. As it can be seen in Step 1 of Algorithm 2, SSAKE does not have a read error correction procedure, but it orders the reads in R (and thus in S) by decreasing frequencies, thus understanding that the coverage guarantees a high frequency to correct reads, whereas reads with sequencing errors will have a low frequency. It is understood that reads with too high frequency correspond to repeats and are removed from R. The prefix tree T of R, usually using the threshold $h = 11$ on the prefix length, has the role of speeding up the searches of the reads starting with a given sequence.

Then, once an initial contig c is chosen (Step 2.1), an overlap length k (for which a minimum min is provided) is fixed for the overlaps search. The rightmost k-mer u of c is identified and searched for in T, starting as usual at the root and spelling $Pref(u, \min\{k, h\})$ along one branch as long as possible. If the spelling ends in a node p, then c may be extended to the right with every read v such that $Pref(v, k) = u$ (such reads are stored in the descendants of p). Now, it is important to notice that if the set Ext of these reads has more than one element, this implies a repeat of u in the target genome. The algorithm may either (this is the *permissive* mode) choose one of the elements in Ext, according to a score function unfortunately not preventing wrong choices, or (this is the *stringent* mode, presented in Algorithm 2) stop the extension of a contig when Ext has several elements, but then the length of the resulting contigs is drastically reduced. The essential repeat solving step is thus absent in the algorithm.

Also note that in its initial version in [57], the algorithm does not provide a scaffolding step and returns thus a set of contigs, not an assembled sequence. SSAKE has been updated since its publication to deal with errors when extending the contig (a consensus of the overlapping reads is used instead of a read) and to scaffold contigs using paired reads.

SHARCGS [15] improves SSAKE by adding pre-assembly read error correction and, in order to avoid false joins, a more strict method for repeat detection before extension. More precisely, when a possible extension with a read e is found, SHARCGS tests whether the extension is ambiguous, in the sense that a repeat

Algorithm 2 SSAKE

Input: Set R of reads of length $l > 0$,
integer min > 0 (minimum length of an overlap),
integer $h > 0$ (prefix length)
Output: An assembled sequence

Step 1: Data preparation
1.1: Frequency based sorting of reads in R

compute the read frequencies $f(r)$
sort R by decreasing value of $f(r)$
compute the prefix tree T of R for length h
1.2: Compute the seed set S and update R

$S \leftarrow R$
Step 2: Build contigs by left-right seed extensions
$C \leftarrow 0$;
while $R \neq \emptyset$ **do**
2.1: choose a seed c from S
$c \leftarrow$ the first element in S
$Remove(c, \overline{c}, R, S, T)$
2.2: repeatedly extend c, then $\overline{c}$, to the right
$extend \leftarrow true$; $switch \leftarrow false$
while ($extend$) **do**
$k \leftarrow l - 1$
while min $\leq k$ **do**
$u \leftarrow$ the rightmost k-mer in c
$Ext \leftarrow \{v \in R \mid Pref(v, k) = u\}$ (uses T)
if $|Ext| = 1$ **then**
let $e \in Ext$
$c \leftarrow$ the extension of c with e
$Remove(e, \overline{e}, R, S, T)$
$k \leftarrow l - 1$
else
if $|Ext| > 1$ **then** $k \leftarrow$ min -1 **else** $k \leftarrow k - 1$ **endif**
end if
end while
if (not $switch$) **then** $switch \leftarrow true$ **else** $extend \leftarrow false$ **endif**;
$c \leftarrow \overline{c}$;
end while
$C \leftarrow C \cup \{c\}$
end while
Step 3: Scaffolding
Use paired reads to solve repeats and connect contigs

exists (not detected by SSAKE since corresponding to a substring of e), and the extension makes a dubious choice, or not. In the first case, the extension is stopped. In the second case, the contig is extended and the extension process continues. **VCAKE [25]** follows the same principles as SSAKE but modifies the read extension method in order to handle errors. In VCAKE, all the possible reads having with c an overlap of length between min and $l - 1$ (see step 2 in Algorithm 2) are considered in order to decide which extension to choose for a contig c. The contig is then successively extended by the character which is the most frequent at its position over all the possible extensions, assuming that the frequencies observed are not too high, in which case a repeat inducing abnormally high frequencies is detected and the extension is canceled.

QSRA [7] modifies VCAKE in the case where a given threshold on the number of extensions is not reached even when min is as low as 11. In this case, QSRA uses so-called quality-values (or q-values) of the reads—statistically estimating the confidence of each read—to decide whether the global quality of the extensions is good enough (depending on another threshold) to nevertheless allow the extension. If so, the extension is done similarly to VCAKE.

PE-Assembler **[2]** has a complex seed construction procedure, where a seed is obtained by the extension of a read similarly to VCAKE, assuming this extension exceeds a given length threshold. Moreover, its contig building procedure—once again similar to VCAKE—takes into account paired reads, which is rare enough to be worth noticing.

12.3.3 Summary

The greedy technique claims simplicity and moderate ambition, both of which are defensible. Recent comparisons [32] show that, in general, greedy assemblers need higher coverage than de Bruijn assemblers (but comparable with OLC assemblers) for a comparable result (in terms of coverage of the target genome by the resulting contigs), but with more assembly errors. Consequently, the time and memory usage—which grows rapidly when the data size increases [61]—are higher than with OLC or de Bruijn assemblers, making the assembly of large genomes not feasible with greedy assemblers. Behind the computational requirements, the generally complex structure of large genomes is a bottleneck for the simple criteria used by greedy approaches. It is worth noticing here that an assembler called SUTTA [41] proposes to replace the greedy extension of a contig by an exact branch-and-bound search of the best extension (as computed using a specific optimization criterion) first to the right and then to the left. Since these directional extensions are independent from each other, the algorithm keeps some greedy features, however.

12.4 Overlap-Layout-Consensus Approach

Introduced for Sanger read assembly [43] and then adapted for NGS data (see for instance [39]), the OLC method proposes a global approach of genome assembly using a graph formulation and following three main steps named overlap, layout, and consensus steps.

12.4.1 General Pattern

Let us first present the principles of the three steps (see Fig. 12.2c for an example).

Overlap Step Given the set R of reads (including their reverse complements), the first aim of the OLC approach is to build a global graph giving an account of the overlaps between reads. This is done during the overlap step, where the set R of input reads (and their reverse complements) is first used to compute a set O of ordered pairs (r_i, r_j) with $r_i, r_j \in R$ $(i \neq j)$ whose overlap (exact or approximate) is significant with respect to some specific criterion defined by a similarity function $s(.,.)$. The choice of O is crucial for downstream procedures since O has an important impact on the assembly result [17]. At the end of the overlap step, the directed graph $G(R) = (R, O, w)$, called the *overlap graph of* R, is built, with node set R and arcs (r_i, r_j) from O. Each arc (r_i, r_j) of it is weighted by the maximum overlap weight $w(r_i, r_j)$ obtained using $s(.,.)$ (see Definition 1). Although computing the overlaps needed by this step and their weights is not algorithmically difficult, with the important amount of data handled by the NGS assemblers, the computational cost of this step is high, even when the most efficient algorithms [53] are used.

Layout Step The goal of this step is to find a path in the overlap graph that uses at least one node from every pair of nodes associated with a read and its complement, meaning that the orientation of the read into the genome is computed at the same time as the path defining the sought assembled sequence. Even when the orientation of the reads is known (which is not the case in practice), this problem is very difficult from an algorithmic viewpoint since it reduces to the Hamiltonian path problem, requiring a path containing each node of a given graph exactly once. The Hamiltonian path problem was proved NP-complete in [19].

Consensus Step The consensus step aims at finding a sequence which best represents all the reads in the path. Given the ordered list $L = r_1, r_2, \ldots, r_n$ of the reads on the path and a weighted edit distance $D(.,.)$ between sequences on Σ, the aim is to find a sequence $s^\star$ and contiguous subsequences $s_1^\star, s_2^\star, \ldots, s_n^\star$ of s^*, in this order from left to right of $s^\star$, such that (1) consecutive sequences $s_i^\star, s_{i+1}^\star, \ldots, s_j^\star$ overlap if and only if the corresponding reads $r_i, r_{i+1}, \ldots, r_j$ overlap and such that (2) $\Sigma_{i=1}^{n} D(s_i^\star, r_i)$ is minimized. Such a sequence $s^\star$ is called a *consensus sequence* for L (and also for the initial path). Since finding a consensus sequence is another NP-complete problem [52], most of the OLC assemblers implement heuristic methods.

As was already the case for greedy algorithms, whereas we seek to solve the GENERAL ASSEMBLY PROBLEM, in practice—and mainly due to the repeats—it is rather the CONTIG ASSEMBLY PROBLEM that the assemblers solve. The OLC assemblers are based on the approach proposed in [26] and are organized as described in Algorithm 3. Once the overlap step is done (Step 2), the layout and the consensus steps are performed once or twice (Steps 3 and 4). In the second case, the contigs are not directly obtained from paths found in the overlap graph $G(R)$ but from paths found in a refined graph $G'(R)$ whose nodes are sequences called

Algorithm 3 Generic OLC

Input: Set R of reads
Output: An assembled sequence

Step 1: Data preparation
  Prepare and correct reads
Step 2: Overlap step
  2.1: Compute the set O of ordered pairs from R
  2.2: Build the overlap graph $G(R)$
  2.3: (optional) Correct errors in $G(R)$
Step 3: (optional) Layout-Consensus 1: unitig building
  3.1: Identify unitigs as paths in $G(R)$
  3.2: Build the refined graph $G'(R)$
Step 4: Layout-Consensus 2: contig building
  Identify contigs as paths in $G'(R)$
Step 5: (optional) Scaffolding step
  Use paired reads, if available, to solve repeats and connect contigs

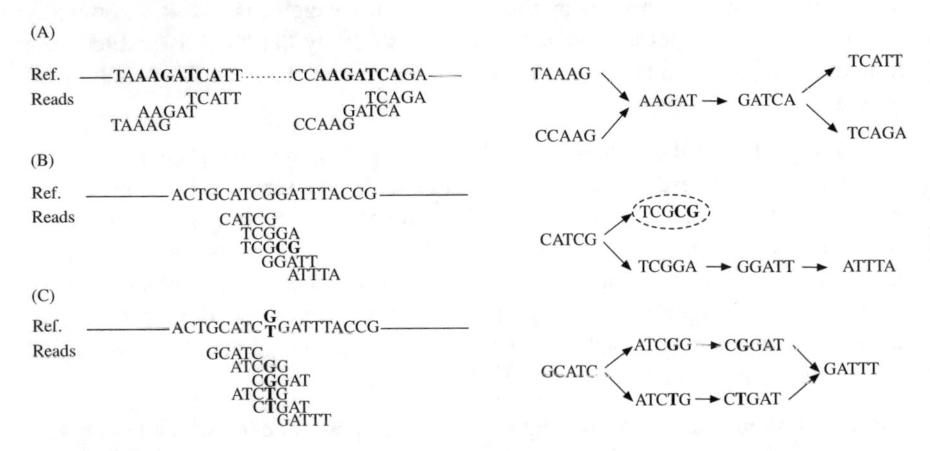

Fig. 12.3 Particular configurations in an overlap graph. *On the left*, a region of the genome and a set of reads (length $l = 5$) are given. *On the right*, the corresponding overlap graph whose nodes are the reads and whose arcs are exact overlaps of length 3. (**a**) A repeat, identified by two paths that converge then diverge. (**b**) A tip (also called a spur): a dead-end path (*dotted circle*) resulting from sequencing errors (CG instead of GA) at the end of a read. (**c**) A bubble (also called a mouth): the change of G into T in some reads results into a pair of paths that converge then diverge

unitigs. It is worth mentioning that the unitig notion is an ambiguous term which does not have the same meaning when talking about different de Bruijn graph-based assemblers. The paths used in $G'(R)$ are usually *linear paths*, i.e., paths whose nodes have in- and out-degree exactly 1 (thus guaranteeing that the reads in the path are involved in no repeat, as repeats introduce branchings into the overlap graph; see Fig. 12.3a). Optionally, a scaffolding step ends the treatment by ordering and orienting the contigs and resolving a number of repeats using paired data.

12.4.2 Refinement

Among the assembly algorithms based on an OLC method, we can observe two different directions: the *classical* approach implementing the overlap graph according to the original definition and the *string graph* method which implements an alternative view of the overlap graph. Classical approaches were usually implemented for Sanger data, and they need huge computational resources to deal with the great amount of data supplied by the NGS experiments. The string graph framework, first introduced by Myers [38], was created in order to reduce the memory space required in the case of large genomes, or of a large input dataset such as NGS data, without losing any information needed for the assembly.

The string graph for a set R of reads can be derived from the overlap graph of R by removing *transitive arcs*. An arc (r_i, r_k) is *transitive* if there exists a third node r_j such that (r_i, r_j) and (r_j, r_k) are arcs of the overlap graph. A non-transitive arc is called *irreducible*. The overlap corresponding to a pair (r_i, r_k) of reads from O is called *irreducible* (*transitive*, respectively) if (r_i, r_k) is an irreducible (transitive, respectively) arc. In the string graph, it is statistically possible to distinguish the arcs which have to be used by the (sought) path defining an assembly of all reads from the arcs which are optional (thus linking non-successive reads on the path) [38]. Solving the GENERAL ASSEMBLY PROBLEM with the string graph consists then in finding the shortest path going through all the required arcs and possibly through optional ones. Medvedev et al. in [36] show that this problem is also NP-hard. In practice, assemblers using the string graph rather solve the CONTIG ASSEMBLY PROBLEM and follow the generic OLC in Algorithm 3 where Step 2.2 is replaced by the construction of the string graph including transitive arcs removal.

We present below an assembler based on the OLC classical approach and its update then discuss two assemblers using the string graph framework.

***Celera*/CABOG** The *Celera* Assembler [39] is one of the first OLC assemblers designed for Sanger data that was updated to deal with Sanger and NGS data. In Algorithm 4 we present the revised pipeline called CABOG [37]. This method needs information about paired reads to build the contigs and the scaffolds and then cannot be correctly used without this kind of information.

To save time during the overlap computation, the CABOG methodology is inspired by a *seed-and-extend* approach. Let S_k be the k-spectrum of R. A filtering procedure is applied to S_k that keeps only k-mers which have at least two occurrences and are identified as non-repetitive, i.e., whose frequency $f(s, R)$ exceeds 2 and does not exceed a computed threshold M. The set of significant overlaps is computed as follows. For each ordered pair of sequences (r_i, r_j) from R such that the number $com(r_i, r_j)$ of k-mers from S_k common to r_i and r_j exceeds a given threshold m, the maximum overlap weight $w(r_i, r_j)$ (see Definition 1) is computed with a dynamic programming approach [29]. Then, the pair (r_i, r_j) is added to the set O if and only if the maximum overlap weight $w(r_i, r_j)$ and the overlap length $len(r_i, r_j)$ are greater than the given thresholds, t and $l_{\min}$.

Algorithm 4 CABOG

Input: Set R of reads,
real t (min. similarity between two reads),
integer $l_{\min}$ (min. overlap length),
integer k (defines the k-mers),
integer m (min. number of shared k-mers)
Output: An assembled sequence

Step 1: Data preparation
$S_k \leftarrow$ the k-spectrum of R
compute $f(s, R)$ of all s in S_k
compute max k-mer frequency threshold M
$S_k \leftarrow \{s \in S_k \,|\, 2 \leq f(s, R) \leq M\}$
Step 2: Overlap step
2.1: Compute the overlap set O from R

$O \leftarrow \emptyset$
for all $(r_i, r_j) \in R^2$ with $com(r_i, r_j) \geq m$ **do**
if $w(r_i, r_j) \geq t$ and $len(r_i, r_j) \geq l_{\min}$ **then** $O \leftarrow O \cup \{(r_i, r_j)\}$ **end if**
end for
2.2: Build the BOG $G(R)$ from O
2.3: (opt., not done) Correct errors in $G(R)$
Step 3: Layout-Consensus 1: unitig building

Break all alternate cycles in $G(R)$
for all read r (together with $\overline{r}$) in R **do**
Score r with $PathLength(r, G(R))$
end for
$U \leftarrow \emptyset$; $C \leftarrow \emptyset$
while $R \neq \emptyset$ **do**
Extract r (and $\overline{r}$) with highest score in R
Build the dovetail paths $p(r)$, $p'(r)$
$L \leftarrow ExtractReadList(p(r), p'(r))$
$u \leftarrow$ a consensus sequence for L
$U \leftarrow U \cup \{u\}$
Remove reads in $p(r)$ and $p'(r)$ from R
end while
Splitting of u, for all $u \in U$
Step 4: Layout-Consensus 2: contig building

Let U' the set of U-unitigs from U
Build unitig graph $G(U')$
Map information about paired reads on $G(U')$
for all confirmed path p in $G(U')$ **do**
Let c be a consensus sequence for p
$C \leftarrow C \cup \{c\}$
end for
Step 5: Scaffolding step
Use paired reads, if available, to solve repeats and connect contigs

To reduce graph complexity before the consensus-layout steps, CABOG constructs the *best overlap graph* or BOG with both directed and undirected edges as follows: a read r together with its reverse complement $\bar{r}$ in R are represented by two nodes, $r.B$ and $r.E$. An (undirected) edge links $r.B$ to $r.E$. In a traversal of the BOG, when $r.B$ is encountered before $r.E$, the edge represents r while it represents $\bar{r}$ in the opposite case. There is at most one output arc for each node. For a read r_i (together with its reverse complement), the node $r_i.E$ is the source of an arc with target $r_j.B$ (resp. $r_j.E$) if and only if the overlap between r_i and r_j (resp. $\bar{r_j}$) is the best overlap using a suffix of r_i. The best overlap is defined as the one with the maximum overlap length, but other criteria can be used. Similarly, the node $r_i.B$ is the source of an arc with target $r_k.E$ (resp. $r_k.B$) if and only if the overlap between $\bar{r_i}$ and $\bar{r_k}$ (resp. r_k) is the best overlap using a suffix of $\bar{r_i}$. An important notion in the BOG is that of the *dovetail path*, defined as an acyclic path that starts with an edge, ends with an edge, and alternates edges and arcs. Such a path represents a series of overlapping reads for which an orientation is chosen.

During the first execution of the layout-consensus steps (in Step 3 of Algorithm 4), a heuristic based on the BOG builds set pre-contigs, so-called *unitigs*, by performing first the *unitig construction* and then the *unitig splitting*.

Unitig Construction First, any cycle that is an alternate series of edges and arcs is broken by arbitrarily removing one arc. The arbitrary choice can impact the results, and CABOG can be a nondeterministic algorithm if the removed arc is chosen randomly. Then, a score is given to each read r as well as to its reverse complement with the function *PathLength*$(r, G(R))$ as follows. If the longest dovetail path with initial node $r.B$ (respectively $r.E$) is called $p(r)$ (respectively $p'(r)$), then the score is equal to $||p(r)|| + ||p'(r)||$, where $||q||$ counts the number of nodes of the path q. Finally, each unitig is built as the consensus sequence for a list L of reads defined by the procedure *ExtractReadList*() as follows. Let L be an empty list of reads. Starting with the read r with the highest score, let L contain r. The reads (or their reverse complement depending on how the undirected edge is traversed) along $p(r)$ are successively added at the end of L until no more reads can be reached or a read already in a unitig is encountered. Similarly, the reads encountered along $p'(r)$ are successively added at the beginning of L. (It is clear that extending $\bar{r}$ to the right is equivalent to extending r to the left.) In CABOG, the heuristic used to build the consensus sequence of L consists in computing the most probable character at conflicting positions of the overlaps between consecutive reads. This step is repeated with the next read with the highest score until each read is included in a unitig.

Unitig Splitting The unitig construction is an aggressive step in which a repeat region is attached to a unique unitig. CABOG performs a second step to break unitigs at locations corresponding to repeat boundaries or to errors following two heuristics, one of which requires paired data. First, two dovetail paths in the BOG used to build the unitigs can intersect (meaning two nodes—one in each path—are connected by an arc in the BOG) and this intersection can be induced by repeats or errors. The *intersection splitting* splits these paths (and the corresponding unitigs) at error-induced intersections based on path length and coverage (not detailed here).

Second, a unitig is broken if the number of *violated read pairs* in this unitig exceeds a given threshold. A read pair is violated if the two reads in the pair lie in the same unitig and the distance between them on the unitig is not consistent with the insert size. As a result, we obtain a set U of unitigs.

In Step 4, CABOG relies once more on information about paired reads, if it is available (otherwise, it outputs the set U as the set of final contigs). For each unitig $u \in U$, CABOG first statistically decides whether u corresponds to a unique sequence of the target genome (in which case u is called a U-unitig) or to a repeat. The set U' of U-unitigs is the basis for the construction of a *unitig graph* $G(U')$, whose nodes are the U-unitigs and whose arcs are given by overlapping U-unitigs. Among these overlaps, some are *confirmed* by information about paired reads, in the sense that the overlap is consistent with the insert distance as witnessed by at least two pairs of paired reads. Then, in $G(U')$, a contig is built from any maximal linear path p of U-unitigs with confirmed overlaps (p is then a *confirmed path*) as a consensus sequence. Finally, all contigs (including repeated unitigs) are oriented and ordered in the scaffolding step, again with the help of information about paired reads.

Edena **[22]** is the first assembler based on the string graph method. It uses efficient data structures to compute exact overlaps as presented by Algorithm 5.

A suffix array [35] SA_R of R is built. The suffix array SA_R is an array of integers containing the starting positions of the sorted suffixes (with respect to the lexicographical order) of $S_R = r_1\$r_2\$r_3\ldots\$r_{|R|}\$$ where the sentinel character \$ separates two reads. Locating every occurrence of a subsequence X within S_R can be efficiently done with a binary search using SA_R. Now, given a read $r \in R$, assume we want to find the exact overlaps of length at least $l_{\min}$ between r and any other read in R. Then, for each suffix sr of r of length at least $l_{\min}$, it is sufficient to search for $\$sr$ in S_R using SA_R. In Algorithm 5, the method *ExactOverlaps*() computes the set O of exact overlaps of length at least $l_{\min}$ for all reads in R with SA_R.

Edena obtains the string graph $G(R)$ from the overlap graph built from R and O by removing the transitive arcs. Then, it corrects particular errors in $G(R)$. These errors are (see Fig. 12.3b, c) the *tips* (also known as *spurs*) which are short dead-end paths resulting from a sequencing error close to the end of a path and the *bubbles* (also known as *mouths*) which are paths that diverge then converge resulting from a sequencing error close to the middle of a read. Tips and bubbles are detected by a local exploration of $G(R)$. The procedure $Extract(G(R), v, md)$ outputs, for each junction node v (i.e., a node with in- or out-degree superior to 1), the set of maximal paths with less than md nodes. They are identified as tips and are marked. Once all junction nodes are visited, the nodes participating only in tips are removed. Bubbles are pruned with the function $Pairs(G(R), v, ms)$ that searches for pairs of paths starting at a junction node v and converging after at most a given number ms of nodes. The procedure *RemovePath*() removes the least covered path meaning the one with less evidence from the reads.

Since *Edena* only considers exact overlaps, the consensus sequence obtained for each maximal linear path $(s_1, s_2), (s_2, s_3), \ldots, (s_{p-1}, s_p)$ of $G(R)$ is in fact obtained by repeatedly extending s_1 with $s_2, s_3, \ldots, s_p$. No scaffolding process is produced.

Algorithm 5 *Edena*

Input: Set R of reads
integer $l_{\min} > 0$ (min. overlap length),
integer $md > 0$ (max. length of a tip)
$ms > 0$ (max. length of a bubble)
Output: A set of contigs

Step 1: Data preparation
index R in a suffix array SA_R
Step 2: Overlap step
2.1: Compute the overlap set O from R

$O \leftarrow ExactOverlaps(R, SA_R, l_{\min})$
2.2: Build string graph $G(R)$

build overlap graph $G(R)$ from O
remove transitive arcs in $G(R)$
2.3: Correct errors in $G(R)$

for all junction node v in $G(R)$ **do**
  for all $p \in Extract(G(R), v, md)$ **do**
    mark all nodes of p
  end for
end for
for all node n used only in tips **do**
  remove n from $G(R)$
end for
for all junction node v in $G(R)$ **do**
  for all $\{p_i, p_j\} \in Pairs(G(R), v, ms)$ **do**
    $RemovePath(p_i, p_j, G(R))$
  end for
end for
Step 3: *(opt., not done) Layout-Consensus 1: unitig building*
Step 4: Layout-Consensus 2: contig building

for all maximal linear path p in $G(R)$ **do**
  let c be a consensus sequence for p
end for

SGA [54] is the practical implementation of the framework presented in [53]. Its originality lies in the use of a compressed data structure, the FM-index [18], all along the assembly process. The FM-index is a compressed substring index, achieving very fast queries and needing a small space storage.

First, SGA implements two procedures to correct errors in reads of R based either on k-mer frequencies or on inexact overlaps that can be both directly computed with the help of the FM-index FM_R of the sequence $S_R = r_1\$r_2\$\ldots r_{|R|}\$$. Second, unlike *Edena*, SGA directly builds the string graph without the need of the overlap graph by avoiding the transitive arc removal step. The algorithm implemented by SGA to output only irreducible (exact or inexact) overlaps is based on the following approach. Let r_i, r_j and r_k be three reads such that r_i overlaps r_j, r_j overlaps r_k and

r_i overlaps r_k (so the latter one is transitive). Then, the maximal suffix x of r_j not overlapping with r_i is a prefix of the maximal suffix y of r_k not overlapping with r_i. To find all the irreducible overlaps between a read r_i and other reads, SGA first searches all the prefixes of reads that overlap a suffix of r_i with FM_R then extends them rightwards letter by letter. For each sequence obtained by this extension, it stops whenever it finds the end of a read (this read is r_j). At this point, the extension sequence represents x, and it is also a prefix of all the suffixes y of all potential reads r_k. Thus, there is an irreducible overlap between r_i and r_j.

Instead of building directly the full string graph, SGA assembles a sequence around each read to get a set A of *pre-assembled reads*. A pre-assembled read is a sequence obtained from a read $r_1 \in R$ by successively extending r_1 with $r_2, r_3, \ldots,$ r_n such that r_i $(1 \leq i < n)$ has only one irreducible overlap, that is, with r_{i+1}. The final string graph $G(A)$ is built from A and the irreducible overlaps computed from the FM-index FM_A of A. After removing tips and bubbles in $G(A)$, contigs are built similarly to the *Edena* algorithm (Step 4 of Algorithm 5), and optionally scaffolding can be done with a strategy similar to classical OLC assemblers if paired data is available.

12.4.3 Summary

The assemblers presented above show that the global view of the overlaps proposed by the OLC method allows to produce assemblies of very good quality, especially when the reads are long [49, 50]. However, when the size of the reads reduces (and particularly for very short reads), the coverage is necessarily increased and results in very important computational requirements. For large genomes, a program failure is therefore possible [27, 49], despite the important efforts made to surpass these problems. Alternative solutions exist, either by using parallelized assemblers, such as *CloudBrush* [12], or by using hybrid assemblers, whose principle is to combine two approaches, among which one usually has the OLC approach. In the hybrid greedy-OLC approach, a seed-and-extend technique is adopted (as in greedy methods), looking for the best extension in a local overlap graph (as in OLC methods) that sometimes is not explicitly built. This is the case of *Taipan* [51], SHORTY[23], and *Telescoper* [6]. Some hybrid methods have been shown to be a successful alternative for greedy and OLC methods on short and very short reads in small genomes with respect to the running time, although the memory requirements are higher [61].

The role of OLC methods in the genome assembly problem will probably increase with the evolution of sequencing technologies. The third-generation sequencing platforms are able to produce much longer reads than the NGS platforms, with lengths between 1 and 20 kbp [9]. The OLC approach is suitable for assembling reads in this interval. However, because of the high error rate of these reads, most of the current assembly methods are based on a hybrid de Bruijn-OLC approach, using both long and short reads (see for instance [28]).

12.5 The de Bruijn Approach

The idea of representing every read as a collection of even smaller subsequences, thus losing long-range connectivity information, was first proposed in 1995 [24]. Applied for Sanger read assembly, this approach allows to transform the search for a Hamiltonian path, corresponding to the assembly of the entire genome in the OLC model, into a search for an Eulerian path (i.e., arc-guided instead of node-guided) with additional read-preserving properties. However, the presence of repeats and errors in the data made this approach impractical for large genomes, until 2001, when the first solutions to these problems are presented in [47]. In this version, the reads are corrected before the graph construction, the read-preserving properties the paths must satisfy are used to disambiguate the passages of the paths into the hubs, and the paired reads are used to scaffold the resulting contigs. Many other assemblers proposed since then use this approach. We first present here the de Bruijn approach in its most general form, and then we point out the algorithmic particularities of specific assemblers.

12.5.1 *General Pattern*

As assumed by convention, let R be the set of reads (including their reverse complements), which are sequences of length l over the alphabet $\Sigma = \{A, C, G, T\}$. Given a positive integer $k \leq l$, any read r of length l may be represented by $l-k+1$ successive k-mers starting respectively at positions $1, 2, \ldots, l-k+1$ of r. Let S_k be the *k-spectrum* of R. See Fig. 12.4 for an illustration. The *de Bruijn* graph $G^k(R)$ is the directed graph defined as follows. Its set of nodes is S_k. It has an arc between the k-mers s and t (in this order) if and only if s and t have an exact overlap of $k-1$ characters, and there exists a read r which contains the $(k+1)$-mer x produced by the extension of s with t, i.e., by the concatenation of s and the last character of t. In the de Bruijn graph, any path $(s_1, s_2), (s_2, s_3), \ldots, (s_{p-1}, s_p)$ *spells* the sequence obtained by successively extending s_1 with $s_2, s_3, \ldots s_p$. As before, a path is *linear* if all its nodes have in-degree and out-degree equal to 1. The *read-path* corresponding to a read r in R is the path $(s_1, s_2), (s_2, s_3), \ldots, (s_{l-k}, s_{l-k+1})$, denoted $p(r)$, such that $s_i = r[i \ldots i+k-1]$. Equivalently, $p(r)$ is the path which spells r. Note that in many assemblers, the de Bruijn graph is represented as a *bidirected* graph, where the nodes containing a k-mer and its complement are stored in the same node.

The GENERAL ASSEMBLY PROBLEM is, in this case, modeled as the research of a path into the de Bruijn graph. The length of this path represents the reconstruction score $rec(S, R)$ of the sequence spelled by the path.

DE BRUIJN H(ALF)-SUPERPATH PROBLEM

Input: An integer $k > 0$, the de Bruijn graph $G^k(R)$ corresponding to the set R of reads, and the set $\mathscr{P}^k(R)$ of read-paths for the reads in R

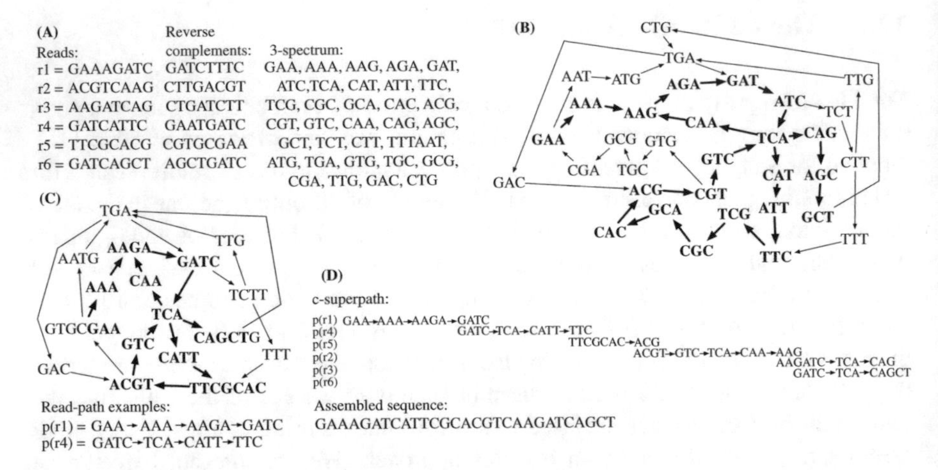

Fig. 12.4 Illustration of the de Bruijn approach, on the same example as in Fig. 12.2. Here, $l = 8$ and $k = 3$. (**a**) The set R of reads and the 3-spectrum $S_3(R)$. (**b**) The resulting de Bruijn graph, where the subgraph in *bold* is the subgraph corresponding to one strand of the genome. (**c**) Paths in (**b**) with no branchings (i.e., linear paths) have been contracted into a node. (**d**) The h-superpath yielding the correct sequence of the genome

Requires: Find in $G^k(R)$ a shortest h-superpath W of $\mathscr{P}^k(R)$, that is, a path of $G^k(R)$ made of overlapping paths from $\mathscr{P}^k(R)$ that contains at least one of $p(r)$ and $p(\overline{r})$, for each $r \in R$, and that is as short as possible.

The NP-completeness of the problem is easily deduced from [36, 40]. The interest of the de Bruijn approach is therefore not given by a lower complexity of the underlying problem but by an easier way to deal with the computational difficulty when the data has some facilitating properties. Note however that most of the algorithms below do not use the read-paths, as the assemblers prefer to avoid the computational requirements associated with them. This information may, up to a point, be replaced with coverage statistics about each arc.

The generic outlines of the de Bruijn approach are given in Algorithm 6. Given that the de Bruijn graph is very sensitive to sequencing errors in reads, these errors are preferably corrected by efficient procedures before the construction of the graph. Additionally, some classes of errors may be detected after the construction of the graph. As in the OLC case, these errors are *tips* (also known as *spurs*) which are short dead-end paths resulting from a sequencing error close to the end of a path and *bubbles* (also known as *mouths*) which are paths that diverge then converge resulting from a sequencing error close to the middle of a read (see Fig. 12.5b, c).

The bottleneck of the contig construction step is, as usual, the presence of repeats (see Fig. 12.5a), and particularly of complex repeats (repeats included in other repeats, tandem repeats etc.). They result into a de Bruijn graph in which solving repeats, i.e., being able to distinguish right the branchings from wrong ones in a configuration as that in Fig. 12.5a, is all the more difficult when shorter sequences like k-mers are used instead of the reads. Consequently, searching for a solution

---

Algorithm 6 Generic de Bruijn

Input: Set R of reads of length $l > 0$, integer $k > 0$ (defines the k-mers),
Output: An assembled sequence

Step 1: Data preparation
 1.1: Compute S_k
 1.2: (optional) Correct errors in reads
Step 2: Construction of the de Bruijn graph $G^k(R)$
 2.1: Define the set of nodes
 2.2: Compute the arcs
 2.3: (optional) Compute the set $\mathscr{P}^k(R)$
Step 3: Cleaning of the de Bruijn graph
 3.1: (optional) Correct errors: bubbles, tips etc.
 3.2: (optional) Resolve repeats
Step 4: Contig building
 Identify contigs as linear paths in $G^k(R)$
Step 5: Scaffolding
 Use paired reads, if available, to solve repeats and connect contigs

---

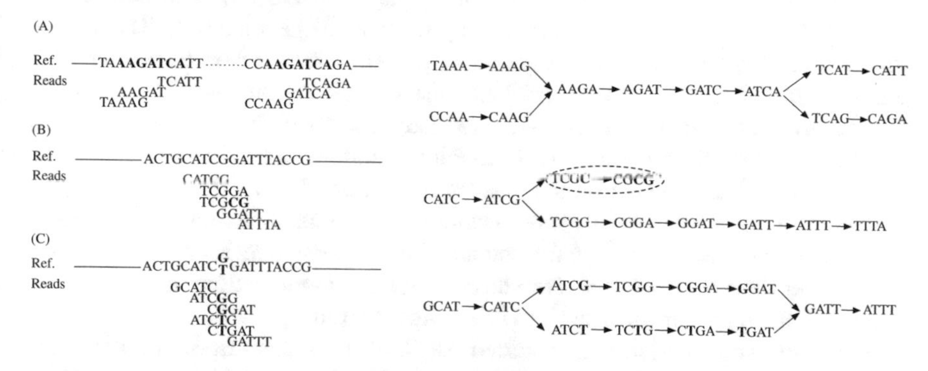

Fig. 12.5 Particular configurations in a de Bruijn graph. *On the left*, a region of the genome and a set of reads (length $l = 5$) are given. *On the right*, the corresponding de Bruijn graph whose nodes are the 4-mers. (**a**) A repeat, identified by two paths that converge then diverge. (**b**) A tip (also called a spur), a dead-end path (*dotted circle*) resulting from sequencing errors (CG instead of GA) at the end of a read. (**c**) A bubble (also called a mouth), the change of G into T in some reads results into a pair of paths that converge then diverge

for the DE BRUIJN H-SUPERPATH PROBLEM goes through attempting to simplify the graph $G^k(R)$ by solving repeats that can be solved (therefore called *resolvable repeats*) and changing the graph accordingly, so as to get as close as possible to the ideal but also rare case where the resulting graph is partitioned into two paths, one corresponding to the target genome and the other one to its reverse complement (note that when the de Bruijn graph is represented as a bidirected graph, the two paths reduce to one). In order to get closer to such a property, linear paths are very important, as they may be contracted into an arc (or, alternatively, a node), without modifying the set of solutions (h-superpaths).

It must be noticed here that such an approach solves the DE BRUIJN H-SUPERPATH PROBLEM only in the case where all the repeats are resolvable, sooner or later according to the order of treatment. In the contrary—and far more frequent—case, unresolvable repeats persist, implying that a part of the graph cannot be disambiguated under the form of a linear path. Consequently, the resolution of the DE BRUIJN H-SUPERPATH PROBLEM is not guaranteed by the proposed algorithm, which finds only parts of the sought h-superpath, thus solving the CONTIG ASSEMBLY PROBLEM rather than the GENERAL ASSEMBLY PROBLEM.

12.5.2 Refinement

We present in detail two algorithms whose contributions to the method are essential. Then, we discuss several other algorithms.

Euler **[47]** (see Algorithm 7) puts the basis of this approach, proposing solutions to the problems encountered by the similar approach on Sanger data [24]. The method for pre-assembly error correction in reads we present here is one of the two methods proposed in [47] that we chose as it has the twofold advantage of showing the principles used for this task and to be more easily described. First, only valid k-mers s with respect to their frequency $f(s, R)$ in R and a bottom threshold m are kept in the k-spectrum S_k. Then, the reads are considered successively and, for each read r, the minimum number $mn(r, S_k)$ of modifications (insertions, deletions, substitutions) needed to ensure that all the k-mers contained in r belong to S_k is computed. When this number does not exceed a given threshold α, the modifications are performed; otherwise, the read is invalidated and it is removed from R.

The graph construction uses a sorted list V of the k-mers in S_k, in which the possible successors of each k-mer with respect to the overlapping requirements are searched for using a binary search. Real successors are identified by searching the $(k + 1)$-mer resulting from the extension of s with t in the reads in R, and the corresponding arcs are built by the procedure $Arc(s, t)$. It is worth noting that linear paths are collapsed so as to reduce the size of the graph, although this is not specifically written in our algorithm.

Repeat solving is an essential part of the method, and in *Euler*, it is based on two operations called *detachments* and *cuts* whose role is to transform a couple $(G, \mathscr{P})$, initially defined as $(G^k(R), \mathscr{P}^k(R))$, into another couple $(G', \mathscr{P}')$ using *equivalent transformations*, that is, transformations preserving the h-superpaths. The specific aim of such a transformation is to carefully untangle the de Bruijn graph by separating paths which have a passage through a common subpath, therefore identified as a repeat. The *(a, b)-detachment* unambiguously couples—based on information issued from the set of paths $\mathscr{P}$—the arc a entering into a repeat with the arc b going out from the same repeat. The *a-cut* removes the arc a from all the paths in $\mathscr{P}$ containing it. Both these transformations need to satisfy strong hypotheses, not given here, in order to ensure the sought equivalence. When these hypotheses are

Algorithm 7 *Euler*

```
Input: Set R of reads of length l > 0,
  integer k > 0 (defines the k-mers),
  integer m > 0 (min frequency),
  integer α ≥ 0 (modifications threshold)
Output: A set of contigs
Step 1: Data preparation
    1.1: Compute S_k and additional information
      compute S_k
      compute frequencies f(s, R), for all s ∈ S_k
    1.2: Correct errors in reads

      S_k ← S_k − {s ∈ S_k | f(s, R) > m}
      while exists r ∈ R with mn(r, S_k) > 0 do
         if mn(r, S_k) ≤ α then modify(r, S_k)
           else R ← R − {r} end if
         update f(s, R) for all s ∈ S_k
         S_k ← S_k − {s ∈ S_k | f(s, R) > m}
      end while
Step 2: Construction of G^k(R)
    2.1: Define a node for each s ∈ S_k
    2.2: Compute the arcs (s, t)

      V ← sorted list of all the k-mers in S_k
      for all s ∈ S_k do
         Candidates ← BinarySearch(V, s[2..k])
         for all t ∈ Candidates do
            x ← the extension of s with t
            if x occurs in R then Arc(s, t) end if
         end for
      end for
    2.3: Compute 𝒫^k(R)
Step 3: Cleaning of the de Bruijn graph
    3.1: (optional, not done) Correct errors
    3.2: Resolve repeats
      while a detachment or a cut is possible do
         modify the couple (G^k(R), 𝒫^k(R))
      end while
Step 4: Contig building

      for all maximal linear path p in G^k(R) do
         let the sequence spelled by p be a contig
      end for
Step 5: (optional, not done) Scaffolding
      Use paired reads, if available, to solve
repeats and connect contigs
```

satisfied for one or another of the transformations, the transformation is performed, allowing to resolve more and more repeats.

It is worth noticing that the repeats shorter than the read length are identified during the identification of the read-paths, and the branching paths representing such repeats are carefully separated in order to form longer unbranched paths. When paired data is available, *Euler* maps the reads onto the de Bruijn graph in order to solve more repeats. As usual, the scaffolding step uses information about paired reads to solve other repeats, as well as to orient and order the contigs.

Euler has several variants: *Euler-DB* [46] uses double-barreled data (particular paired reads) to further untangle the de Bruijn graph, *Euler-SR* [10] is optimized for high-coverage short-read assemblies, and *Euler-USR* [11] corrects reads using a so-called repeat graph of the target genome, which can be approximated by simplifying the de Bruijn graph.

***Velvet* [59]** In order to minimize memory requirements, the graph $C^k(R)$ built by *Velvet* compresses in one node n a series of successive k-mers from the same read, of which only the first and the last k-mer may overlap another read (equivalently, the k-mers form a linear path in the standard de Bruijn graph). Moreover, the node n contains, in this case, both the representation of the sequence and the representation of its reverse complement. Arcs are defined on the same principle as in the standard de Bruijn graph between two nodes, n_1 and n_2, such that the last k-mer in the series of n_1 overlaps on $k - 1$ characters the first k-mer in the series of n_2. Note that in *Velvet*, the value k is odd by convention, thus preventing a k-mer from becoming its own reverse complement (which is unsuitable for the bidirectional nodes).

Velvet (see Algorithm 8) orders the reads by giving them identity numbers and computes for each read r the set $O(r)$ of *original* k-mers (not occurring in preceding reads) as well as the set $P(r)$ of original k-mers also occurring in succeeding reads (sets $P(r)$ are stored in a hash-table H_{read}). A series of consecutive original k-mers from a read is then a node if and only if none of the k-mers belongs to $P(r)$. Procedures *FindSeries* and *FindAdj* respectively compute the nodes and the arcs of the graph using information from $P(r)$ and $O(r)$. To speed-up the computations, two hash-tables are used.

Velvet does not correct sequencing errors before the graph construction but corrects three types of errors in the de Bruijn graph (see Fig. 12.5b, c): the tips, the bubbles, and the *erroneous connections* that are due to false overlaps created by sequencing errors. The algorithm will remove the tips shorter than $2k$ and whose first arcs have the so-called *minority count property*, that is, its multiplicity (defined as the number of reads inducing that arc, which may be computed at the same time as the de Bruijn graph) does not reach the maximum value over all arcs outgoing from the same node. These conditions are required in order to avoid removing paths due to reads without sequencing errors. In order to remove bubbles, *Velvet* uses a breadth-first search algorithm called the "Tour bus" algorithm similar to Dijkstra's well-known algorithm. Each time a bubble is discovered, the algorithm merges the two alternative paths into a consensus path. Once the "Tour bus" algorithm is finished, the erroneous connections are identified as nodes with a low coverage, according to a chosen threshold.

After the correction of the graph, contigs are generated with a method similar to Step 4 in Algorithm 7. A noticeable difference with *Euler* is that in *Velvet*, repeat solving is postponed after contig generation and is thus a part of the scaffolding step.

Algorithm 8 *Velvet*

Input: Set R of reads of length $l > 0$,
integer $k > 0$ (defines the k-mers),
Output: An assembled sequence

Step 1: Data preparation
  1.1: Compute S_k and additional information
    compute S_k
    compute an id number $ID(r)$, for $r \in R$

    for all $s \in S_k$ **do**
      $a(s) \leftarrow \min\{ID(r) \mid r \text{ contains } s\}$
    end for
    for all $r \in R$ **do**
      $O(r) \leftarrow \{s \in S_k \mid a(s) = ID(r)\}$
      $P(r) \leftarrow \{s \in O(r) \mid \exists r' \neq r, \text{containing } s\}$
      insert $(r, P(r))$ into H_{read}
    end for
  1.2: (not done) Correct errors in reads
Step 2: Construction of $C^k(R)$
  2.1: Compute the nodes

    for all $r \in R$ **do**
      $FindSeries(r, O(r), H_{read})$
    end for
  2.2: Compute the arcs

    for all node n **do**
      $FindAdj(n, H_{read})$
    end for
  2.3: (optional, not done) Compute $\mathscr{P}^k(R)$
Step 3: Cleaning of the de Bruijn graph
  3.1: Correct errors: bubbles, tips etc.

    remove frequent tips of length less than $2k$
    Dijkstra-like breadth-first search
    for all bubble b identified **do**
      merge the two paths of b into one
    end for
    remove nodes with low coverage
  3.2: (optional, not done) Resolve repeats
Step 4: Contig building

    for all maximal linear path p in $C^k(R)$ **do**
      let the sequence spelled by p be a contig
    end for
Step 5: Scaffolding
  Use paired reads, if available, to solve repeats and connect contigs

---

The second version of *Velvet* improves the scaffolding step by including two algorithms called *Pebble* and *Rock Band* [60]. These two algorithms have the role of connecting and filling the gap between contigs by using paired-end reads in the case of *Pebble* and long reads in the case of *Rock Band*.

***AllPaths* [8]** marks a breakthrough with the other assemblers, proposing several original approaches. First, it starts with a read correction method similar to the one used in *Euler* (Step 1, Algorithm 7) but applies it simultaneously to three sets of k-mers, for several values of k. Once the reads are corrected, the assembler continues with a unique value of k. Second, it uses paired reads even in the contig building step. These reads are used to compute groups of reads from R; these groups correspond to overlapping regions of the target genome. Third, it performs a local assembly before the whole-genome assembly. Indeed, for each group of reads corresponding to a region, a local de Bruijn-like graph (called a *sequence graph*) is computed, analyzed, and cleaned before being merged into a global sequence graph. The contigs are deduced after a series of editing steps, whose aim is to correct errors and resolve repeats. As usual, the scaffolding step finishes the assembly. The improved version of *AllPaths* proposed in [20] is called *AllPaths-LG*. It increases the performances of the pre-assembly read error correction procedure of the repeat solving step and of the memory usage and is very efficient [49].

***SOAPdenovo* [31]** closely follows the generic de Bruijn algorithm (Algorithm 6). However, it aims at assembling large genomes (specifically, human genomes) and thus proposes a fine modularization of each step, with sequential execution of the modules under the form of a pipeline, in order to load in the computer memory a limited amount of data at each step. As a consequence of this goal, each step is optimized (and even simplified) for such a modularization. In order to still reduce the needed memory space, the second version of the software, *SOAPdenovo2* [34], implements a sparse graph structure [58] (see also the alternative approaches below) and improves the choice of the k-mers' length.

IDBA [44, 45] iteratively adapts the de Bruijn graph for different values of k in an accumulated de Bruijn graph, thus avoiding the difficulty of choosing a unique value of k. A small value for k allows a more accurate correction of the reads, while a larger value favors the detection of small repeats (which remain inside the same k-mer). IDBA takes advantage of small and large values of k by using an interval $[k_{\min}, k_{\max}]$ instead of a fixed value for k. At each iteration, the de Bruijn graph G^k corresponding to a value k is corrected using techniques similar to *Velvet*, the potential contigs are built, and then G^k is transformed into G^{k+1} by converting each edge of G^k into a vertex of G^{k+1}. Two vertices in G^{k+1} are connected by an arc if the previously obtained contigs verify this connection. At the end, the contigs are connected using information about paired reads to build scaffolds.

12.5.3 Alternative Approaches

Lowering the computational requirements is the major concern for all the assemblers, and those based on a de Bruijn graph propose a number of solutions that we skim through below. In order to reduce memory requirements, the design and implementation of highly efficient data structures are investigated.

***SparseAssembler* [58]** uses a so-called *sparse de Bruijn graph* which replaces the storage of $g + 1$ consecutive k-mers (i.e., a $(k + g)$-mer) in a read with the storage of one *representative* k-mer, thus allowing to reduce the k-mer storage to $1/g$ of its initial size but increasing the memory needed to store the adjacencies of each k-mer. Tests [58] show that the memory savings are increasing when the difference between the length l of the reads and k increases and range from 50% to 90% with respect to the assemblers using a classical de Bruijn graph.

SPAdes [3] is based on a special case of de Bruijn graph called *paired de Bruijn graph* which includes the paired read information directly into its structure. Instead of using simple k-mers as nodes, the paired de Bruijn graph associates a unique pair of k-mers with each node such that the distance on the target genome between the two k-mers from a pair is equal to the insert length of the paired reads. Since in practice the paired reads are not separated by an exact insert length, the paired de Bruijn graph used in SPAdes adds to each node an integer representing the estimated distance between the two k-mers of the node. This distance is computed using an accumulated de Bruijn graph similar to the one used in IDBA. Thus, the first step of SPAdes consists in building and correcting an accumulated de Bruijn graph. During the second step, an estimated distance on the target genome is computed between k-mers based on paths in the accumulated de Bruijn graph. Then the assembler builds the paired de Bruijn graph and constructs the contigs.

***Minia* [14]** uses a probabilistic representation of the de Bruijn graph, proposed in [42] and based on a Bloom filter [4]. A *Bloom filter* is a compact structure storing a set of elements in a bit array with the help of several functions indicating the precise bits that have to be set to 1 when a given element is in the filter. Searching for an element in a Bloom filter thus may have false positives, whose probability of apparition increases with the number of elements inserted in the filter. However, the Bloom filter also has undeniable advantages in terms of space requirements. *Minia* solves the false nodes problem by storing false positives in a supplementary table, thus compromising on the memory requirements. Given the little information stored in a Bloom filter, another additional structure is needed, allowing a breadth-first search of the graph in order to build contigs. The resulting global data structure has a larger size than the initial Bloom filter but still allows important memory savings [27]. Note that there is no scaffolding step in *Minia*.

In order to exploit the full potential of the computer systems, some assemblers are able to distribute both the set of k-mers and the computation of the de Bruijn graph over different computers working in parallel, to recollect the results, and to put them together in order to output a unique assembled sequence. In this way, they benefit from the important memory space offered by the computer cluster and make important running time savings.

ABySS [55] is the first parallelized assembler which distributes the representation of the de Bruijn graph over a network of commodity computers. Two main problems need efficient solutions when such a distributed representation is proposed. First, the sequence of a k-mer must uniquely and efficiently determine the location of the k-mer, i.e., the computer to which the k-mer is affected. Second, the k-mers adjacent to a given k-mer must be stored in a way independent from the location of the given

k-mer. ABySS solves both problems using a base-4 representation for each k-mer (coupled with the use of a hash function), and eight bits per node to indicate which of its up to eight incident arcs are present in the de Bruijn graph.
PASHA [33] is another parallelized de Bruijn graph-based assembler, which is moreover able to exploit the hybrid computing architectures composed of both multiprocessor machines with shared memory and distributed-memory computer clusters. Like ABySS, the main steps are the distribution of the k-mers over the computers, the construction of the de Bruijn graph, and the contig generation. In order to reduce the running time, PASHA implements a different hash function and uses two different threads on each computer: one which performs local computations for the assembly and another for the communications between the different computers. The construction and correction of the de Bruijn graph, as well as the generation of the contigs and scaffolds are similar to the ones used in *Velvet*.

12.5.4 Summary

Assemblers based on a de Bruijn graph have the undeniable advantage of producing a global view of the genome without needing to perform all-to-all comparisons between reads. Moreover, using k-mers instead of reads, the de Bruijn assemblers have no difficulty in mixing reads of different lengths. However, finding the best value for k is another difficult task. A large k allows to store more information about the short repeats (which remain inside the same k-mer) but misses overlaps shorter than the (large) $k - 1$ value. A smaller k loses information about short repeats but captures more information about overlaps between reads. Choosing k is thus making a trade-off between specificity (large k) and sensitivity (small k). Possible solutions include checking different values of k, using a dedicated tool [13], and adopting a multi-k approach (like in IDBA [44] and SPAdes [3]).

Progressively, the quality of the assemblies produced by de Bruijn methods for short reads improves and sometimes equals that produced with Sanger data, for large genomes [16]. However, it must also be noticed that there is no universal supremacy of de Bruijn assemblers with respect to the other types of methods, in the sense that depending on many parameters (sequencing technology, criteria used for comparison, size of the genomes etc.) de Bruijn assemblers may behave very differently from each other, sometimes occupying the both ends of a ranking comprising also greedy-based and OLC assemblers [49]. Generally, de Bruijn based assemblers are suitable for assembling large sequences with short reads, especially when they are parallelized or when they benefit from multithreaded parallelization, as is the case for *AllPaths*, *SOAPdenovo*, and *Velvet*.

12.6 Perspectives and Conclusion

De novo assembly algorithms for NGS data have made a lot of progress recently. Their performances get closer and closer to that of assemblers for Sanger data, in terms of coverage and accuracy [16, 20], and this is mainly due to the de Bruijn approach. Still, the assembly of a genome—and especially of large genomes—remains a very important challenge. There are several reasons for this. First, no assembler is able to guarantee a minimum quantitative or qualitative performance for a given dataset [61]. Second, no assembler performs better (according to the usual criteria) than all the others on all types of datasets [5, 16, 49]. Third, the evaluation protocols for assemblers use numerous criteria [16], which are able altogether to give insights about the behavior of assemblers on general types of genomes, but imply no predictive model for evaluating the expectations one may have from a given assembler on a given data (the Lander–Waterman model [30] successfully used for Sanger data shows inadequate for NGS data [50]). Various research directions are inferable from these observations, in order to improve the result of an assembly process. Here are three of them.

The quality and quantity of the data are essential for the performances of any assembler. Progresses in this direction are systematically made since the emerging third-generation sequencing produces, with much lower cost and a slightly higher error rate, reads of 150–400 bp and, with a larger cost and a much higher error rate, reads of thousands of bp [21].

The major bottleneck of the general assembly problem is the detection and the process of repeats. Repeats already got a lot of attention but are still responsible for many mis-assemblies and missing repetitive parts in the genome [61]. To handle them, the current solutions must be completed with coverage statistics helping to identify repeats, the use of different sizes for the reads or for the inserts between paired reads, and possibly a more global view of the assembly construction.

Another major bottleneck is related to the computational resources, i.e., memory storage and running time. de Bruijn assemblers obviously improved over greedy and OLC assemblers concerning the running time, but their memory requirements are still important in general [27, 61]. New solutions for storing the de Bruijn graph [14, 58] are promising with respect to memory savings, although they seem to compensate with a sometimes very important running time for large genomes [27].

In this chapter, we proposed a detailed algorithmic view of the existing approaches, including both the classical ones and the more recent ones. We preferred rigorous formulations of the problems to (only) intuitive formulations, following the underlying idea that improving the assemblers requires bringing the practice and the theory closer. Practice is the ultimate, true evaluation environment. Theory may highly improve the algorithms, and thus the results, by proposing new approaches to the numerous problems risen by genome assembly.

References

1. Ansorge, W.J.: Next-generation DNA sequencing techniques. New Biotechnol. **25**(4), 195–203 (2009)
2. Ariyaratne, P.N., Sung, W.K.: PE-Assembler: de novo assembler using short paired-end reads. Bioinformatics **27**(2), 167–174 (2011)
3. Bankevich, A., Nurk, S., Antipov, D., Gurevich, A.A., Dvorkin, M., Kulikov, A.S., Lesin, V.M., Nikolenko, S.I., Pham, S., Prjibelski, A.D., et al.: SPAdes: a new genome assembly algorithm and its applications to single-cell sequencing. J. Comput. Biol. **19**(5), 455–477 (2012)
4. Bloom, B.H.: Space/time trade-offs in hash coding with allowable errors. Commun. ACM **13**(7), 422–426 (1970)
5. Bradnam, K.R., Fass, J.N., Alexandrov, A., Baranay, P., Bechner, M., Birol, I., Boisvert, S., Chapman, J.A., Chapuis, G., Chikhi, R., et al.: Assemblathon 2: evaluating de novo methods of genome assembly in three vertebrate species. GigaScience **2**(1), 1–31 (2013)
6. Bresler, M., Sheehan, S., Chan, A.H., Song, Y.S.: Telescoper: de novo assembly of highly repetitive regions. Bioinformatics **28**(18), i311–i317 (2012)
7. Bryant, D.W., Wong, W.K., Mockler, T.C.: QSRA–a quality-value guided de novo short read assembler. BMC Bioinf. **10**(1), 69 (2009)
8. Butler, J., MacCallum, I., Kleber, M., Shlyakhter, I.A., Belmonte, M.K., Lander, E.S., Nusbaum, C., Jaffe, D.B.: ALLPATHS: de novo assembly of whole-genome shotgun microreads. Genome Res. **18**(5), 810–820 (2008)
9. Carneiro, M.O., Russ, C., Ross, M.G., Gabriel, S.B., Nusbaum, C., DePristo, M.A.: Pacific biosciences sequencing technology for genotyping and variation discovery in human data. BMC Genomics **13**(1), 375 (2012)
10. Chaisson, M.J., Pevzner, P.A.: Short read fragment assembly of bacterial genomes. Genome Res. **18**(2), 324–330 (2008)
11. Chaisson, M.J., Brinza, D., Pevzner, P.A.: De novo fragment assembly with short mate-paired reads: Does the read length matter? Genome Res. **19**(2), 336–346 (2009)
12. Chang, Y.J., Chen, C.C., Chen, C.L., Ho, J.M.: A de novo next generation genomic sequence assembler based on string graph and MapReduce cloud computing framework. BMC Genomics **13**(Suppl. 7), S28 (2012)
13. Chikhi, R., Medvedev, P.: Informed and automated k-mer size selection for genome assembly. Bioinformatics **30**, 31–37 (2014)
14. Chikhi, R., Rizk, G.: Space-efficient and exact de Bruijn graph representation based on a Bloom filter. In: Proceedings of WABI 2012. LNBI, vol. 7534, pp. 236–248. Springer, Heidelberg (2012)
15. Dohm, J.C., Lottaz, C., Borodina, T., Himmelbauer, H.: SHARCGS, a fast and highly accurate short-read assembly algorithm for de novo genomic sequencing. Genome Res. **17**(11), 1697–1706 (2007)
16. Earl, D., Bradnam, K., John, J.S., Darling, A., Lin, D., Fass, J., Yu, H.O.K., Buffalo, V., Zerbino, D.R., Diekhans, M., et al.: Assemblathon 1: a competitive assessment of de novo short read assembly methods. Genome Res. **21**(12), 2224–2241 (2011)
17. El-Metwally, S., Hamza, T., Zakaria, M., Helmy, M.: Next-generation sequence assembly: four stages of data processing and computational challenges. PLoS Comput. Biol. **9**(12), e1003,345+ (2013). doi:10.1371/journal.pcbi.1003345
18. Ferragina, P., Manzini, G.: Opportunistic data structures with applications. In: Proceedings of FOCS'00, pp. 390–398 (2000)
19. Garey, M.R., Johnson, D.S.: Computers and Intractability: A Guide to NP-Completeness. Freeman, New York (1979)
20. Gnerre, S., MacCallum, I., Przybylski, D., Ribeiro, F.J., Burton, J.N., Walker, B.J., Sharpe, T., Hall, G., Shea, T.P., Sykes, S., et al.: High-quality draft assemblies of mammalian genomes from massively parallel sequence data. Proc. Natl. Acad. Sci. USA **108**(4), 1513–1518 (2011)

21. Henson, J., Tischler, G., Ning, Z.: Next-generation sequencing and large genome assemblies. Pharmacogenomics **13**(8), 901–915 (2012)
22. Hernandez, D., François, P., Farinelli, L., Østerås, M., Schrenzel, J.: De novo bacterial genome sequencing: millions of very short reads assembled on a desktop computer. Genome Res. **18**(5), 802–809 (2008)
23. Hossain, M.S., Azimi, N., Skiena, S.: Crystallizing short-read assemblies around seeds. BMC Bioinform. **10**(Suppl. 1), S16 (2009)
24. Idury, R.M., Waterman, M.S.: A new algorithm for DNA sequence assembly. J. Comput. Biol. **2**(2), 291–306 (1995)
25. Jeck, W.R., Reinhardt, J.A., Baltrus, D.A., Hickenbotham, M.T., Magrini, V., Mardis, E.R., Dangl, J.L., Jones, C.D.: Extending assembly of short DNA sequences to handle error. Bioinformatics **23**(21), 2942–2944 (2007)
26. Kececioglu, J.D., Myers, E.W.: Combinatorial algorithms for DNA sequence assembly. Algorithmica **13**, 7–51 (1993)
27. Kleftogiannis, D., Kalnis, P., Bajic, V.B.: Comparing memory-efficient genome assemblers on stand-alone and cloud infrastructures. PLoS One **8**(9), e75505 (2013)
28. Koren, S., Harhay, G.P., Smith, T., Bono, J.L., Harhay, D.M., Mcvey, S.D., Radune, D., Bergman, N.H., Phillippy, A.M.: Reducing assembly complexity of microbial genomes with single-molecule sequencing. Genome Biol. **14**(9), R101 (2013)
29. Landau, G.M., Vishkin, U.: Introducing efficient parallelism into approximate string matching and a new serial algorithm. In: Proceedings of the Eighteenth Annual ACM Symposium on Theory of Computing, STOC '86, pp. 220–230. ACM, New York (1986). doi:10.1145/12130.12152. http://doi.acm.org/10.1145/12130.12152
30. Lander, E.S., Waterman, M.S.: Genomic mapping by fingerprinting random clones: a mathematical analysis. Genomics **2**(3), 231–239 (1988)
31. Li, R., Zhu, H., Ruan, J., Qian, W., Fang, X., Shi, Z., Li, Y., Li, S., Shan, G., Kristiansen, K., et al.: De novo assembly of human genomes with massively parallel short read sequencing. Genome Res. **20**(2), 265–272 (2010)
32. Lin, Y., Li, J., Shen, H., Zhang, L., Papasian, C.J., et al.: Comparative studies of de novo assembly tools for next-generation sequencing technologies. Bioinformatics **27**(15), 2031–2037 (2011)
33. Liu, Y., Schmidt, B., Maskell, D.: Parallelized short read assembly of large genomes using de Bruijn graphs. BMC Bioinform. **12**, 354 (2011)
34. Luo, R., Liu, B., Xie, Y., Li, Z., Huang, W., Yuan, J., He, G., Chen, Y., Pan, Q., Liu, Y., et al.: SOAPdenovo2: an empirically improved memory-efficient short-read de novo assembler. GigaScience **1**(1), 1–6 (2012)
35. Manber, U., Myers, G.: Suffix arrays: a new method for on-line string searches. In: Proceedings of the First Annual ACM-SIAM Symposium on Discrete Algorithms, SODA '90, pp. 319–327. Society for Industrial and Applied Mathematics, Philadelphia, PA (1990). http://dl.acm.org/citation.cfm?id=320176.320218
36. Medvedev, P., Georgiou, K., Myers, G., Brudno, M.: Computability of models for sequence assembly. In: Proceedings of WABI 2007. LNCS, vol. 4645, pp. 289–301. Springer, Heidelberg (2007)
37. Miller, J.R., Delcher, A.L., Koren, S., Venter, E., Walenz, B.P., Brownley, A., Johnson, J., Li, K., Mobarry, C., Sutton, G.: Aggressive assembly of pyrosequencing reads with mates. Bioinformatics **24**(24), 2818–2824 (2008)
38. Myers, E.W.: The fragment assembly string graph. Bioinformatics **21**(Suppl. 2), ii79–ii85 (2005)
39. Myers, E.W., Sutton, G.G., Delcher, A.L., Dew, I.M., Fasulo, D.P., Flanigan, M.J., Kravitz, S.A., Mobarry, C.M., Reinert, K.H., Remington, K.A., et al.: A whole-genome assembly of Drosophila. Science **287**(5461), 2196–2204 (2000)
40. Nagarajan, N., Pop, M.: Parametric complexity of sequence assembly: theory and applications to next generation sequencing. J. Comput. Biol. **16**(7), 897–908 (2009)

41. Narzisi, G., Mishra, B.: Scoring-and-unfolding trimmed tree assembler: concepts, constructs and comparisons. Bioinformatics **27**(2), 153–160 (2011)
42. Pell, J., Hintze, A., Canino-Koning, R., Howe, A., Tiedje, J.M., Brown, C.T.: Scaling metagenome sequence assembly with probabilistic de Bruijn graphs. Proc. Natl. Acad. Sci. USA **109**(33), 13272–13277 (2012)
43. Peltola, H., Söderlund, H., Ukkonen, E.: SEQAID: a DNA sequence assembling program based on a mathematical model. Nucleic Acids Res. **12**(1), 307–321 (1984)
44. Peng, Y., Leung, H.C., Yiu, S.M., Chin, F.Y.: IDBA–a practical iterative de Bruijn graph de novo assembler. In: Research in Computational Molecular Biology, pp. 426–440. Springer, Heidelberg (2010)
45. Peng, Y., Leung, H.C., Yiu, S.M., Chin, F.Y.: Idba-ud: a de novo assembler for single-cell and metagenomic sequencing data with highly uneven depth. Bioinformatics **28**(11), 1420–1428 (2012)
46. Pevzner, P.A., Tang, H.: Fragment assembly with double-barreled data. Bioinformatics **17**(Suppl. 1), S225–S233 (2001)
47. Pevzner, P.A., Tang, H., Waterman, M.S.: An Eulerian path approach to DNA fragment assembly. Proc. Natl. Acad. Sci. USA **98**(17), 9748–9753 (2001)
48. Pop, M., Phillippy, A., Delcher, A.L., Salzberg, S.L.: Comparative genome assembly. Brief. Bioinform. **5**(3), 237–248 (2004)
49. Salzberg, S.L., Phillippy, A.M., Zimin, A., Puiu, D., Magoc, T., Koren, S., Treangen, T.J., Schatz, M.C., Delcher, A.L., Roberts, M., et al.: GAGE: a critical evaluation of genome assemblies and assembly algorithms. Genome Res. **22**(3), 557–567 (2012)
50. Schatz, M.C., Delcher, A.L., Salzberg, S.L.: Assembly of large genomes using second-generation sequencing. Genome Res. **20**(9), 1165–1173 (2010)
51. Schmidt, B., Sinha, R., Beresford-Smith, B., Puglisi, S.J.: A fast hybrid short read fragment assembly algorithm. Bioinformatics **25**(17), 2279–2280 (2009)
52. Sim, J.S., Park, K.: The consensus string problem for a metric is NP-complete. J. Discrete Algorithms **1**(1), 111–117 (2003). doi:10.1016/S1570–8667(03)00011-X
53. Simpson, J.T., Durbin, R.: Efficient construction of an assembly string graph using the FM-index. Bioinformatics **26**(12), i367–i373 (2010)
54. Simpson, J.T., Durbin, R.: Efficient de novo assembly of large genomes using compressed data structures. Genome Res. **22**(3), 549–556 (2012)
55. Simpson, J.T., Wong, K., Jackman, S.D., Schein, J.E., Jones, S.J., Birol, I.: ABySS: a parallel assembler for short read sequence data. Genome Res. **19**(6), 1117–1123 (2009)
56. Smith, T.F., Waterman, M.S.: Identification of common molecular subsequences. J. Mol. Biol. **147**(1), 195–197 (1981)
57. Warren, R.L., Sutton, G.G., Jones, S.J., Holt, R.A.: Assembling millions of short DNA sequences using SSAKE. Bioinformatics **23**(4), 500–501 (2007)
58. Ye, C., Ma, Z.S., Cannon, C.H., Pop, M., Yu, D.W.: Exploiting sparseness in de novo genome assembly. BMC Bioinform. **13**(Suppl. 6), S1 (2012)
59. Zerbino, D.R., Birney, E.: Velvet: algorithms for de novo short read assembly using de Bruijn graphs. Genome Res. **18**(5), 821–829 (2008)
60. Zerbino, D.R., McEwen, G.K., Margulies, E.H., Birney, E.: Pebble and rock band: heuristic resolution of repeats and scaffolding in the velvet short-read de Novo assembler. PLoS One **4**(12), e8407 (2009)
61. Zhang, W., Chen, J., Yang, Y., Tang, Y., Shang, J., Shen, B.: A practical comparison of de novo genome assembly software tools for next-generation sequencing technologies. PLoS One **6**(3), e17,915 (2011)

Chapter 13
An Efficient Approach to Merging Paired-End Reads and Incorporation of Uncertainties

Tomáš Flouri, Jiajie Zhang, Lucas Czech, Kassian Kobert, and Alexandros Stamatakis

13.1 Introduction

Next-Generation Sequencing (NGS) technologies have reshaped the landscape of life sciences. The massive amount of data generated by NGS is rapidly transforming biological research from traditional wet-lab work into a data-intensive analytical discipline [1]. The *Illumina* "sequencing by synthesis" technique [2] is one of the most popular and widely used NGS technologies. The sequences produced by the *Illumina* platform are short reads, typically in the range of 75- to 300-bp. The short length of these reads poses various challenges for data analysis. Note that, the *Illumina* platforms also have the ability to produce paired-end reads, which are sequenced from both ends of the target DNA fragment. With proper experimental design, these pairs of paired-end reads can be merged into longer reads, which may potentially almost double the read length.

There already exist state-of-art production-level paired-end read mergers, such as PEAR [3], PANDAseq [4], FLASH [5], and VSEARCH [6]. However, one challenge for merging the paired-end reads is the sheer size of the data. For instance, the *Illumina HiSeq* platform can produce eight billion paired-end reads in a single run, which amounts to 1Tb of data (as of October 2014). Because merging the paired-end reads is often the first step in most data analysis pipelines, users usually

T. Flouri (✉) • J. Zhang • L. Czech • K. Kobert
Heidelberg Institute for Theoretical Studies, Heidelberg, Germany
e-mail: Tomas.Flouri@h-its.org; bestzhangjiajie@gmail.com; Lucas.Czech@h-its.org; Kassian.Kobert@h-its.org

A. Stamatakis
Heidelberg Institute for Theoretical Studies, Heidelberg, Germany

Institute for Theoretical Informatics, Karlsruhe Institute of Technology, Karlsruhe, Germany
e-mail: Alexandros.Stamatakis@h-its.org

M. Elloumi (ed.), *Algorithms for Next-Generation Sequencing Data*,
DOI 10.1007/978-3-319-59826-0_13

run paired-end read mergers multiple times with different parameters to improve the quality of the results. Thus, merger performance represents a bottleneck.

Once the paired-end reads have been merged, one can assemble the reads with a de novo assembler, for shotgun sequenced genomes [7]. De novo assembly is based on computing overlaps among short reads to build longer target sequences called *contigs*. A naïve but accurate approach is to compute overlaps between every pair of reads and merge the pair with the best overlap. However, this involves $\mathcal{O}(n^2)$ overlap computations between reads pairs, where n is the number of reads. Thus, this approach is rarely used for large data sets due to the lack of efficient overlap computation algorithms.

This chapter is divided into two parts. We first present an algorithm for finding overlapping regions between two short reads without considering gaps (as gaps are infrequent on *Illumina* platforms [8]). It follows the same principles of *Dynamic Programming* (DP) as algorithms for pairwise sequence alignment [9–11] or *edit distance* (*Levenshtein distance*) computation [12]. In our case, insertions and deletions (gaps) are not taken into account, and hence, the algorithm is closely related to computing the *Hamming distance* between two equally long segments of sequences (see [13]). The difference to the *Hamming distance* is that we use a scoring matrix to assign a score to a pair of bases, instead of the value 1 for a match and 0 for a mismatch. Moreover, we devise a novel, vectorization scheme for speeding up the computation which, as we will show, achieves optimal speed-ups (i.e., speed-up proportional to the vector length) compared to the non-vectorized approach. Our scheme is an adapted version of the vectorization used for computing the DP table of Smith and Waterman for calculating local pair-wise sequence alignments [14].

In the second part, we focus on quantifying uncertainties in overlap computations. One characteristic of the *Illumina* platform is that the quality of base calls decreases toward the end of the sequence. Therefore, the overlapping region of the paired-end reads often contains the low-quality ends of the reads. The raw *Illumina* data come with a quality score for each base call, representing the error probability. We will discuss how these quality scores can be used to improve merging accuracy.

The remainder of this chapter is organized as follows: In Sect. 13.2, we introduce basic definitions that will be used throughout the chapter. In Sect. 13.3, we first formally define the problem of finding overlaps between two sequences; we then describe the sliding window approach and show how to convert it into a DP formulation. Finally, we extend the DP approach to a vectorized scheme which yields optimal speed-up. In Sect. 13.4, we discuss how to utilize the quality scores of the *Illumina* reads to improve merging accuracy. Finally, Sect. 13.5 provides a summary of the topics discussed in this chapter.

13.2 Preliminaries

A *string* is a finite, possibly empty, sequence of symbols taken from a finite non-empty set Σ, which we call *alphabet*. We denote the empty string by ε. Further, we denote the length of a string x by $|x|$ and the concatenation of two strings x and y, by xy. Now, let x, y, w and z be strings such that $y = wxz$. String x is called a *substring* of y. If $w \neq \varepsilon$, then w is a *prefix* of y. On the other hand, if $z \neq \varepsilon$, then z is a *suffix* of y. We use $Pref(x)$, $Suff(x)$, and $Subs(x)$ to denote the set of prefixes, suffixes, and substrings of x, respectively. The i-th character of a string x is denoted by x_i, starting from 0. Substrings of x are denoted by $x[i \ldots j]$, for $0 \leq i \leq j < |x|$, where i and j are the positions of the first and last symbol of a substring within x. To avoid duplicate computations that would arise had we considered a string as prefix (resp. suffix) of itself, we define the notion of *proper* prefix (and suffix), which in addition requires that $xz \neq \varepsilon$ ($wx \neq \varepsilon$). We implicitly assume prefixes and suffixes as proper, unless otherwise stated.

In the rest of the text, we work with biological sequences (strings), in particular nucleotide data, and hence, we assume that the alphabet Σ consists of four symbols (`A`,`C`,`G`, and `T`) which we refer to as nucleotides. We also use the term *sequence* and *string* interchangeably. All algorithms are presented for the four nucleotides, although they can be extended to arbitrary finite alphabets, such as the alphabet for all 15 degenerate nucleotide characters (see Sect. 13.4).

We distinguish among three types of overlaps among strings x and y and describe them using the following three definitions:

Definition 1 (Suffix–Prefix Overlap) A *Suffix–Prefix* overlap between strings x and y is a pair $(u, v) \in Suff(x) \times Pref(y)$ such that $|u| = |v|$. The set of all overlapping regions between suffixes of x and prefixes of y is defined by:

$$\overrightarrow{\mathscr{C}}_{x,y} = \{(u, v) \mid (u, v) \in Suff(x) \times Pref(y) \land |u| = |v|\}.$$

Definition 2 (Prefix–Suffix Overlap) A *Prefix–Suffix* overlap between strings x and y is a pair $(u, v) \in Pref(x) \times Suff(y)$ such that $|u| = |v|$. The set of all overlapping regions between prefixes of x and suffixes of y is defined by:

$$\overleftarrow{\mathscr{C}}_{x,y} = \{(u, v) \mid (u, v) \in Pref(x) \times Suff(y) \land |u| = |v|\}.$$

Definition 3 (Substring Overlap) A *Substring* overlap for two strings x and y is a pair $(x, u) \in \{x\} \times Subs(y)$ (resp. $(u, y) \in Subs(x) \times \{y\}$) such that $|x| = |u|$ (resp. $|y| = |u|$), given that $|x| > |y|$ resp. $|x| \leq |y|$. The set of all substring overlaps between x and y is defined by:

$$\overline{\mathscr{C}}_{x,y} = \begin{cases} \{(x, z) \mid z \in Subs(y) \land |z| = |x|\} : |x| \leq |y| \\ \{(z, y) \mid z \in Subs(x) \land |z| = |y|\} : |y| < |x|. \end{cases}$$

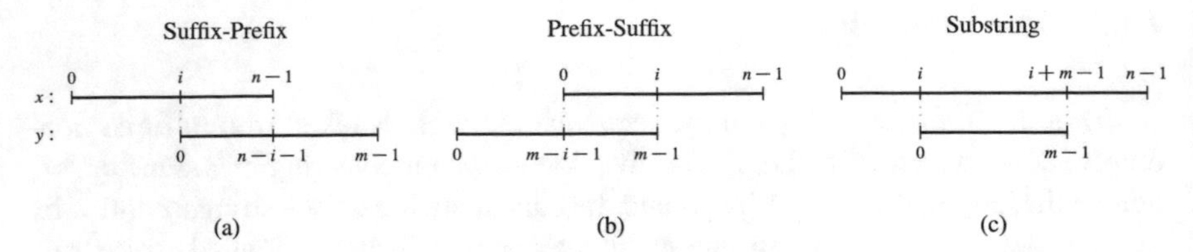

Fig. 13.1 The three types of overlapping regions between two strings x and y of lengths n and m. (**a**) illustrates the Suffix–Prefix case. There are exactly $n - \max(0, n-m) - 1$ overlapping regions $(x[i \ldots n-1], y[0 \ldots n-i-1])$ for $1 + \max(0, n-m) \leq i \leq n-1$. (**b**) illustrates the Prefix–Suffix case. There are exactly $\min(m, n) - 1$ overlapping regions of the form $(x[0 \ldots i], y[m-i-1 \ldots m-1])$, for $0 \leq i < \min(m, n) - 1$. (**c**) illustrates the substring overlap case. There are exactly $\max(n-m, m-n) + 1$ overlapping regions of the form $(x[i \ldots i+m0-1], y[0 \ldots m-1])$ if $n > m$ or $(x[0 \ldots n-1], y[i \ldots i+n-1])$ if $n < m$, and (x, y) if $n = m$, for $0 \leq i \leq n-m$

In the rest of the text, we use the term *overlap* to refer to elements of $\mathscr{C}$, which we formally define as the union of the three previously defined sets, that is:

$$\mathscr{C}_{x,y} = \overrightarrow{\mathscr{C}}_{x,y} \cup \overleftarrow{\mathscr{C}}_{x,y} \cup \overline{\mathscr{C}}_{x,y}$$

Figure 13.1 illustrates the three types of overlaps. If strings x and y are already given, we may use $\mathscr{C}$ instead of $\mathscr{C}_{x,y}$. Moreover, for simplicity, we define the operation $|c|$ for an element $c \in \mathscr{C}$ as the length of the overlap, that is, the length of one of its two components.

13.3 Finding Overlaps Between Two Sequences

Given two sequences and a scoring matrix, our goal is to find the overlap that yields the highest score. Typically, a scoring matrix rewards matches and penalizes mismatches. Let

$$s : \Sigma \times \Sigma \rightarrow \mathbb{Z}$$

be a mapping between pairs of symbols and scores, and let $s(u, v)$ be the value of the scoring matrix for symbols u and v. We evaluate the *Alignment Score* (AS) of each possible overlap $C = (x, y) \in \mathscr{C}$ by:

$$AS(C) = \sum_{i=0..|C|-1} s(x_i, y_i). \tag{13.1}$$

We then consider the *best* overlap to be the overlap that maximizes the AS. The two sequences we want to overlap are referred to as the reference sequence r and the query sequence q. Note that the choice of the scoring matrix influences the probability of finding the *true* overlap, that is, the real merging region for the two sequences. Also, the best overlap might not necessarily be the true overlap. For instance, a scoring matrix where matches are rewarded by positives scores and mismatches are assigned a zero value will have a high false-positive rate of finding long overlaps between two random sequences.

In what follows, an example of an overlap between two sequences is provided:

```
r = G T T C G C T A A T A C
                | | |  .  | |
q =             T A A G A C T  A A C
```

For simplicity, assume our scoring function rewards matches with a score of 1 and penalizes mismatches with a penalty of -1, that is $s(u, v) = -1$ if and only if $u \neq v$, and $s(u, u) = 1$ otherwise. As there are five matches and one mismatch in the example of overlap above, the AS is 4. This is the highest AS for any possible overlap between the two sequences (given the current scoring matrix), and therefore, we consider it as the best overlap.

Using a scoring matrix for evaluating a sequence alignment (in our case, the overlap) is a standard approach that is widely used, for instance, in BLAST [15], Bowtie2 [16], and PEAR [3].

Keep in mind that the best overlap does not necessarily correspond to the true overlap (as the two sequences may not overlap at all and the best overlap found by the scoring matrix is just a random phenomenon; sequencing errors may also prevent us from finding the true overlap [3]). Therefore, the significance of the overlap identified by this approach can be determined by a user-specified threshold for the minimal overlap size (such as in FLASH [5]) or via a statistical test (such as in PEAR [3]).

The following subsections present three methods for finding the best overlap. We first present a simple method that uses a sliding window to compute the AS of every overlapping region. We then extend it to a DP formulation. Based on this DP method, we subsequently present a novel vectorized algorithm.

13.3.1 Sliding Window Approach

The sliding window approach directly follows Definitions 1–3 to evaluate the AS of all overlaps $\mathscr{C}_{x,y}$ of sequences x and y of lengths n and m, respectively. First, the Suffix–Prefix overlaps $(u, v) \in \overrightarrow{\mathscr{C}}_{x,y}$ are considered, which correspond to $u = x[i \ldots n-1]$ and $v = y[0 \ldots n-i-1]$, for $1 + \max(0, n-m) \leq i \leq n-1$, by sliding i through the values of its domain, and thus forming a "sliding window" that indicates the overlap (see Fig. 13.1a). Then all substring overlaps are evaluated by iterating over all elements $(u, v) \in \overline{\mathscr{C}}_{x,y}$, which depend on whether $|x| > |y|$. If the inequality holds, then $u = x[i \ldots i+m-1]$ and $v = y[0 \ldots m-1]$, for

$0 \leq i \leq |x| - |y|$. In the other case, $u = x[0 \ldots n-1]$ and $v = y[i \ldots i+n-1]$, for $0 \leq i \leq |y| - |y|$ (Fig. 13.1c). Finally, we evaluate the set of Prefix–Suffix overlaps $(u, v) \in \overrightarrow{\mathscr{C}}_{x,y}$, where u and v correspond to $x[0 \ldots i]$ and $y[m-i-1 \ldots m-1]$, such that $0 \leq i < \min(m, n) - 1$ (Fig. 13.1b).

Theorem 1 (Complexity) *The sliding window approach runs in $\Theta(nm)$ time.*

Proof The sliding window approach progressively evaluates all overlaps in order to determine the one that yields the highest score. Set $\overrightarrow{\mathscr{C}}_{x,y}$ consists of $n - \max(0, n - m) - 1$ elements that represent Suffix–Prefix overlaps of sizes $1, 2, \ldots, n - \max(0, n - m) - 1$. This requires $(n - \max(0, n-m))(n - \max(0, n-m) - 1)/2$ base pair comparisons. If we assume $n > m$, it amounts to $m(m-1)/2$ base pair comparisons. On the other hand, if $n \leq m$, the number of comparisons is $n(n-1)/2$. The set $\overleftarrow{\mathscr{C}}_{x,y}$ consists of $\min(m, n) - 1$ elements that represent Prefix–Suffix overlaps of sizes $1, 2, \ldots, \min(m, n) - 1$. This leads to $\min(m, n)(\min(m, n) - 1)/2$ comparisons of base pairs. Again, the number of comparisons can be rewritten as $m(m-1)/2$ if $n > m$, or $n(n-1)/2$ otherwise. Finally, $\overline{\mathscr{C}}_{x,y}$ comprises of $\max(n - m, m - n) + 1$ elements of size $\min(m, n)$ and amounts to $(\max(n - m, m - n) + 1)\min(m, n)$ comparisons of base pairs, that is, $(n - m)m$ comparisons if $n > m$, $(m - n)n$ if $n < m$, or no comparisons if $m = n$. The sum of performed comparisons over all cases gives the total number of exactly nm comparisons. Therefore, we have proven that the asymptotic complexity of the sliding window approach is $\Theta(nm)$. □

13.3.2 Dynamic Programming Approach

There are two approaches to parallelize the overlap computation of two sequences on a single CPU. One way is to evaluate several base pairs of a single overlapping region simultaneously. The other method is to simultaneously evaluate base pairs of multiple overlapping regions. To maximize the efficiency of the parallelization scheme, it is necessary to maximize the number of elements being processed in `x86` vector units per CPU cycle. Hence, an efficient memory layout is also important for attaining optimal performance.

Here, we focus on designing a memory layout for evaluating base pairs of multiple overlapping regions simultaneously. We focus on this vectorization method for two reasons. First, overlapping regions of the same type (e.g., Prefix–Suffix) have varying lengths, which complicates the design of an efficient memory layout. Second, the AS is the sum of scores of every base pair in a single overlapping region. Since the base pair scores are computed in parallel, additional instructions are required after this initial computation to obtain the AS of the overlapping region by computing the sum over those scores. We resort to a DP algorithm that uses a matrix to compute the scores of all overlaps, in analogy to DP algorithms for computing the *Hamming* and *edit* distances of two strings (see [17] for a review).

Given two sequences x and y of lengths n and m, we construct a matrix D of size $(n+1)(m+1)$. Each cell $D(i,j)$ stores the score of the corresponding base pair, that is, $s(x_{i-1}, y_{j-1})$. Note that we initialize the first row and first column with zeroes to define a recurrent DP formula. As proved in the previous section, there exist $\min(m,n) - 1$ prefix–Suffix overlap regions. Their AS can be computed by summing the cells of diagonals of the matrix ending at cells $D(i,m)$, where $0 < i \leq \min(n,m) - 1$. The lengths of those diagonals are $1, 2, \ldots, \min(n,m) - 1$, and each diagonal ending at $D(i,m)$ corresponds to the scores of base pairs in the Prefix–Suffix overlap $(x[0 \ldots i-1], y[m-i \ldots m-1])$. Given the cell $D(i,m)$, we can compute the score of its respective overlap C by summing over the elements of the diagonal, that is,

$$AS(C) = \sum_{k=0}^{i-1} D(i-k, m-k) \tag{13.2}$$

Analogously, cells $D(n,j)$ for $0 < j \leq \min(n,m) - 1$ denote Suffix–Prefix overlaps, and their diagonals contain the scores of $(x[n-j \ldots n-1], y[0 \ldots j-1])$. Their AS for an overlap C represented by cell (n,h) is computed as

$$AS(C) = \sum_{k=0}^{j-1} D(n-k, j-k) \tag{13.3}$$

Finally, cells $D(i,m)$ and $D(n,j)$ for $\min(n,m) \leq i \leq n$ and $\min(n,m) \leq j \leq m$ represent substring overlaps, and the AS of the respective overlapping regions can be computed using Eqs. (13.2) and (13.3).

To avoid a second traversal of the DP matrix, we can compute the AS on the fly by storing in each cell $D(i,j)$ the sum of all previously computed diagonal cells $D(i-k, j-k)$, $0 < k \leq \min(i,j)$ plus the score $s(x_{i-1}, y_{j-1})$. We may then formulate the matrix cell computation in the following recurrent way:

$$D(i,j) = \begin{cases} 0 & : \quad i = 0,\ j \leq m \\ 0 & : \quad j = 0,\ i \leq n \\ D(i-1, j-1) + s(x_{i-1}, y_{j-1}) & : \quad 0 < i \leq n,\ 0 < j \leq m \end{cases}$$

Figure 13.2 illustrates the resulting matrix for two strings, $x =$ TAAAGACTAAC and $y =$ GTTCGCTAATAC.

Theorem 2 (Complexity) *Given two sequences x and y of lengths n and m, the DP algorithm runs in $\Theta(nm)$ time and requires $\Theta(nm)$ space.*

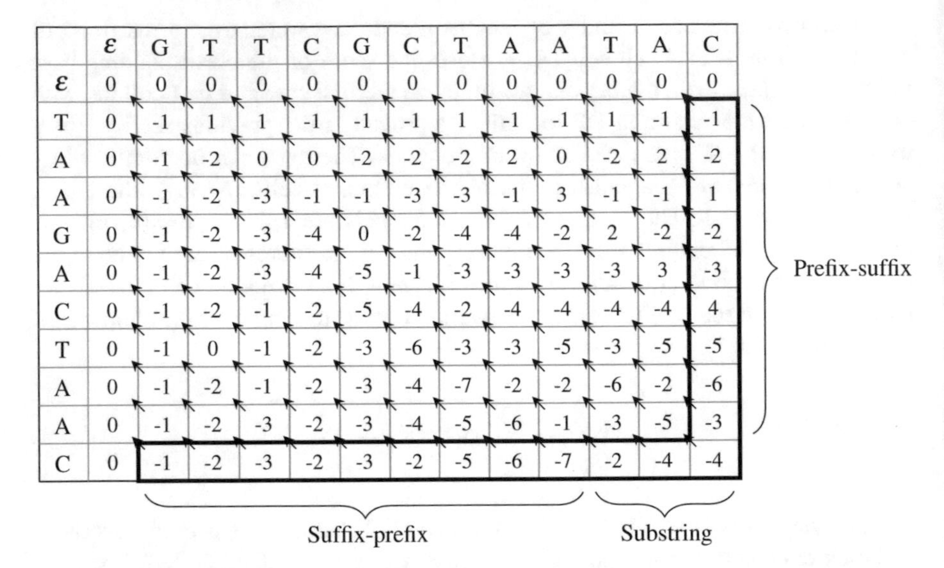

	ε	G	T	T	C	G	C	T	A	A	T	A	C
ε	0	0	0	0	0	0	0	0	0	0	0	0	0
T	0	-1	1	1	-1	-1	-1	1	-1	-1	1	-1	-1
A	0	-1	-2	0	0	-2	-2	-2	2	0	-2	2	-2
A	0	-1	-2	-3	-1	-1	-3	-3	-1	3	-1	-1	1
G	0	-1	-2	-3	-4	0	-2	-4	-4	-2	2	-2	-2
A	0	-1	-2	-3	-4	-5	-1	-3	-3	-3	-3	3	-3
C	0	-1	-2	-1	-2	-5	-4	-2	-4	-4	-4	-4	4
T	0	-1	0	-1	-2	-3	-6	-3	-3	-5	-3	-5	-5
A	0	-1	-2	-1	-2	-3	-4	-7	-2	-2	-6	-2	-6
A	0	-1	-2	-3	-2	-3	-4	-5	-6	-1	-3	-5	-3
C	0	-1	-2	-3	-2	-3	-2	-5	-6	-7	-2	-4	-4

Fig. 13.2 The DP solution for computing overlaps between sequences $x =$ TAAAGACTAAC and $y =$ GTTCGCTAATAC. The *arrows* indicate the dependency of each cell with its upper left diagonal cell. Each cell $D(i,j)$ of the matrix corresponds to the AS of a specific overlapping region of $\mathscr{C}_{x[0...i-1],y[0...j-1]}$, depending on the concrete values of i and j. If $i = |x| - 1$ and $0 \leq j < \min(|x|, |y|) - 1$, then it is a Suffix–Prefix; otherwise, if $j \geq \min(|x|, |y|) - 1$ it is a substring overlap. Similarly, if $j = |y| - 1$ and $0 \leq i < |x| - \max(0, |x| - |y|) - 1$, then the value represents the AS of a Prefix–Suffix overlap; otherwise if $i > |x| - \max(0, |x| - |y|) - 1$, then it is a substring overlap. The three cases for x and y are illustrated in the figure

Proof Each cell $D(i,j)$ is the sum of two values: the score $s(x_{i-1}, y_{j-1})$ and the value of cell $D(i-1, j-1)$. We have $n + m + 1$ initialization steps (first row and first column). Then, the computation of the $n \times m$ cells requires $\mathscr{O}(1)$ time per cell, and hence, the asymptotic complexity is $\Theta(nm)$. Moreover, since we store the complete matrix in memory, the space requirements are also $\Theta(nm)$. □

We can further reduce the memory footprint from $\Theta(nm)$ to $\Theta(m + n)$ by computing the matrix on a column-by-column basis. At each iteration, we only store the current column in memory. Computation of the next column then uses the results from the previous column which is stored in memory. The new values overwrite the previous values in the storage column. This reduces the memory footprint to $\Theta(n)$. Since we need to store the bottom value of each column in memory as the final result (the bottom row of the matrix) along with the last column (rightmost column of matrix), we require a total space of $\Theta(n + m)$.

Figure 13.3 illustrates the approach in detail. Lines 2–5 initialize the array `HH` of size `QLEN` (query sequence length). The outer loop at line 6 iterates over the nucleotides of the reference sequence `RSEQ`. The inner loop iterates over the nucleotides of the query sequence `QSEQ` and fills the corresponding column at lines 9–15 according to the score $s(i,j)$ of the nucleotides `QSEQ[i]` and `RSEQ[j]`. The

```
OVERLAP(RSEQ,QSEQ,RLEN,QLEN)
                                               RSEQ is the reference sequence
                                               QSEQ is the query sequence
                                               RLEN is the length of the RSEQ
                                               QLEN is the length of the QSEQ

 1. INTEGER i,j                                Loop indices
 2. FOR i=0 TO QLEN-1 DO
 3. {
 4.     HH[i] = 0                              Initialize HH-array of H-values
 5. }

 6. FOR j=0 TO RLEN-1 DO                       For each nucleotide in the reference sequence
 7. {
 8.     H = 0                                  Initialize HH-array of H-values
 9.     FOR i=0 TO QLEN-1 DO                   For each nucleotide along the query sequence
10.     {
11.         X = HH[i]                          Save old H for next round
12.         H = H + score[QSEQ[i]][RSEQ[j]]    Add score of two nucleotide to value from diagonal

13.         HH[i] = H                          Store the computer value in HH-array
14.         H = X                              Retrieve saved value to act as diagonal value
15.     }
16.     EE[j] = HH[qlen-1]                     Store last value of HH-array in EE
17. }
```

Fig. 13.3 Pseudo-code for the non-vectorized approach of computing the scores of all overlapping regions

following operations are performed at each iteration j of the outer loop and i of the inner loop:

1. Store the value of HH[i] from the previous (inner loop) iteration $i - 1$ in a temporary variable X (line 11). This value will be needed in the next inner loop iteration since it is the upper-left diagonal cell of the next cell (below the current one) that will be computed.
2. Add the score value of the two nucleotides QSEQ[i] and RSEQ[j] to the value of H (which contains the value of the upper-left diagonal cell).
3. Store H at HH[i].
4. Retrieve the value saved in X, and place it to H. Now, H holds the value of the upper-left diagonal cell for the next cell that will be computed and stored in our column.

After a column has been computed, its last value is stored in array EE of size RLEN (line 16) which corresponds to the bottom row of the matrix.

This algorithm forms the cornerstone of the vectorized approach. In fact, we directly build upon this algorithm to devise the vectorization.

13.3.3 Vectorized Algorithm

In this section, we present a vectorized algorithm that can be executed on *Single Instruction, Multiple Data* (SIMD) computers and achieves substantially better performance than the non-vectorized approach. This is because we can process multiple matrix cells with a single SIMD instruction. On `x86` architectures, SIMD instructions operate on specific vector *registers* (storage space within the [co]processor) with a length of 128, 256, and 512 bits. These registers can hold multiple independent data values of smaller size (e.g., a 256-bit register can hold 32 8-bit integer values, or 16 16-bit integer values). If two registers contain 16 values each, for instance, we can perform an element-wise addition of the vector elements using a single instruction, instead of executing 16 separate instructions. Typically, Intel-based architectures employ 128-bit registers using *Streaming SIMD Extensions* (SSE) and 256-bit registers using *Advanced Vector Extensions* (AVX) instructions. The AVX-2 instruction set implements some additional vector operations, such as a bitwise AND on 256-bit integer data (`vpand` instruction). It also offers the `vpalignr` instruction, which concatenates two registers into an intermediate composite, shifts the composite to the right by a given fixed number of bytes, and extracts the right-aligned result. These instructions can be leveraged by our implementation. See Fig. 13.5 for some examples of typical SSE-3 instructions.

The description of the vectorization has two parts: first, we describe why the vectorized algorithm is a generalization of the non-vectorized approach, and then we explain the vector operations in more detail.

13.3.3.1 Matrix Computation and the Query Profile

We build upon the algorithm presented in Fig. 13.3 by extending the computation of a single cell to multiple cells. The underlying idea is to process cells simultaneously along the query sequence (in a vertical direction along the DP matrix), as shown in Fig. 13.4a. The advantage of this approach is that the corresponding substitution scores can be loaded more efficiently from memory.

Moreover, as we assume a finite alphabet, the same look-ups $s(x_i, y_i)$ (16 different types for nucleotides) are repeated to calculate the matrix. Therefore, we can improve performance by creating a so-called score profile for the query sequence. This profile, which is a query-specific substitution score matrix, is computed only once for the entire DP procedure and saves one memory access in the inner loop of the algorithm as the query sequence no longer needs to be accessed. Instead of indexing the substitution score matrix s by the query *and* the reference nucleotide, the profile is indexed by the query sequence position and the reference nucleotide. The scores for matching nucleotide `A` in the reference with each nucleotide in the query are stored sequentially in the first query profile row, followed by the scores for matching nucleotide `C` in the next row and so on. This query sequence profile is

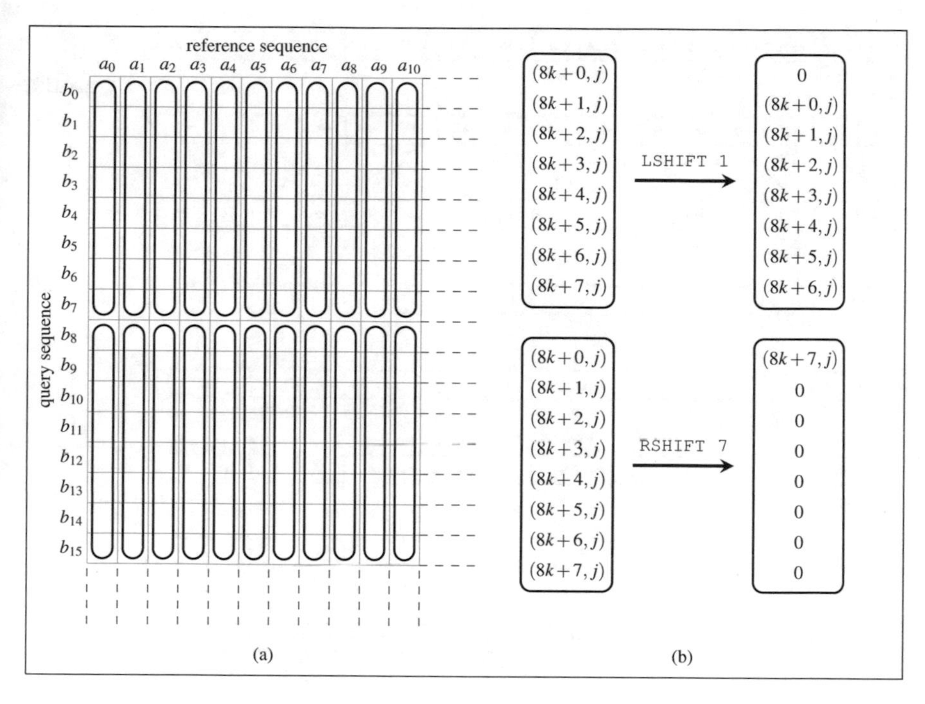

Fig. 13.4 Vector arrangements in the SIMD implementation of the DP approach for computing the AS of overlapping regions

accessed frequently in the inner loop of the algorithm and is usually small enough to fit the L-1 cache (Fig. 13.5).

The query profile can be precomputed by iterating over the query sequence once. For each nucleotide q_i of the query q, we set the i-th value of rows representing $\sigma \in \Sigma$ (for each character of the alphabet) to $s(\sigma, q_i)$, for all $0 \leq i < |q|$.

Before describing the vectorization scheme in more detail, we introduce the notation we use for our algorithm. In our description, we use vectors of size 8. This corresponds, for instance, to a 128-bit SSE vector register, with 16-bit long integers. Figure 13.4b shows the two types of shift operations that we use in our method. LSHIFT 1 is used to shift the contents of a SIMD register by one element to the left. We use the SSE2 `pslldq` (`vpslldq` for AVX2) instruction for this. Similarly, RSHIFT 7 shifts seven elements to the right (SSE2 `psrldq`; AVX-2 `vpsrldq` instructions). The new elements of the register resulting from the shifts are automatically filled with zeroes. Although our notation might seem counter-intuitive given the vertical representation of vectors as columns, the first bit in the register (topmost in our illustration) is the *Least Significant Bit* (LSB) and the last bit (bottom in our illustration) is the *Most Significant Bit* (MSB). In CPU registers, the LSB appears in the rightmost slot, while the MSB in the leftmost.

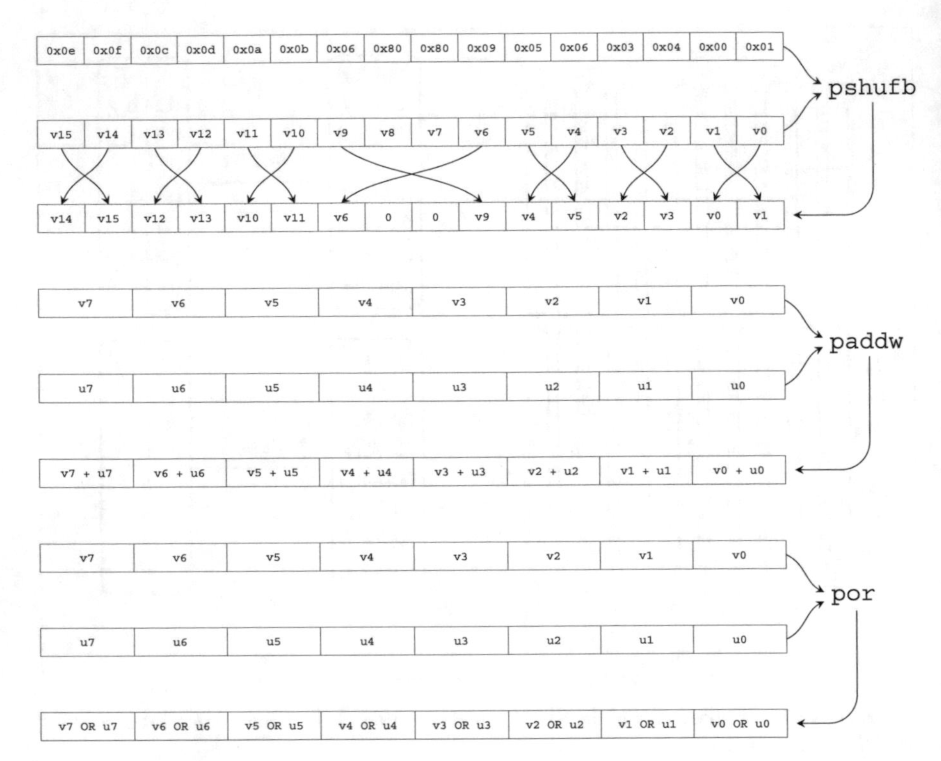

Fig. 13.5 Illustration of typical vector instructions. The `pshufb` instruction accepts two arguments, an index register and a shuffle register. The values of the shuffled register are permuted according to the index register and stored in an output register. The slots of an index register indicate the slot number where the value at the current position of the shuffle register should be moved to in the output register. The `paddw` instruction treats cells as 16-bit values and performs a pairwise addition among cells of the two input registers. Finally, `por` performs a 128-bit logical OR on the two input registers

The vectorization scheme is presented in Fig. 13.6. The outer loop iterates horizontally over the reference sequence and is similar to the non-vectorized algorithm in Fig. 13.3 apart from a small difference in the inner loop. While the non-vectorized version processes one query nucleotide per inner loop iteration, the vectorized version processes a complete query vector in the vertical direction along a column of the DP matrix. Hence, vectors are always oriented vertically. To compute such a query vector along a matrix column, we first copy the values of the adjacent vector (part of the previous column) to the left into a vector register `H` (line 16). Then, we left-shift (shift values down the column) by one element to vertically align the cells of the vector in the current column we want to update with their diagonal counterparts in the previous column. However, because of the left-shift (shift down by one element in the part of the column the vector spans), we are now missing the uppermost value in the present vector. Therefore, we copy the bottom value of the vector above it and one column to the left first, and place it to the first

```
OVERLAP-SSE(RSEQ,QPROFILE,n,m)
                                          RSEQ is the reference sequence
                                          QPROFILE is a scoring profile based on the query
                                          n is the length of the reference
                                          m is the length of the query

1. INTEGER SCORE = 0                      Score of the best overlap
2. INTEGER i,j                            Loop indices
3. CONST INTEGER y = (m+7)/8              Number of vectors along query sequence (y)
4. VECT_SET HH, EE                        HH is the set of vectors in a column
                                          EE is the set of vectors for the horizontal last row
5. VECTOR H, X, T1

6. FOR i=0 TO y-1 DO
7. {
8.     HH[i] = [0, 0, 0, 0, 0, 0, 0, 0]   Initialize HH-array of H-values
9. }

10. FOR j=0 TO n-1 DO                     For each nucleotide in the reference sequence
11. {
12.     X = [0, 0, 0, 0, 0, 0, 0, 0]      Initialize X-vector for 1. round
13.     CHAR c = RSEQ[j]                  Current nucleotide from reference

14.     FOR i=0 TO y-1 DO                 For each vector along query sequence
15.     {
16.         H = HH[i]                     Load previous H-vector from HH-array

17.         T1 = H RSHIFT 7               Save H[7] for use below
18.         H = (H LSHIFT 1) OR X         Shift H-vector and OR with H[7] from previous round
19.         X = T1                        Save old H[7] in X for next round

20.         H = H + QPROFILE[c][i]        Add query profile vector to H

21.         HH[i] = H                     Store H-vector in HH-array
22.     }
23.     EE[j] = H[(m-1) MOD 8]            Store last value of H in EE-array
24. }
25. FOR i=0 TO y-1 DO
26. {
27.     SCORE = MAX(HH[i],SCORE)          Store the maximal value
28. }
29. FOR i=0 TO n DO
30. {
31.     SCORE = MAX(EE[i],SCORE)          Store the maximal value
32. }
```

Fig. 13.6 Pseudo-code for the vectorized approach. The LSHIFT and RSHIFT operations shift the elements of a vector the specified number of times to the left or right. The OR operation returns the bitwise or of the vector elements. The query sequence is assumed to be padded to the closest multiple of 8 that is greater or equal to the sequence length

entry of the current vector via a bitwise OR with the vector X. Thus, the vector X contains the last value of the vector above the present one before this vector above was updated. Hence, this corresponds to the last element of the vector above and one column to the left. Then, the elements of the query profile for the current vector (current fraction of the column) that correspond to the match/mismatch scores for reference nucleotide c with the query nucleotide vector are added to the elements of the current vector H (line 20).

At the beginning of iteration j of the outer loop, array HH holds the computed vectors from the previous iteration $j - 1$, that is, the values from column $j - 1$ to the left. At the end of each inner loop iteration i, vector HH[i] will have been overwritten with the updated vector i for the current column j (line 21). Before overwriting the vector entry i in HH, its last value is stored in vector X (lines 17 and 19). This vector X will be used in the next inner loop iteration $i + 1$ to set the first element of the first vector of column $i + 1$ (line 18).

13.3.3.2 Query Profile Vectorization

Typically, NGS platforms generate short reads of length between 100 and 300 bp. Given that the query profile must be computed from scratch for each pair of reads we want to overlap and that the query profile for two sequences x and y over Σ consists of $|\Sigma|$ rows of size $|x|$, its computation amounts to $\frac{|\Sigma|}{|y|}$ of the time required to compute the matrix values. This might seem to be a tiny fraction of the run-time. However, if we consider that the matrix computation is vectorized, then it amounts to $\frac{s \cdot |\Sigma|}{|y|}$, where s is the speed-up we obtain from vectorizing the matrix computation. Assuming a speed-up of 8, an alphabet of all degenerate nucleotides (16), and reads of length 200, the non-vectorized query profile computation accounts for 64% of total run-time. Thus, we need to design a vectorization scheme for computing the query profile given the scores are encoded as 16-bit values. The concept is illustrated in Fig. 13.7 for a vector of 8 16-bit elements and for $|\Sigma| = 4$.

Briefly, the underlying idea is to transform the value of each nucleotide q_i of the query sequence q into $|\Sigma|$ indices pointing to the values $s(\sigma, q_i)$, for all $\sigma \in \Sigma$. The scoring matrix is stored in vector registers, and hence, we use each of the $|\Sigma|$ sets of $|q|$ indices pointing to $s(\sigma, q_i)$, $0 \leq i \leq |q| - 1$, as index registers for shuffling the register holding the values $\{s(\sigma, \sigma') \mid \forall \sigma' \in \Sigma\}$. This operation is repeated $|\Sigma|$ times to create the $|\Sigma|$ columns of the query score profile.

Figure 13.8 illustrates the process. We iterate through the query sequence and read 16 (8-bit) nucleotide values at each iteration, using the movdqa instruction (step 1). Note that the scoring matrix values are loaded into two registers, which have the layout illustrated in Fig. 13.7 and/or Fig. 13.8, and that the values are 16-bit integers (8 values per register). In our concrete example, the first scoring matrix register holds the scores for aligning any nucleotide with nucleotides A and C, and the second register holds the scores for aligning any nucleotide with G and T. Initially, we transform the value of each actual nucleotide q_i stored in a vector to indices of the entries of the two scoring matrix registers that contain the values $s(q_i, \texttt{A})$, $s(q_i, \texttt{C})$, $s(q_i, \texttt{G})$, and $s(q_i, \texttt{T})$. The nucleotides are encoded as A $\mapsto 0$, C $\mapsto 1$, G $\mapsto 2$, and T $\mapsto 3$. Hence, we multiply these encoded q_i values by four to index the A resp. G scores of the first resp. second scoring matrix register. We proceed analogously for the C resp. T scores of the two scoring matrix registers. The multiplication of the q_i values by four is carried out by adding the vector register twice to itself (step 2). Because we use 16-bit integer scores while the nucleotides

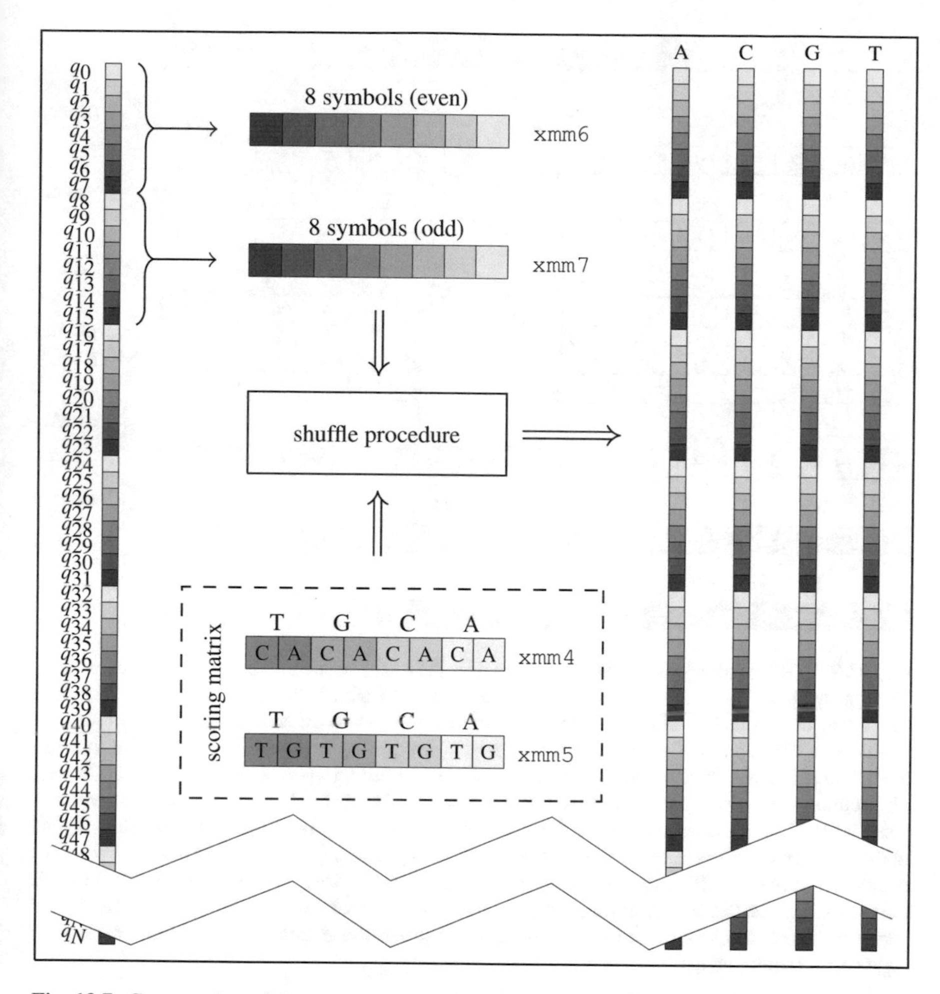

Fig. 13.7 Computation of the query score profile. The query sequence q is assumed to be padded to the closest multiple of 16 that is greater or equal to the sequence length, and the profile is constructed in $|q|/16$ steps. At each step k, $0 \leq k < |q|/16$, 16 nucleotides from the query are loaded and split into two registers (xmm6 and xmm7). These registers operate as the bitmasks for shuffling the 4×4 scoring matrix of 16-bit scores, to finally obtain the values $s(q_i, \mathtt{A})$, $s(q_i, \mathtt{C})$, $s(q_i, \mathtt{G})$, and $s(q_i, \mathtt{T})$ for each nucleotide q_i, $16 \times k \leq i < 16 \times (k + 1)$. These values are stored in four arrays A, C , G, and T of the same length as q such that $A[i] = s(q_i, \mathtt{A})$, $C[i] = s(q_i, \mathtt{C})$, $G[i] = s(q_i, \mathtt{G})$, $T[i] = s(q_i, \mathtt{T})$. The four arrays represent the query score profile

are stored as 8-bit values, we need to use two indices for each score of a single nucleotide q_i, one pointing to the lower 8 bits and one pointing to the upper 8 bits of the score. Therefore, we split the 16 loaded 8-bit nucleotides into two vector registers using the pshufb command and an appropriate mask (step 3 only shows one of the resulting two registers). In those two vector registers, the 8 nucleotide values are stored in the odd bytes of the vector and the even bytes are filled with

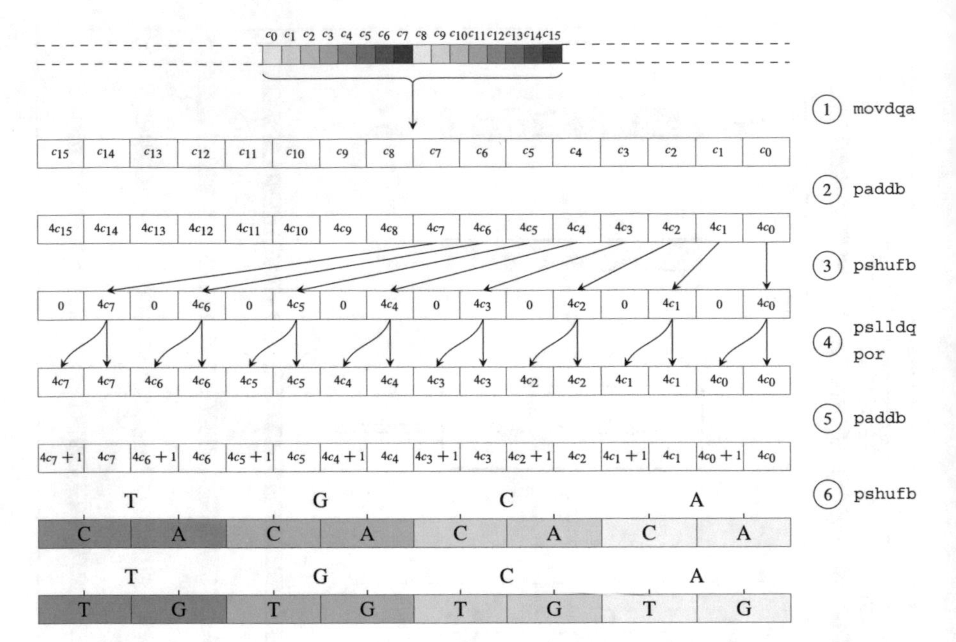

Fig. 13.8 The shuffling procedure. We iterate over the query sequence in steps of 16 nucleotides. At each iteration, 16 nucleotides are loaded into a register using one load (movdqa) instruction (1). Then, each value is multiplied by 4. Multiplication is carried out using two add (paddb) instructions (2). The 16 8-bit values in the register are then split into two registers, one holding the first eight values (in the illustration), the second one holding the last eight values (not shown). The actual 8-bit values are then moved to the odd (starting from 1) bytes of the register using a shuffle (pshufb) instruction with the appropriate mask (3). The 8-bit values are then copied to the even bytes of the vector register using a left-shift (pslldq) instruction and a bitwise OR (por) instruction (4). Finally, using one add (paddb) instruction, we add 1 to the even bytes in the vector register (5). The resulting register serves as the index register for shuffling the scoring matrix registers, in our case the AC component, to obtain the A and C score profile vectors for eight characters in the query sequence

zeroes (as shown in step 3 of Fig. 13.8). At this point we have already transformed the values of each nucleotide q_i into indices pointing to the lower 8-bits of the scores $s(q_i, \mathtt{A})$ resp. $s(q_i, \mathtt{G})$ of the first resp. second scoring matrix register. The next step is to transform the zero value of the even slots to an index pointing to the upper 8 bits of the corresponding score. This is achieved by using an additional vector register where the eight indices representing q_i are shifted one byte to the left each (pslldq instruction). Then, we perform a logical OR (por instruction) on the shifted and unshifted vector registers (see step 4). Finally, we use the paddb instruction to increment the even bytes of the resulting register by one, for they must point to the corresponding upper 8-bit part of the score values (recall that their adjacent bytes in the direction of the LSB index the lower 8-bit part of the score values). Thereby, each adjacent pair of bytes starting at odd positions (we enumerate byte positions from 1 to 16 in this concrete example) in this modified index vector register holds indices to the corresponding lower and upper 8-bit score value. We use this index to shuffle the two scoring matrix registers (`pshufb` instruction) and obtain for each loaded nucleotide the 16-bit score of aligning it with an `A` or a `G` (step 6). We then reuse the results of step 5 but now add two to each byte of the index register, in order to shift the indices to the `C` resp. `T` component of the two scoring matrix registers. Then, we execute the `pshufb` instruction again on the original scoring matrix registers to get the scores for aligning the nucleotides with `C` or `T`. The same operations are conducted with the register holding the next eight nucleotide values (not shown in the figure). Therefore, we can construct the query score profile as shown in Fig. 13.7, in a vectorized way.

13.3.3.3 Experimental Results

We implemented four variations of the vectorized algorithm. First, an 8-bit SSE3 version, which handles 16 elements at a time, and the individual DP matrix cells are represented by 8-bit values. This limitation to 8 bits works when the scoring matrix contains relatively small values, typically with a distance of not more than 2 units from 0, and when the short reads are up to 150-bp long. Second, a 16-bit SSE3 version that handles eight elements at a time. This is useful for more diverse scoring matrices or reads that are longer than 150-bp. We have also ported the two implementations to the AVX-2 instruction set.

We tested our implementations on a simulated paired-end with reads comprising 150, 250, and 300-bp, which are typical sizes for *Illumina* platforms. We generated two random reads for each length and measured the performance of the vectorized code by executing it 1,000,000 times. Note that neither the overlap size nor the nucleotide composition of the reads affects the run-time. The same amount of computations is performed for any pair of reads of the same size, regardless of their overlap sizes and nucleotide composition.

We ran the tests on an Intel Core i5-4200U CPU with 4 GB of memory. We included both the times for the construction of the query score profile and the overlapping algorithm into the total run-time. Table 13.1 lists the speed-ups for the vectorized implementation compared the non-vectorized implementation. Figure 13.9 provides a speed-up comparison between the four implementations.

Table 13.1 Speed-ups of the three implementations for different paired-end read lengths

	SSE		AVX-2	
Length (bp)	8-bit	16-bit	8-bit	16-bit
100	11.26	6.50	11.66	6.94
150	13.01	7.11	17.10	8.29
250	16.36	7.42	18.40	10.57
300	16.58	6.59	19.29	10.55

The run-times of the naive algorithm are 27.8, 75.7, and 108.4 s for read lengths of 150, 250, and 300-bp, respectively.

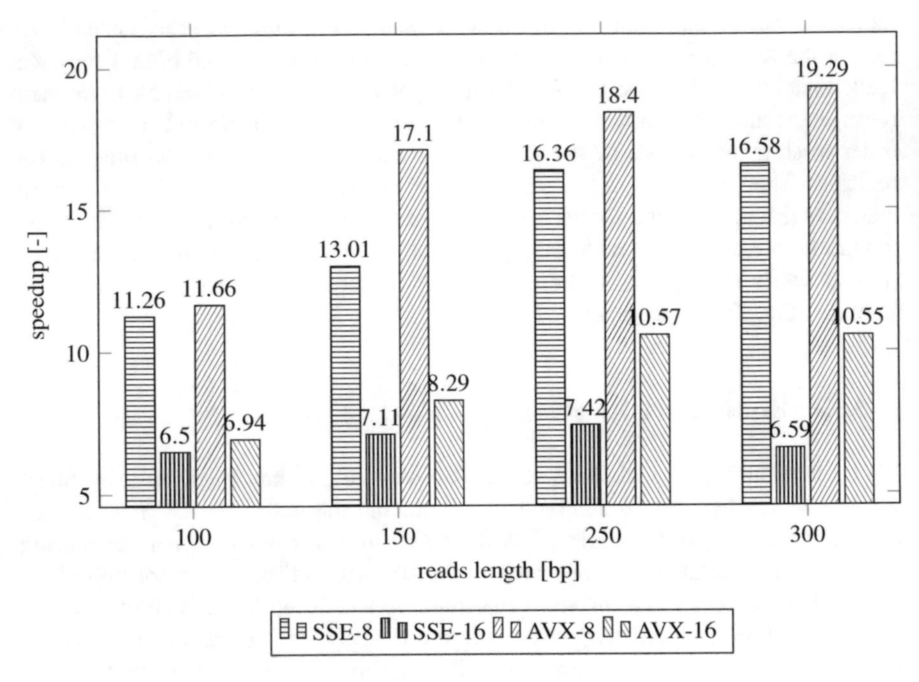

Fig. 13.9 Speed-up comparison of the four implementations

The results show that we obtain an optimal speed-up over the non-vectorized version of approximately 16× when using the 8-bit SSE implementation given that the read length is sufficiently large. The 16-bit SSE implementation yields a near-optimal speed-up of approximately 7×. Finally, the AVX-2 implementations only yield a run-time improvement of approximately 16% over the corresponding SSE implementations, instead of the expected twofold speed-up. This is because of the limited availability of certain integer arithmetic and logic instructions in the AVX-2 instruction set. For instance, shifting integer values from the lower 128-bit bank to the upper 128-bit bank of a 256-bit vector register is not possible using a single instruction. This is implemented using more expensive shuffle and permutation operations.

13.4 Integrating Uncertainties Using Quality Scores and Degenerated Characters

In this section we deal with uncertainties for base calls and provide methods for computing the best overlap given these uncertainties. We handle two levels of uncertainty. First, the observed base may not be the true base (Sect. 13.4.1). Second, a base call may be represented as a degenerate character, which stands for a set of nucleotides instead of a single one (Sect. 13.4.2). Note that we would like to point out that the vectorization scheme described in Sect. 13.3.3.1 can directly be applied to degenerate characters. The query profile simply needs to be precomputed for an alphabet of size 16 (instead of 4) for nucleotide data. Incorporating quality scores renders the application impractical due to the large size of query profile. A unique entry in the query profile must be precomputed for each possible nucleotide (16 when considering degenerate characters) and quality score combination (40 for *Illumina* data). Hence, the size of the query profile (precomputation) can become as large as (or larger than) the size of the matrix that needs to be computed. Thus, this method is less efficient, especially in the case of shorter reads (e.g., 75- to 200-bp). To alleviate this limitation, it is possible to compute the entries in the query profile on the fly as needed. However, such an approach is appropriate when comparing one (fixed) sequence against multiple query sequences. In this case, we can dynamically build a query profile for the fixed sequence. For merging paired-end reads, it is unlikely that an entry in the query profile will be used more than once.

Finally, we show how to determine the consensus base for the overlapping region and how to derive the error estimate for the consensus base in Sect. 13.4.3.

13.4.1 Computing the Best Overlap Using Quality Scores

The *Illumina* platform is one of the most accurate NGS platforms [18]. However, it still has a typical average error rate of 1% [18], and the quality of base calling decreases toward the end of these reads (Fig. 13.10). The raw *Illumina* reads are accompanied with a quality score (*Q*-score) for each base call. Quality scores are

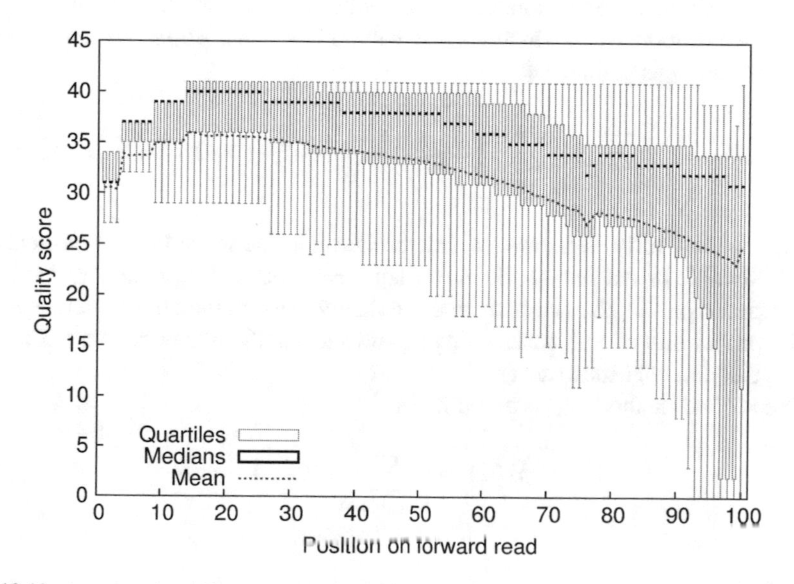

Fig. 13.10 An example of *Illumina* average read quality scores

expressed on a logarithmic scale known as the *Phred scale* [19]. More specifically, the quality score Q is defined as:

$$Q = -10 \times \log_{10} e \,, \tag{13.4}$$

where e is the estimated error probability for a base call. Q-scores are rounded to the nearest integer and usually range between 1 and 40. It is easy to see that a higher Q-score corresponds to a more reliable base call (lower error probability), and vice versa. For instance, $Q = 30$ suggests the estimated accuracy of the base call is 99.9% and $Q = 20$ indicates an accuracy of 99%.

For visual convenience, the Q-scores are encoded in ASCII by adding a base number b (typically $b = 33$ or $b = 64$). Thus, given a Q-score, we can easily convert it back to e with:

$$e = 10^{\frac{b-Q}{10}} \,. \tag{13.5}$$

These quality scores introduce uncertainties when computing the *Alignment Score* (AS) as described in Sect. 13.3. Here, we present three related but distinct methods, ϕ_E, ϕ_A, and ϕ_B, for integrating the Q-scores into the AS computation.

Let $C = (x, y) \in \mathscr{C}$ be an overlapping region, and e_x resp. e_y, a vector of quality score values such that the i-th element of e_x resp. e_y indicates the error probability of observing x_i resp. y_i. We use the notation e_{x_i} (resp. e_{y_i}) to index such elements. Also, let $\boldsymbol{\pi} = (\pi_{\mathtt{A}}, \pi_{\mathtt{C}}, \pi_{\mathtt{G}}, \pi_{\mathtt{T}})$ be the vector of empirical frequencies for nucleotides A, C, G, and T ($\boldsymbol{\pi}$ can be estimated using the entire input data set). We usually consider e_{x_i} and e_{y_i} to be inherent properties of x_i and y_i. Therefore, we omit them from the equations. We also use the notation $\Sigma_{\overline{x}}$ to denote $\Sigma \setminus \{x\}$ (set Σ without x).

First we introduce ϕ_E, which we also call the *Expected Alignment Score* (EAS). The function ϕ_E is defined as:

$$\phi_E(C) = \sum_{i=0\ldots|C|-1} \sum_{(x,y)\in\Sigma^2} \Big(\Pr[x'_i = x \mid x_i] \Pr[y'_i = y \mid y_i] s(x, y) \Big), \tag{13.6}$$

The EAS has a valid mathematical interpretation in the sense that it represents the expected value of AS for the given overlap, given the scoring matrix s, that is, for each ordered pair of characters from the alphabet, we compute the AS and weight it by the probability of true bases being equal to a specific nucleotide pair. The EAS is then the sum over these terms.

The second method, ϕ_A, is computed as:

$$\phi_A(C) = \sum_{i=0\ldots|C|-1} \delta(x_i, y_i) \tag{13.7}$$

where

$$\delta(x_i, y_i) = \begin{cases} s(x_i, y_i)\Pr[x_i' = y_i' \mid x_i, y_i] & : \quad x_i = y_i \\ s(x_i, y_i)\Pr[x_i' \neq y_i' \mid x_i, y_i] & : \quad x_i \neq y_i \end{cases}$$

For the sake of convenience, we drop the subscript i from x_i, y_i, x_i', and y_i'. Expanding $\Pr[x' = y' \mid x, y]$, we get:

$$\Pr[x' = y' \mid x, y] = \sum_{z \in \Sigma} \Pr[x' = y' = z \mid x, y] \tag{13.8}$$

A priori, we assume independence between the two sequences. Thus, the probability for x' only depends on x (and e_x, as well as π). With this, we can rewrite the equation as:

$$\Pr[x' = y' \mid x, y] = \sum_{z \in \Sigma} \Pr[x' = z \mid x]\Pr[y' = z \mid y] \tag{13.9}$$

For the DNA alphabet, we obtain:

$$\begin{aligned} \Pr[x' = y' \mid x, y] = \Pr[x' = \mathtt{A} \mid x] \cdot \Pr[y' = \mathtt{A} \mid y] \\ + \Pr[x' = \mathtt{T} \mid x] \cdot \Pr[y' = \mathtt{T} \mid y] \\ + \Pr[x' = \mathtt{G} \mid x] \cdot \Pr[y' = \mathtt{G} \mid y] \\ + \Pr[x' = \mathtt{C} \mid x] \cdot \Pr[y' = \mathtt{C} \mid y]. \end{aligned} \tag{13.10}$$

Note that

$$\Pr[x' \neq y' \mid x, y] = 1 - \Pr[x' = y' \mid x, y]. \tag{13.11}$$

Thus, for computing ϕ_A, it is sufficient to only calculate the probability of a match, given the two observed bases and quality scores.

The calculation of ϕ_A assumes that if a match (resp. mismatch) is observed, all other possible matches (resp. mismatches) at this position should contribute to the score. We now introduce a third method ϕ_B, which only considers the observed pairs of bases. For an overlap C, it is defined as:

$$\begin{aligned} \phi_B(C) &= \sum_{i=0\ldots|C|-1} \Pr[x_i' = x_i \mid x_i]\Pr[y_i' = y_i \mid y_i]s(x_i, y_i) \\ &= \sum_{i=0\ldots|C|-1} (1 - e_{x_i})(1 - e_{y_i})s(x_i, y_i). \end{aligned} \tag{13.12}$$

The common element in Eqs. (13.6), (13.9), and (13.12) is $\Pr[x' = z \mid x]$. We can compute it as:

$$\Pr[x' = z \mid x] = \begin{cases} 1 - e_x & : \quad z = x \\ e_x \cdot \frac{\pi_z}{\sum\limits_{q \in \Sigma_{\overline{x}}} \pi_q} & : \quad z \neq x \end{cases} \tag{13.13}$$

We now outline Eq. (13.9) using two concrete examples. The first one shows the case where the two observed characters are identical. We assume $x = y = C$ with error probabilities e_x and e_y:

$$\begin{aligned} \Pr[x' = y' \mid x, y] = (1 - e_x)(1 - e_y) \\ + e_x \frac{\pi_{\mathrm{A}}}{\pi_{\mathrm{A}} + \pi_{\mathrm{T}} + \pi_{\mathrm{G}}} e_y \frac{\pi_{\mathrm{A}}}{\pi_{\mathrm{A}} + \pi_{\mathrm{T}} + \pi_{\mathrm{G}}} \\ + e_x \frac{\pi_{\mathrm{T}}}{\pi_{\mathrm{A}} + \pi_{\mathrm{T}} + \pi_{\mathrm{G}}} e_y \frac{\pi_{\mathrm{T}}}{\pi_{\mathrm{A}} + \pi_{\mathrm{T}} + \pi_{\mathrm{G}}} \\ + e_x \frac{\pi_{\mathrm{G}}}{\pi_{\mathrm{A}} + \pi_{\mathrm{T}} + \pi_{\mathrm{G}}} e_y \frac{\pi_{\mathrm{G}}}{\pi_{\mathrm{A}} + \pi_{\mathrm{T}} + \pi_{\mathrm{G}}} \end{aligned} \tag{13.14}$$

The second example illustrates the case where the two observed characters are different. We assume $x = \mathtt{C}$ and $y = \mathtt{A}$ by Eq. (13.10):

$$\begin{aligned} \Pr[x' = y' \mid x, y] = \; & e_x \frac{\pi_{\mathrm{A}}}{\pi_{\mathrm{A}} + \pi_{\mathrm{T}} + \pi_{\mathrm{G}}} (1 - e_y) \\ & + (1 - e_x) e_y \frac{\pi_{\mathrm{C}}}{\pi_{\mathrm{C}} + \pi_{\mathrm{T}} + \pi_{\mathrm{G}}} \\ & + e_x \frac{\pi_{\mathrm{T}}}{\pi_{\mathrm{A}} + \pi_{\mathrm{T}} + \pi_{\mathrm{G}}} e_y \frac{\pi_{\mathrm{T}}}{\pi_{\mathrm{C}} + \pi_{\mathrm{T}} + \pi_{\mathrm{G}}} \\ & + e_x \frac{\pi_{\mathrm{G}}}{\pi_{\mathrm{A}} + \pi_{\mathrm{T}} + \pi_{\mathrm{G}}} e_y \frac{\pi_{\mathrm{G}}}{\pi_{\mathrm{C}} + \pi_{\mathrm{T}} + \pi_{\mathrm{G}}}. \end{aligned} \tag{13.15}$$

The function ϕ_A has been used in PEAR [3] with a simplified scoring matrix such that $s(a, b) = +1$ if $a = b$ and $s(a, b) = -1$ if $a \neq b$. For this simple case, we deduce the following correlation between ϕ_E and ϕ_A.

Proposition 1 *Given $\alpha = +1$, $\beta = -1$, it follows that $\phi_E(C) = 2 \cdot \phi_A(C) + \alpha \cdot n_m + \beta \cdot n_h$, where n_m and n_h are the numbers of mismatches and matches of bases in C, irrespective of quality scores.*

Proof The proof is straightforward and is therefore omitted. □

Functions ϕ_E, ϕ_A, and ϕ_B provide three different ways of integrating uncertainties into computing the AS. The function ϕ_E sums over the expected score for each position in the overlapping region. Thus, as mentioned above, it represents the expected AS for an overlap. The second function ϕ_A explicitly considers whether the two observed bases are the same, and the score for each position is scaled by the probability of the true underlying base match (or mismatch), given the observed bases and error probabilities. The last function ϕ_B scales the value of each pair of nucleotides by the confidence we have in those values, that is, how likely they are to be true. Function ϕ_B represents a general formula from which ϕ_A is deduced. If the scoring for ϕ_A is such that $s(a, a) = s(b, b)$ for all a and b and $s(a, b) = c$ for all $a \neq b$ and some constant c, then ϕ_A is a simplification of ϕ_B. In this case, only two events can occur—either both values are identical, irrespective of whether both are A, C, G or T, or they are different.

The scoring functions ϕ_E and ϕ_A have been implemented in PEAR [3]. Empirical tests suggest that ϕ_A merges more reads correctly than ϕ_E. The performance of ϕ_B remains to be tested.

13.4.2 Degenerate Characters

Another approach to expressing uncertainties for a base call is to use degenerate characters. Degenerate characters denote that the actual nucleotide at a given position in a read is not uniquely defined but lies within a set of characters with a certain probability. The common IUPAC notation for these degenerate characters is shown in Table 13.2.

We can classify the IUPAC nucleotide symbols from Table 13.2 into four categories: Category 1 consists of *A*, *T*, *G*, and *C*, which represent a single nucleotide; category 2 includes *W*, *S*, *M*, *K*, *R*, and *Y*, where each stands for the combination of two nucleotides; category 3 comprises the symbols *B*, *D*, *H*, and *V*,

Table 13.2 IUPAC nucleotide encoding for degenerate/ambiguous characters

Symbol	A	C	G	T
A	A			
C		C		
G			G	
T				T
W	A			T
S		C	G	
M	A	C		
K			G	T
R	A		G	
Y		C		T
B		C	G	T
D	A		G	T
H	A	C		T
V	A	C	G	
N	A	C	G	T
–				

each of which stands for three ambiguous nucleotides; and finally, category 4 is the N character that represents an unknown nucleotide (A, T, G or C) and a gap "-." We employ a set notation for those symbols. For example, $S=$ {C, G} and $N=$ {A, C, G, T}.

First, consider the case of calculating the EAS (ϕ_E) as described in Eq. (13.6). The same formula can be used for degenerate characters as well; we only need to be able to compute $\Pr[x' = z \mid X]$ for a degenerate character X.

For this, we consider the observed characters in the four categories discussed above. Category 1 only comprises non-degenerate characters; thus, the formulas are the same as in Eq. (13.13). For categories 2 and 3, we differentiate between z being an element of X and z being different to any element in X. For N, no further distinction is needed as x' will be in N.

13.4.2.1 Case: Categories 2 and 3

For z in X, we get

$$\Pr[x' = z \mid X] = (1 - e_X) \cdot \frac{\pi_z}{\sum_{q \in X} \pi_q}, \tag{13.16}$$

and

$$\Pr[x' = z \mid X] = e_X \cdot \frac{\pi_z}{\sum_{q \notin X} \pi_q} \tag{13.17}$$

for any z not in X.

13.4.2.2 Case: N

The probability that the true underlying character is z can be calculated if we observe that the N character is simply given by the base frequencies $\boldsymbol{\pi}$ (recall that $\boldsymbol{\pi}$ can be estimated from the input data set), that is:

$$\Pr[x' = z | X = \mathrm{N}] = \pi_z. \tag{13.18}$$

Next, for the case of Eq. (13.7) (ϕ_A), we employ a slightly different approach:

$$\phi_A(C) = \sum_{i=0 \ldots |C|-1} \delta(X_i, Y_i) \tag{13.19}$$

where

$$\delta(X_i, Y_i) = \begin{cases} \alpha \cdot \Pr[x_i' = y_i' \mid X_i, Y_i] & : \quad X_i \cap Y_i \neq \emptyset \\ \beta \cdot \Pr[x_i' \neq y_i' \mid X_i, Y_i] & : \quad X_i \cap Y_i = \emptyset \end{cases}$$

Here, α and β correspond to the simplified scoring matrix s where $s(a, b) = \alpha$ if $X_i \cap Y_i \neq \emptyset$ and $s(a, b) = \beta$ if $X_i \cap Y_i = \emptyset$. Again, notice that we only need to calculate $\Pr[x' = z \mid X]$ for Eq. (13.19).

Finally, for evaluating ϕ_B, we introduce a scoring matrix s' for the degenerate characters based on the scoring matrix s defined for the non-degenerate characters. We define $s'(X, Y)$, where either X or Y is degenerate, as:

$$s'(X, Y) = \sum_{(\hat{x},\hat{y}) \in X \times Y} \left(\frac{\pi_{\hat{x}}}{\sum_{z \in X} \pi_z} \cdot \frac{\pi_{\hat{y}}}{\sum_{z \in Y} \pi_z} \cdot s(\hat{x}, \hat{y}) \right). \tag{13.20}$$

With $s'(X, Y)$, Eq. (13.12) can be easily applied to compute ϕ_B for degenerate characters. The interpretation of s' is that it is the expected score of a given pair of degenerate characters with respect to the scoring matrix s.

13.4.3 Computing the Consensus Quality Score

The Q-score for the two characters prior to the merging step measures the probability of errors for observing those two characters in each read, respectively. After merging the two reads, we need to compute a new Q-score for each position in the overlapping region that represents the error probability given the corresponding characters in the merged read. Note that we implicitly assume the two reads to have the same characters in the overlapping region. Bayes' theorem gives us the probability of observing any of the four DNA states (A, C, G, and T) in the merged read by:

$$\begin{aligned}
&\Pr[x' = y' = x^* \mid x' = y', X, Y] \\
&= \frac{\Pr[x' = y' \mid x' = y' = x^*, X, Y] \cdot \Pr[x' = y' = x^* \mid X, Y]}{\Pr[x' = y' \mid X, Y]} \\
&= \frac{1 \cdot \Pr[x' = y' = x^* \mid X, Y]}{\Pr[x' = y' \mid X, Y]} \\
&= \frac{\Pr[x' = x^* \mid X]\Pr[y' = x^* \mid Y]}{\Pr[x' = y' \mid X, Y]} \\
&= \frac{\Pr[x' = x^* \mid X]\Pr[y' = x^* \mid Y]}{\sum_{z \in \{A,C,G,T\}} \Pr[x' = z \mid X]\Pr[y' = z \mid Y]}.
\end{aligned} \tag{13.21}$$

where the variable x^* is the character for this position in the merged string.

The probabilities required to calculate the new Q-score are the same as explained in Sect. 13.4.2. The character in the overlapping region should be the character with the highest value obtained, thereby. The Q-score is computed using Eq. (13.4), where $e = 1 - \Pr[x' = y' = x^* | x' = y', X, Y]$. An explicit formulation for a subset of the posterior probabilities has been developed independently and simultaneously in [20].

13.5 Conclusions

In the first part of this chapter, we discussed a discrete algorithm for finding overlapping regions between two short reads. We showed how to convert the classical sliding window approach into a DP problem. Using this DP formulation, we then implemented a vectorization scheme and showed that the speed up over the non-vectorized implementation is near-optimal. In the second part of this chapter, we presented three functions that allow to incorporate quality score information into the overlap score calculation. These quality scores represent the error probabilities of base calls from the sequencing platforms. Furthermore, we presented solutions for handling reads that contain degenerate characters. Finally, we discussed how to compute the posterior error probabilities for a consensus base in a Bayesian setting.

Acknowledgements T.F is supported by DFG project STA/860-4. L.C, K.K and J.Z are funded by a HITS scholarship.

References

1. Koboldt, D.C., Steinberg, K.M., Larson, D.E., Wilson, R.K., Mardis, E.R.: The next-generation sequencing revolution and its impact on genomics. Cell **155**(1), 27–38 (2013)
2. Mardis, E.R.: Next-generation DNA sequencing methods. Annu. Rev. Genomics Hum. Genet. **9**, 387–402 (2008)
3. Zhang, J., Kobert, K., Flouri, T., Stamatakis, A.: PEAR: a fast and accurate Illumina Paired-End reAd mergeR. Bioinformatics (Oxford, England) **30**(5), 614–620 (2014)
4. Masella, A.P., Bartram, A.K., Truszkowski, J.M., Brown, D.G., Neufeld, J.D.: PANDAseq: paired-end assembler for illumina sequences. BMC Bioinf. **13**(1), 31 (2012)
5. Magoč, T., Salzberg, S.L.: FLASH: fast length adjustment of short reads to improve genome assemblies. Bioinformatics (Oxford, England) **27**(21), 2957–2963 (2011)
6. Rognes, T., Flouri, T., Nichols, B., Quince, C., Mahé, F.: VSEARCH: a versatile open source tool for metagenomics. PeerJ **4**, e2584 (2016)
7. Paszkiewicz, K., Studholme, D.J.: De novo assembly of short sequence reads. Brief. Bioinform. **11**(5), 457–472 (2010). [Online] Available: http://bib.oxfordjournals.org/content/11/5/457.abstract
8. Nakamura, K., Oshima, T., Morimoto, T., Ikeda, S., Yoshikawa, H., Shiwa, Y., Ishikawa, S., Linak, M.C., Hirai, A., Takahashi, H., Altaf-Ul-Amin, M., Ogasawara, N., Kanaya, S.: Sequence-specific error profile of Illumina sequencers. Nucleic Acids Res. **39**(13), e90 (2011)
9. Needleman, S.B., Wunsch, C.D.: A general method applicable to the search for similarities in the amino acid sequence of two proteins. J. Mol. Biol. **48**(3), 443–453 (1970)
10. Gotoh, O.: An improved algorithm for matching biological sequences. J. Mol. Biol. **162**(3), 705–708 (1982)

11. Smith, T., Waterman, M.: Identification of common molecular subsequences. J. Mol. Biol. **147**(1), 195–197 (1981)
12. Levenshtein, V.I.: Binary codes capable of correcting deletions, insertions and reversals. Dokl. Akad. Nauk SSSR **163**(4), 845–848 (1965)
13. Hamming, R.: Error detecting and error correcting codes. Bell Syst. Tech. J. **29**(2), 147–160 (1950)
14. Rognes, T., Seeberg, E.: Six-fold speed-up of smith-waterman sequence database searches using parallel processing on common microprocessors. Bioinformatics **16**(8), 699–706 (2000)
15. Altschul, S., Gish, W.: Local alignment statistics. Methods Enzymol. **266**, 460–480 (1996)
16. Langmead, B., Salzberg, S.L.: Fast gapped-read alignment with Bowtie 2. Nat. Methods **9**(4), 357–359 (2012)
17. Gusfield, D.: Algorithms on Strings, Trees, and Sequences – Computer Science and Computational Biology. Cambridge University Press, Cambridge (1997)
18. Quail, M.A., Smith, M., Coupland, P., Otto, T.D., Harris, S.R., Connor, T.R., Bertoni, A., Swerdlow, H.P., Gu, Y.: A tale of three next generation sequencing platforms: comparison of Ion Torrent, Pacific Biosciences and Illumina MiSeq sequencers. BMC Genomics **13**(1), 341 (2012)
19. Ewing, B., Green, P.: Base-calling of automated sequencer traces using Phred. II. Error probabilities. Genome Res. **8**(3), 186–194 (1998)
20. Edgar, R.C., Flyvbjerg, H.: Error filtering, pair assembly and error correction for next-generation sequencing reads. Bioinformatics **31**(21), 3476 (2015)

Chapter 14
Assembly-Free Techniques for NGS Data

Matteo Comin and Michele Schimd

14.1 A Brief History of Sequencing Technologies

Sequencing technologies have undergone a considerable evolution in the last decades; the first expensive machines (appearing in the late 70s) have today been substituted by cheaper and more effective one. At the same time, data processing evolved concurrently to face new challenges and problems posed by the new type of sequencing records. In this first section, we briefly outline how such an evolution of sequencing technologies developed and how new challenges were posed by each new generation.

14.1.1 Shotgun Sequencing

As computational power became cheaper and massively available, scientific communities of various fields began to take full advantage of it by developing innovative approaches and techniques to exploit the technological advances within their own research field. Life sciences were no exception in this *gold rush*; as a result, ambitious projects, such as the reconstruction of the whole human genome, become more feasible than ever before. What before was only theoretically possible suddenly became achievable, and many techniques started to gain interest and importance. In an early work [46], Rodger Staden first introduced the *shotgun*

M. Comin (✉) • M. Schimd
Department of Information Engineering, University of Padova, Padova, Italy
e-mail: comin@dei.unipd.it; schimdmi@dei.unipd.it

M. Elloumi (ed.), *Algorithms for Next-Generation Sequencing Data*,
DOI 10.1007/978-3-319-59826-0_14

sequencing methodology, which uses biochemical reactions to read nucleotides of small fragments (often called *reads*) obtained from the original sequence. What made this protocol really appealing was the development of ad hoc algorithms that were able to process many such fragments in a short time. As shotgun sequencing began to be extensively investigated in even more detail, it was observed that, despite the increasing power of the available supercomputers, the presence of repetitive and "complex" zones on genomic sequences made *brute-force* approaches too complex if not impossible to implement. A possible solution to this problem relies on biology experts who, by spending many hours on sequence analysis, *map* zones of the genome that may be of particular interest to *classify* and *tag* interesting parts of the sequence under investigation. Although still useful and successful, such an approach cannot keep up with the ever increasing amount of available sequencing data. Moreover, it has recently been shown that human DNA contains important and essential information in almost all of its parts (including those previously called *junk DNA*) [16], making the task of classifying different portions of the genomic sequence no longer a possible way to reduce the amount of data to be processed (while still maintaining the same information content). It should by now be clear that *efficient*, *scalable* and *effective* algorithms to reconstruct the reference sequence from a shotgun experiment are fundamental in the field of life science; this has been even more the case in recent years, during which we observed an unprecedented explosion of shotgun sequence experiments thanks to the advent of *Next Generation Sequencing* (NGS) technologies able to produce billions of fragments in a short amount of time and with relatively low costs. As new data become available, the request for automated analysis tools increases. The bioinformatics community is continuously facing the thrilling challenge of developing and designing new methods to process data in a faster and more effective way than ever before. Because of this constant evolution, shotgun sequencing (like many other techniques) has been modified and improved even to suit sequencing methods that at the time of its first introduction by Staden were not available. This section aims to give a brief overview of some of the most important and studied challenges that bioinformaticians need to solve, focusing on the application to NGS data. Before doing this, however, it is useful to provide a short introduction to the different sequencing technologies currently available.

14.1.2 The Sanger Sequencer, the First Generation

The genomic era started around the mid-1970s with the development of the first effective sequencing technologies. Among them, the most successful has been *chain termination methods*, usually referred to as *Sanger methods* after one of the authors of the original paper [39]. The process starts by separating the two DNA strands by means of heating; afterwards, a *primer* is attached to one of the strands. The primed template is then inserted into a mixture containing several reagents that allow the *chain reaction* to start; the result is excited using *gel capillary electrophoresis*, and, by means of either dyes or radio labeling, the sample is read and translated into a sequence of bases. Despite being one of the first available methods for DNA sequencing, Sanger technology possesses several desirable properties. Thanks to the relatively long reads (between 600 and 900 bases) and high reliability (1% error or below), this technology is still the only choice when high-quality data are needed, for example, for the completion of complex parts of genomic sequences. The Sanger methods' main problem is the high cost (a few thousand dollars per megabase), especially when compared with NGS methods.

14.1.3 Next Generation Sequencing

Sanger technology remained the de facto choice for sequencing for more than 30 years. During this period, huge projects were successfully completed. Among them, the *whole human genome* project deserves particular attention; it culminated in the milestone paper [26] that announced the first *non-draft* version of the human (*Homo sapiens*) genome as the result of a huge collaboration involving many different research groups worldwide. At the beginning of the 2000s, new sequencing methods started to be developed by several life science and bioengineering companies. As a result of this competitive market, several novel methods for whole shotgun sequencing became available; they were (and still are) all called *Next Generation Sequencing* (NGS).

Such new technologies became very popular in the scientific community thanks to their low cost. Using NGS sequencers, it is currently possible to produce one million bases for as little as 0.1$, which is five times cheaper than the Sanger sequencers. Such new methods have changed the way the scientific community looks at genomic projects [30]. Since their introduction, the number of massive sequencing projects initiated has increased day by day, and, as groups all over the world have started to share their experiments, the amount of data available has exponentially increased, as shown in Fig. 14.1. Of all the projects, the *1000 Genomes Project*[1] is worth mentioning, which aims to sequence a total of 1000 human individuals belonging to different races. The project's site reports *[...] the goal of the 1000 Genomes Project is to find the most genetic variants that have frequencies of at least 1% in the populations studied [...]*. Finding all such variants is much harder than the reconstruction of 1000 individuals' genomes, and tasks of this complexity were not feasible until the NGS technologies appeared.

One of the key features of NGS methods is the possibility of producing really huge amounts of data during a single experiment. This reduces the total cost of both the reagents and experts needed to supervise the entire sequencing process. This combination of cost-lowering factors allows NGS sequencers to attain competitive costs in terms of dollars per base.

[1] http://www.1000genomes.org/.

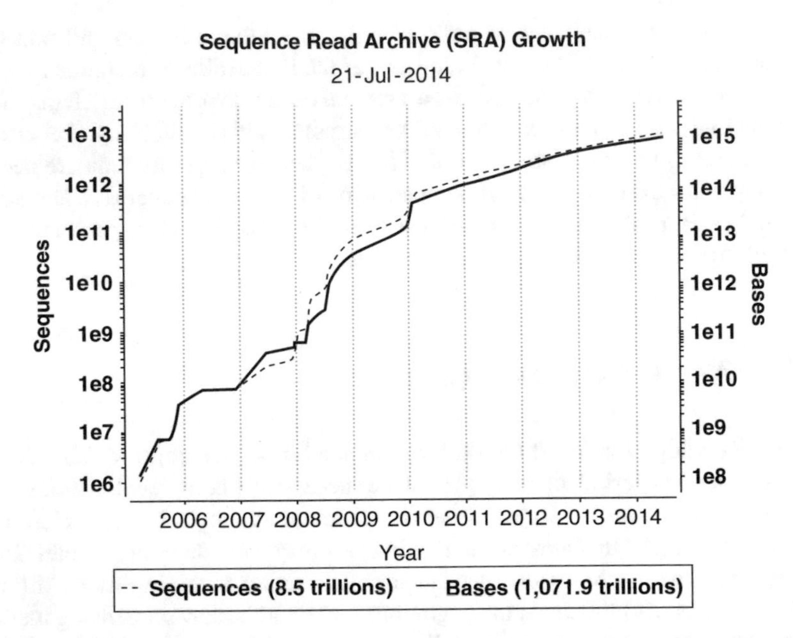

Fig. 14.1 Growth rate for the *Sequence Read Archive* (SRA). http://www.ebi.ac.uk/ena/about/statistics, accessed 21 July 2014

Although different technologies apply different chemistries and principles to obtain reads from sample(s), they all share the same high-level procedure, which is summarized here.

1. A *library* is prepared by ligating *adapters* to ends of a DNA sample. Different technologies use different adapters; this is one of the main sources of difference on error models and *biases* between technologies.
2. The sample is then immobilized into a solid surface to produce massive copies of the original fragments. The fact that the samples are immobilized is the key feature allowing NGS machines to produce millions of copies per single experiment.
3. Finally, many copies of the same sample are sequenced using platform-specific methods, and the final data set is then created. During this step, the machine must assess the actual bases of the sample. This operation is usually performed by dying or radio tagging, and it is another discriminant between the different sequencing technologies.

If on the one hand NGS technologies supply researchers with overwhelming amounts of data, on the other hand, this does not come for free. Next-generation sequencers have several drawbacks that make data analysis and processing difficult. When NGS sequencers were first introduced, the available methods allowed fragment lengths of at most few hundreds bases; back then, this was a huge disadvantage compared with the Sanger methods, which were (and still are) able to produce fragments of lengths up to 700 bases. These days, however, NGS methods can compete with the Sanger ones in terms of read length even though the data sets produced with previous versions of the same technologies represent a big fraction of the total available data on public databases. A second aspect related to NGS data is their reliability; next generation reads are known to be more noisy than Sanger ones. The error rate on NGS data is approximately one order of magnitude higher than the average error rate of Sanger methods (which is usually under 1%); on average, NGS data contain one erroneous base every hundred. With such an error rate, the chance of miscalled bases within a single read becomes very high, and while designing and developing algorithms for NGS data, researchers must take this into account.

Most of the NGS sequencers are able to produce *mate pairs*, which are pairs of reads sequenced using the same sample but starting from the opposite ends. The advantage of mate pairs is the possibility of using different *libraries* with different sample sizes. This allows the sequencers to produce pair of reads with an approximately known spacing called *insert size*, which is much larger than the actual read length and (in most of the cases) can be set to span repeats, allowing their resolution.

14.1.4 Future Generation Sequencing: PacBio

NGS methods are now the standard in sequencing technologies; however, further advances are being revealed. The most promising *future generation sequencer* was introduced by *Pacific Biosciences* and is based on a new sequencing procedure known as *single molecule real time* (SMRT) sequencing [18]. The commercial product (usually referred to as *PacBio*), which has recently been presented and discussed [4], produces reads with characteristics that are considerably different from both NGS and Sanger reads. PacBio reads can be as long as 10,000 bases, a size never available before, but the error rate on such reads is as high as 15% with 12% insertions, 2% deletions and (just) 1% substitutions. This combination of factors makes the algorithms developed for all previous technologies not feasible or not effective for this type of data.

In this chapter, we present a parameter-free alignment-free method, called $Under_2$, based on variable-length patterns. We will define a similarity measure using variable-length patterns along with their statistical and syntactical properties, so that "uninformative" patterns will be discarded. The rest of the chapter is organized as follows: in Sect. 14.2, we present the main challenges the NGS data pose, in Sect. 14.3 we describe how NGS data are used to solve problems of comparative genomics, and finally Sect. 14.4 gives our conclusions.

14.2 Algorithmic Challenges of NGS Data

The available variety of sequencing technologies, each having its own *reads profile* (in terms of error, length, biases, ...), makes the task of designing and developing efficient and effective algorithms very challenging. In this scenario, it is not true that *one size fits all*, and approaches designed for one technology may be completely useless for another one. Each feature of the produced reads has an impact on the development of algorithms, and it is important to be aware of all such aspects before starting to work on new approaches.

14.2.1 High Coverage

As already mentioned, a big role in the success of NGS technologies can be attributed to their ability to perform massive sequencing at a relatively low cost. Although desirable, high coverage generates huge numbers of reads that need to be stored and processed. The choice of data structures and algorithms used is crucial when it comes to NGS data. Even when the most efficient approaches are adopted, the data sets may still be too big, and the only remaining solution relies on either *filtering* high- and low-level data or reducing the size of the set by computing a new data set that somehow maintains the important information of the original one while reducing the amount of data to be processed.

14.2.2 Short Reads

Genomic sequences are the result of the evolution process that the corresponding species experienced. While some organisms (like humans) have a relatively short story, many others (like plants) have gone through several mutations. Each time the DNA replicates, there is a small chance that the process creates mutations that were not previously present. Several types of mutations can happen during the replication process, but many of them have the net effect of replicating portions of the sequence in other positions on the genome itself. This creates complicated structures generally called *repeats*. The characteristics of repeated structures on a real genomic sequence are quite variable. The same genomic sequence (like the human one) can contain repeats of a few bases to thousand of bases long. The number of copies of the same repeated sequence is also a variable factor of all genomes; we may have structures that are repeated a few times as well as sequences that appear thousands of times in the same genome. Finally, the relative position of repeated regions can be adjacent (like *tandem repeats*) or far apart from each other. It is easy to see how short reads make the process of discovery and resolving repeats very difficult. For example, a read sequenced "inside" a repeat appearing in many different places of the sequence can never be reliably mapped into either of these positions. Even though NGS technologies can currently be used to resolve short to medium-long repeats, there are still genomic structures that need to be addressed with different data (like, for example, mate pairs or PacBio reads).

14.2.3 High Error Rate

Although noisy data may look like a minor problem, combining this with the problems we already discussed (short reads, repetitive structures, ...) amplifies them. Approaches to error correction are often part of a proposed approach and may rely on pre-processing (e.g., filtering) or "online" procedures that try to identify errors as "unexpected" behaviors of the algorithms.

Although every single challenge may not seem that difficult to cope with (with the possible exception of repeats that in some situations make the solution not unique), the combined effect of all of them makes the bioinformatics problem difficult if not impossible to solve without using heuristics and approximate algorithms [33, 47].

14.2.4 Assembly

As already mentioned, a huge achievement obtained so far in the bioinformatics field is the reconstruction of the whole human genome sequence, the process of reconstructing a genetic sequence starting from fragments of thereof (i.e., reads) is called *assembly*. Assembly can be carried out *without any knowledge of the sequence* (e.g., sequencing a newly discovered organism), in which case the problem is named *de novo assembly*. When the sequence is reconstructed using another (possibly related) sequence as guide, the problem is called *comparative assembly*. Since the former is a more difficult problem, and also is a prerequisite to the latter, we will focus only on de novo assembly.

Giving a formal definition of the problem of assembly is not straightforward at all. We can easily (and correctly) state that the assembly problem is *the problem of reconstructing the (unknown) reference sequence from which a collection of (known) reads are drawn from.* This definition, although correct, is useless because of its recursive nature arising from the reference to the solution itself (i.e., the reference sequence). Based on this definition, the only way to assess the effectiveness of the assembly algorithms is to compare its result with the reference sequence (which is what we are trying to obtain with de novo assembly). In other words, we need a way to model the reference sequence, but the only model sufficiently precise to describe it is the entire sequence of the nucleotide that is the output of the assembly process. Consequently we cannot have a precise definition of the problem, but we rather must deal with the fact that *the assembly problem can only be described using heuristic definitions*. Despite these difficulties, assembly is a well-defined problem in the field of biology. In fact, it is probably the most important problem that needs to be solved before starting the analysis of the genetic material of an organism. Reconstructing a sequence with millions or even billions of bases using fragments no longer than 1000 bases is a hard task regardless of the sequencing technology used. The problem is even worse because of the complex structure that genomic sequences contain (e.g., repeats, mutations,...). NGS data make the problem even harder, especially when performed using the very short reads currently available. Even with the advent of future generation sequencing technologies, the assembly problem remains challenging because the read lengths remain much shorter than the genomic sequence we aim to reconstruct.

When NGS data started to become massively available, existing assembly algorithms were not suitable any more. With billions of reads, it was no longer practical to perform pairwise overlaps between reads; therefore, the community started to develop new approaches to the assembly problem with the main goal of developing algorithms and methods able to process NGS data efficiently. A survey on current most successful approaches to assembly using NGS data can be found in [31]; we give here a brief introduction and discussion of the most important aspect of these methods.

According to [31], assembly algorithms for NGS can be roughly classified into three different macro-categories: *greedy* [17, 51] algorithms, *overlap layout consensus* [22, 23] and *De Bruijn graphs* [32, 40, 52].

14.2.4.1 Greedy

Greedy algorithms [17, 51] start by selecting a *seed* read (or set of reads) that represents the initial assembly and proceed by *finding the best alignment* to extend it. At each iteration, the "best read" is chosen and removed from the original set. The process continues until either the set is empty or no good enough alignment can be obtained with all the remaining reads. This approach is clearly greedy since, once a read has been used to extend the current assembly, there is no possible "regret" for such a decision. The advantage of greedy strategies is that they are fast and easy to implement, but, on the other hand, it is very likely (especially with big and noisy sets) that the algorithm stops at a *suboptimal* solution.

14.2.4.2 Overlap Layout Consensus

The *overlap layout consensus* (OLC) [22, 23] algorithms divide the task of producing an assembly into three subsequent phases. During the *overlap* phase, an *all-against-all* pairwise read comparison is performed to determine the best possible overlaps. Since this phase could require lot of computational time, usually *seed* overlap is detected using k-mer sharing (i.e., an overlap is actually tested between reads that share at least a certain amount of k-mers), which is easier to compute than an actual alignment; only for those pairs of reads that satisfy the k-mer requirement is the actual alignment computed. The alignment relation between reads is *laid out* on a graph where each node represents a read and an arc between nodes exists if a valid alignment has been detected between the corresponding reads. After the graph has been simplified, a path detection algorithm is run to identify a *consensus* sequence that is then output as the candidate assembly.

14.2.4.3 De Bruijn Graphs

There are two major problems with OLC algorithms: the first one is related to all-against-all pairwise alignment, which is a time-consuming operation; the second problem is the phase of path discovery that requires finding a *Hamiltonian path*, which is known to be an NP-hard problem.

The latter has been the main motivation triggering the development of assembly algorithms based on *De Bruijn* graphs [32, 40, 52]. The idea is to create what is called *k-mers* graph, which is a simplified version of the De Bruijn graph.

On k-mer graphs, each node corresponds to a k-mer (but, as opposed to De Bruin graphs, not all possible k-mers have an associated node). Edges represent an observed $(k - 1)$-mer. The graph is constructed by scanning the read set so that each k-mer that is present in at least one of the reads has a corresponding node on the graph. If the read is sequenced without error and with *perfect* coverage, then the graph constructed from the read set and the graph constructed from the reference sequence are identical. A key observation is that the k-mer graph, by construction, contains an *Eulerian path* that represents the reference sequence [32].

14.2.5 Comparative Genomics

Comparative genomics is the biology field that studies genetic material with the goal of identifying and classifying biological features shared by different organisms or by different individuals of the same species. Practically, this can be achieved using genome analysis techniques to test the correlation between sequences. There are many known correlation structures for which specific algorithms exist: *variant detection*, *rare variation and burden testing*, and *identification of de novo mutations* are a few of them, and a detailed survey can be found in [25]. With the introduction of NGS data, the number of organisms that can be simultaneously compared has dramatically increased. As a the consequence of this, the overall complexity of the problems has increased as well. Moreover, using *High Performance Computing* techniques, we can now perform comparative genomics analysis at the genome scale, and *genome-wise association studies* (GWAS) are now becoming fundamental steps in all comparative genomics projects. For many years, genome assembly and sequence alignment have been essential primitives to perform comparative genomics studies; however, with the advent of massive sequencing, many of the approaches based on these primitives are no longer practical. When looking for variations between sequences, (short) reads need to be aligned to a reference allowing non-perfect matching; these are, indeed, the variation to be discovered, on the form, for example, of *single nucleotide polymorphisms* (SNPs). When performed on millions (or even billions) of reads, this task becomes computing intensive, and standard alignment tools may not be the best choice.

14.2.5.1 Alignment of Sequences

Since the beginning of the genomic era, researchers have focused their interests on finding coding sections of the human genome. More generally, one of the main tasks has been (and still is) finding recurrent *patterns* and classifying them based on the role they have in the regulation of human biology. Pattern identification, reconstruction and many other problems at the heart of bioinformatics all resort to the fundamental algorithmic primitive of *alignment*. Informally, *alignment* is the process of superimposing two different sequences to obtain the best possible match between them; more technically, the problem of alignment does not have a unique definition. Alignment can be performed with or without mismatches, which means that a certain degree of difference between superimposed sequences is or is not tolerated. Due to mutations induced by evolutionary events, sequence alignment is (almost) always performed allowing mismatches. How the mismatches are treated during the process is a matter of specific algorithms and may lead to different results on the same input data.

One of the most popular ways to perform sequence alignment relies on a dynamic programming approach called the *Smith–Waterman* after the authors of the original paper [42], which was published more than 30 years ago. The idea is to define a recurrence that assigns *scores* to: matches, mismatches, insertions and deletions. Given two sequences x and y with size n and m, respectively, the Smith–Waterman algorithm computes an $n \times m$ matrix. This matrix is then used to derive the optimal sequence of string operations (i.e., substitutions, insertions and deletions) that transforms x into y. The complexity of this approach is $O(nm)$ for time, while the space (still $O(nm)$ with a naïve implementation) can be kept at $O(m+n)$ using some tricks. It is important to note, however, that dynamic programming approaches are able to find the optimal solution once the recurrence has been given. Another popular approach, first proposed and implemented in a tool called BLAST [1], relies on hash maps to perform fast alignment of sequences based on their k long subsequence (i.e. k-mers). Based on the specific necessities, the alignment obtained using hash maps can be refined using dynamic programming algorithms. With the advent of NGS data, the dynamic programming approach started to become impractical. The main reason is not the complexity of a single alignment itself, but the number of alignments required when the read set contains millions of reads. Moreover, the shorter the reads are, the more likely they can fit in more than a single position of the reference sequence for any given score. This means that multiple positions can give the same (optimal) score for one given read.

Consequently, new alignment algorithms have been devised to specifically work on NGS data, and, at the same time, *alignment-free* approaches to sequence comparison and pattern discovery have started to gain interest in the community [27, 45].

14.2.5.2 Alignment-Free Sequence Comparison

The comparison of sequences is fundamental for the analysis of many biological processes. The use of alignment tools such as BLAST [1] to assess the degree of similarity between two sequences is a dominant approach. Alignment-based methods produce good results only if the biological sequences under investigation share a reliable alignment. However, there are cases where traditional alignment-based methods cannot be applied, for example, when the sequences being compared do not share any statistical significant alignment. This is the case when the sequences come from distant related organisms or they are functionally related but not orthologous. Moreover, as discussed above, another drawback is that alignment methods are usually time consuming. Thus, they cannot be applied to large-scale sequence data produced by NGS technologies.

Alignment-free sequence comparison methodology was introduced during the mid 1980s with the seminal paper of Edwin Blaisdell [3] where the D_2 statistic was introduced as a way of correlating different sequences based on the frequency of their constituent k-mers, with k being an adjustable parameter of the algorithms. The idea, although simple, proved to be effective, especially after several improvements were developed [35, 50]. Recently, alignment-free techniques have been used to perform *assembly-free* sequence comparison using NGS data [8, 44]. A good survey on the most recent advances in alignment-free techniques can be found in [45].

14.3 Assembly-Free Comparative Genomics

With the advent of NGS technologies, a many short read data have been generated. These data are used to study many biological problems, such as transcription factor binding site identification, de novo sequencing, alternative splicing, and so on. The first step of most studies is to map the reads onto known genomes. However, if a reference genome is not available, the assembly of a template sequence is usually challenging, as we have already discussed in previous sections.

When the NGS reads cannot be mapped onto a reference genome, alignment-based methods are not applicable. Moreover, the size of NGS data demands the use of very efficient algorithms. For these reasons, the comparison of genomes based on the direct comparison of NGS reads has been investigated only recently using alignment-free methods [44].

The use of alignment-free methods for comparing sequences has proved useful in different applications. Some alignment-free measures use the pattern distribution to study evolutionary relationships among different organisms [20, 34, 41]. In [15], researchers have shown that the use of k-mer frequencies can improve the construction of phylogenetic trees traditionally based on a multiple-sequence alignment, especially for distantly related species. The efficiency of alignment-free measures also allows the reconstruction of phylogenies for whole genomes [11, 12, 41]. Several alignment-free methods have been devised for the detection of enhancers in ChIP-Seq data [13, 21, 24, 29] and also of entropic profiles [6, 7]. Another application is the classification of remotely related proteins, which can be addressed with sophisticated word counting procedures [9, 10]. For a comprehensive review of alignment-free measures and applications, we refer the reader to [49].

To the best of our knowledge, so far only one group of researchers has compared sets of NGS reads using alignment-free measures based on k-mer counting [44]. In this chapter, we intend to follow the same approach by adapting the alignment-free pairwise dissimilarity, called $Under_2$ [11], for the comparison of two sets of NGS reads. The current chapter differs from our previous studies [11, 12, 14] in the following aspects. First, instead of considering two long sequences as input, we need to modify $Under_2$ so that it can be applicable for genome comparison based on two sets of NGS reads. Another important aspect is the way patterns are weighted in our similarity score, where we need to consider the expected number of occurrences of a pattern in a set of reads.

Almost all other methods are based on statistics of patterns with a fixed-length k, where the performance depends dramatically on the choice of the resolution k [41]. Finally, one the most important contributions is the use of reverse and reverse-complement patterns, as well as variable-length patterns, to mimic the exchange of genetic material.

14.3.1 Previous Work on Alignment-Free Methods

Historically, one of the first papers introducing an alignment-free method was by Blaisdell in 1986 [3]. He proposed a statistic called D_2 to study the correlation between two sequences. The D_2 similarity is the correlation between the number of occurrences of all k-mers appearing in two sequences. Let A and B be two sequences from an alphabet Σ. The value A_w is the number of times w appears in A, with possible overlaps. Then, the D_2 statistic is:

$$D_2 = \sum_{w \in \Sigma^k} A_w B_w$$

This is the inner product of the word vectors A_w and B_w, each one representing the number of occurrences of words of length k, i.e., k-mers, in the two sequences. However, Lippert et al. [28] showed that the D_2 statistic can be biased by the stochastic noise in each sequence. To address this issue, two other popular statistics, called D_2^S and D_2^*, were introduced respectively in [24] and [35]. These measures were proposed to standardize the D_2 in the following manner. Let $\tilde{A}_w = A_w - (n - k + 1) * p_w$ and $\tilde{B}_w = B_w - (n - k + 1) * p_w$ where p_w is the probability of w under the null model and n is the length of the strings A and B. Then, D_2^* and D_2^s can be defined as follows:

$$D_2^* = \sum_{w \in \Sigma^k} \frac{\tilde{A}_w \tilde{B}_w}{(n - k + 1) p_w}$$

and,

$$D_2^s = \sum_{w \in \Sigma^k} \frac{\tilde{A}_w \tilde{B}_w}{\sqrt{\tilde{A}_w^2 + \tilde{B}_w^2}}$$

These similarity measures respond to the need for normalization of D_2. All these statistics have been studied by Reinert et al. [35] and Wan et al. [50] for the detection of regulatory sequences. In [44], the authors extended these statistics for genome comparison based on NGS data and defined d_2, d_2^s and d_2^*. The major difficulties are the random sampling of reads from the genomes and the consideration of double strands of the genome. They tested the performance of d_2, d_2^s and d_2^* on synthetic and real data sets. In particular, the common motif model, introduced by Reinert et al. [35], is used to mimic the exchange of genetic material between two genomes, and *MetaSim* [36] is used to simulate the sequencing. We describe the common motif model in the next subsections and propose a more realistic formulation. In this chapter, we will follow the same experimental setup as [44] and compare our results with these statistics.

14.3.2 $\overline{Under_2}$ an Assembly-Free Genome Comparison Based on Next-Generation Sequencing Reads and Variable Length Patterns

In this section, we describe our parameter-free, alignment-free dissimilarity measure, called $\overline{Under_2}$, which extends our previous work [11] to the case of NGS reads. The dissimilarity $\overline{Under_2}$ is based on two concepts: irredundancy and underlying positioning.

Let us consider two sets of reads R_1 and R_2, sampled from two genomes. Every set is composed by M reads of length β in the $\Sigma = \{A, C, G, T\}$ alphabet. We say that a pattern in Σ^* is shared between the two sets of reads if it appears at least once in some read of R_1 and once in some other read of R_2. For example, consider the following sets:

$$R_1 = \{TGCG, CGAA, TAAC\}$$
$$R_2 = \{AACC, GGTC, TAGG\}$$

representing two sets of $M = 3$ with reads each having length $\beta = 4$; an example of a pattern occurring in both sets is *AAC* or *TA*. Note that the set of all patterns (with length at least 2) *common* to both sets is:

$$\{AA, TA, AAC, AC\}.$$

The notion of irredundancy is meant to remove the redundant patterns, i.e., those patterns that do not convey extra information for the similarity measure. The second driving principle is the fact that, in previous approaches, every position of a read contributes a multiple number of times to the final score.

In the following, we address these two issues separately. The goal is to build a similarity measure between the two sets of reads R_1 and R_2 using all exact patterns of any length, Σ^*, that are shared between the two sets.

14.3.2.1 Removing Redundant Patterns

One can easily show that most sequences share an unusually large number of common patterns that do not convey extra information about the input. To keep the article self-contained, here we summarize the basic facts already proved in [10] and extend the notion of an irredundant common pattern to the case of two sets of reads. If the occurrence of a pattern in a read completely overlaps with the occurrence of another longer pattern, we say that the occurrence of the first pattern is covered by the second one.

Definition 1 (Irredundant/Redundant Common Patterns) A pattern w is *irredundant* if and only if at least an occurrence of w in R_1 or R_2 is not covered and by other patterns. A pattern that does not satisfy this condition is called a *redundant common pattern*.

Let us consider the two read sets R_1 and R_2 defined above. In the third read of set R_1, i.e., *TAAC*, the common pattern *AC* occurs; however, it is covered by the other common pattern *AAC*. We highlight in bold the pattern occurrences that are not covered by other patterns. Following the definition above, *AC* is a redundant common pattern, and the only irredundant common patterns are *AA*, *TA* and *AAC*.

	R_1			R_2	
TGCG	CGAA	TAAC	AACC	GGTC	TAGG
	AA	AA	AA		
		TA			**TA**
		AC	AC		
		AAC	**AAC**		

Note that the in the first set R_1, these patterns partially overlap. This observation will be important later. We observe again that the set of irredundant common patterns $\mathscr{I}_{R_1,R_2}$ is a subset of the well-known linear set of maximal patterns [2]; therefore, the number of irredundant common patterns is bounded by $|R_1| + |R_2|$, where $|R_1| = |R_2| = M\beta$.

A simple algorithm that can discover all such patterns has already been described in [11], and it employs a generalized suffix tree of two sequences. To extend this algorithm to the new input R_1 and R_2, it is sufficient to use the two sets of reads, while keeping the occurrences that belong to the two sets separate. The construction of the generalized suffix tree and the subsequent extraction of the irredundant common patterns can be completed in time and linear space in the size of sequences [11]. In summary, the notion of irredundancy is useful for removing non-informative patterns and thus for drastically reducing the number of candidates to be analyzed to estimate the sequence similarity between R_1 and R_2.

14.3.2.2 Selecting Underlying Patterns

The basic idea behind our approach is that a position on the sequences should contribute to the final similarity only once. Traditionally, alignment-free statistics fail to comply with this simple rule. In fact, every position, apart from the borders, belongs to k different k-mers and thus contributes k times to the similarity.

In previous works on whole-genome comparison, to solve this problem we used the notions of pattern priority and of underlying pattern [11]. The pattern priority rule is mainly based on the idea of selecting, for each position, those patterns that represent the largest number of matching sites between sequences and that are thus more likely to be conserved patterns Here we recall the definition of pattern priority and of the underlying pattern from [11], and we adapt these concepts to the new settings.

Let us consider the set of irredundant common patterns $\mathscr{I}_{R_1,R_2}$ as input. Given two patterns, w and w', we say that w has priority over w', denoted $w \rightarrow w'$, if and only if either $|w| > |w'|$ or $|w| = |w'|$ and w are less likely to appear in the sequences than w', or if w and w' have the same length and probability to appear, but the first occurrence of w appears before the first occurrence of w'; that is, w appears on the same read of w' but at a lower index or w appears on a read that precedes (using an arbitrary order) the reads where w' appears. We say that an occurrence l of w is *tied* to an occurrence l' of another pattern w' if these occurrences (partially) overlap each other, $[l, l + |w| - 1] \cap [l', l' + |w'| - 1]) \neq \emptyset$, and $w' \rightarrow w$. Otherwise, we say that l is *untied* from l'. As observed above the patterns, *TA* and *AAC* partially overlap on the third read *TAAC*; in other words, *TA* is tied with *AAC* (because the former is longer than the latter even though it occurs later on the same read).

Definition 2 (Underlying Patterns) A set of patterns $\mathscr{U}_{R_1,R_2} \subseteq \mathscr{I}_{R_1,R_2}$ is said to be *Underlying set* of $\{R_1, R_2\}$ if and only if:

(i) every pattern w in $\mathscr{U}_{R_1,R_2}$, called *underlying pattern*, has at least one occurrence in both sets of reads, that is untied from all the untied occurrences of other patterns in $\mathscr{U}_{R_1,R_2} \setminus w$, and
(ii) a pattern $w \in \mathscr{I}_{R_1,R_2} \setminus \mathscr{U}_{R_1,R_2}$ does not exist such that w has at least two untied occurrences, one per set of reads, from all the untied occurrences of patterns in $\mathscr{U}_{R_1,R_2}$.

Let us consider the two sets of reads introduced above and the set of irredundant patterns $\mathscr{I}_{R_1,R_2} = \{AA, TA, AAC\}$.

	R_1			R_2	
TGCG	CGAA	TAAC	AACC	GGTC	TAGG
	AA	AA	AA		
		TA			**TA**
		AAC	**AAC**		

The relative priority is by definition $AAC \rightarrow TA$ and $AAC \rightarrow AA$ since the first is strictly longer than the other two patterns. In the third read of set R_1 (i.e., *TAAC*), the three patterns partially overlap, and because of the relative priority, the occurrences of *AA* and *TA* are tied by *AAC*. All untied occurrences are highlighted in bold. According to Definition 2, *TA* and *AA* are not in $\mathscr{U}_{R_1,R_2}$, because they do not have one untied occurrence per set of reads. Therefore, the only underlying pattern is *AAC*.

The objective Definition 2 is to select the most important patterns in $\mathscr{I}_{R_1,R_2}$ for each location of the reads in the two sets, according to the pattern priority rule. If a pattern w is selected, we filter out all occurrences of patterns with less priority than w that lay on the untied locations of w in a simple combinatorial fashion. The complete procedure to discover the set $\mathscr{U}_{R_1,R_2}$ can be found in [11]. Below we give an overview of the algorithm.

Algorithm *Underlying Pattern Extraction* (**Input**: R_1, R_2; **Output:** $\mathscr{U}_{R_1,R_2}$)

Compute the set of irredundant common patterns $\mathscr{I}_{R_1,R_2}$.
Rank all patterns in $\mathscr{I}_{R_1,R_2}$ using the pattern priority rule.
for Select the top pattern, w, from $\mathscr{I}_{R_1,R_2}$: **do**
  if Check in Γ if w has at least one untied occurrence per sequence that is not covered by some other patterns already in $\mathscr{U}_{R_1,R_2}$ **then**
    Add w to $\mathscr{U}_{R_1,R_2}$ and update the location vector, Γ, in which w appears as untied.
  else
    Discard w.
  end if
end for

An auxiliary vector Γ of length L is used to represent all locations of R_1 and R_2. For a pattern w in $\mathscr{I}_{R_1,R_2}$, we can check whether its occurrences are tied to other patterns by looking at the vector Γ. If some untied occurrences are found, then we can add the new underlying pattern w to $\mathscr{U}_{R_1,R_2}$ and update the vector Γ accordingly using all the untied occurrences of w. In total, the extraction of all underlying patterns, using this scheme, takes $O(L^2)$ time. A more advanced algorithm with a better complexity, $O(L \log L \log \log L)$ time and $O(L)$ space can be found in [11].

14.3.2.3 Building the $\overline{Under_2}$ Similarity Measure

Our similarity is inspired by the *average common subword* (ACS) approach [48], where the scores of common patterns found are averaged over the length of sequences. Here we follow the same approach, but, instead of counting all common patterns, we use just the untied occurrences of the underlying patterns, which by definition do not overlap [11]. We can note that the set of underlying patterns $\mathscr{U}_{R_1,R_2}$ is not symmetric, in general $\mathscr{U}_{R_1,R_2} \neq \mathscr{U}_{R_2,R_1}$. Thus, to build a symmetric measure, we need to consider both sets.

In ACS, the contribution of each position is given by the length of the pattern covering that position. In our approach, we use instead the ratio of the number of occurrences for an underlying pattern w and the expected number of occurrences for that pattern. Let us define occ_w as the number of occurrences of w in all reads of R_1 and R_2 and $untied^1_w$ as the number of untied occurrences of w in R_1. First, we compute the score:

$$Score(R_1, R_2) = \frac{\sum_{w \in \mathscr{U}_{R_1,R_2}} |w| * untied^1_w * \frac{occ_w}{E[occ_w]}}{|R_1|}.$$

Recalling that the untied occurrences do not overlap with each other, we notice that the term $|w| * untied^1_w$ counts the positions where w appears without overlapping any other pattern. For each such position, we sum the score $\frac{occ_w}{E[occ_w]}$, where $E[occ_w]$ is the expected number of occurrences. Note that the expectation of this ratio is exactly 1. This sum is then averaged over the length of the first sequence under examination, R_1. This score is large when the two sequences are similar; therefore, we take its inverse. Another observation is that the number of occurrences of a pattern increases logarithmically with the length of R_2. Thus, we consider the measure $\log_4(|R_2|)/Score(R_1, R_2)$, where a base-4 logarithm is used to represent the four DNA bases.

To center the formula, such that it goes to zero when $R_1 = R_2$, we subtract the term $\log_4 |R_1|$. If $R_1 = R_2$ there will be just one underlying pattern that is equal to the sequence itself. In this case, $Score(R_1, R_1)$ will be 1, and the term $\log_4 |R_1|$ makes sure that $\widehat{Under_2}(R_1, R_1) = 0$. These observations are implemented in the general formula of $\widehat{Under_2}(R_1, R_2)$.

$$\widehat{Under_2}(R_1, R_2) = \frac{\log_4 |R_2|}{Score(R_1, R_2)} - \log_4 |R_1|$$

$$\overline{Under_2}(r_1, R_2) = \frac{\widehat{Under_2}(R_1, R_2) + \widehat{Under_2}(R_2, R_1)}{2}$$

Finally, to correct the asymmetry, our similarity measure called $\overline{Under_2}$ is the average of the two statistics $\widehat{Under_2}(R_1, R_2)$ and $\widehat{Under_2}(R_2, R_1)$.

An important aspect of this formula is the computation of the expected number of occurrences of a pattern w. A Markov model usually outperforms the Bernoulli model on biological sequences. In our case, the length of reads is relatively short. Thus, to avoid overfitting, we will rely on a first order Markov model. In summary, the expectation is computed as $E[occ_w] = p_w M(\beta - |w| + 1)$, where p_w is the probability of w using the Markov model, M is the number of Reads, and $(\beta - |w| + 1)$ are the possible occurrences of w. Finally, we extend our approach to account for untied occurrences that are present in the reverse, complement and reverse-complement of each sequence to simulate the DNA strand and the evolution of sequences. For more details about this extension, we refer to [11].

14.3.3 Experimental Results on Synthetic and Real Data

To compare the performance of $\overline{Under_2}$ and all d-type statistics proposed in [44], we performed several experiments using both simulated and real data.

14.3.3.1 The Common Motif Model Revised

We start from background sequences (20 in our experiment), which can be either synthetic or a real genomic reference; we call such sequences negative to indicate that no correlation exists between any two of them. For each negative sequence, we created a positive one using three different correlation models. The first is the *Common Motif* (CM) model introduced in [35]. In the CM model, a pattern of length five is inserted at position j with probability λ, while the background is left unchanged with probability $1 - \lambda$; we chose the same pattern and the same length used in [35, 44]. In the *CM* model, the pattern inserted is always the same. The second model we adopted is the *simple multiple motifs* (SMM). In this model, five patterns with length varying from four to six bases are considered. Note that the five patterns are all different now. Moreover, we also consider their reverse complement in this model. For each position j, a pattern is inserted with probability λ; the pattern to be inserted is chosen so that all five patterns and their reverse complements are inserted with the same probability as follows: first The last model introduced is the *full multiple motifs* (FMM) model, which is a slight variation of the SMM where, for each pattern, not only the reverse complement is considered, but also the reverse is inserted. The introduction of these two models SMM and FMM tries to mimic the exchange of genetic material between genomes, where regions of variable length as well as reverse and reverse complements are important.

14.3.3.2 Experimental Setup

We test the performance of the different statistics by assessing if sequences from the positive set score higher than those from the negative set. We compute the similarity scores for all pairs of sequences in the positive set and all pairs of sequences in the negative set. Then, we sort all scores in one combined list. We consider as *positive predictive value* (PPV) the percentage of pairs from the positive set that are in the top half of this list. A PPV of 1 means perfect separation between positive and negative sequences, while a PPV of 0.5 means no statistical power.

Following the experimental setup of [44], during all the experiments we maintained a constant pattern intensity $\lambda = 0.001$. For each sequence (either positive or negative), we used *MetaSim*[2] [36] to generate M reads with $\beta = 200$ length and with 0 standard deviation (i.e., all reads have a length of exactly β) to obtain an overall coverage $\gamma = (M\beta)/N$ (where N is the length of the sequence) equals 5. We will use these parameters for most of the experiments. Except where indicated, exact (i.e., no errors) sequencing has been simulated; when errors are considered, the *MetaSim* preset for 454 model is used with all parameters set to their default values. For each experimental setup, we compute the average score over five runs of $\overline{Under_2}$ and of all d-type statistics.[3] During all simulations, parameters of different algorithms have been maintained fixed; more specifically, we used $k = 5$ for d-type statistics because this is the best value measured in [44] as well as the best value we observed in a set of preliminary tests.

14.3.3.3 Simulations with Random Background

In this first test, we use random sequences as background. Although real data sets are always more desirable than simulations, the use of random sequences is very useful to establish the behavior of alignment-free statistics. Moreover, random background sequences can be used to formally prove the statistical power of the d-type statistics (see [35, 50]).

To simulate data, we used the same setup of [44]. We considered two different i.i.d. models for negative sequences, uniform background with $p_A = p_C = p_G =$

[2]http://ab.inf.uni-tuebingen.de/software/metasim/.

[3]http://www-rcf.usc.edu/~fsun/Programs/D2_NGS/D2NGSmain.html.

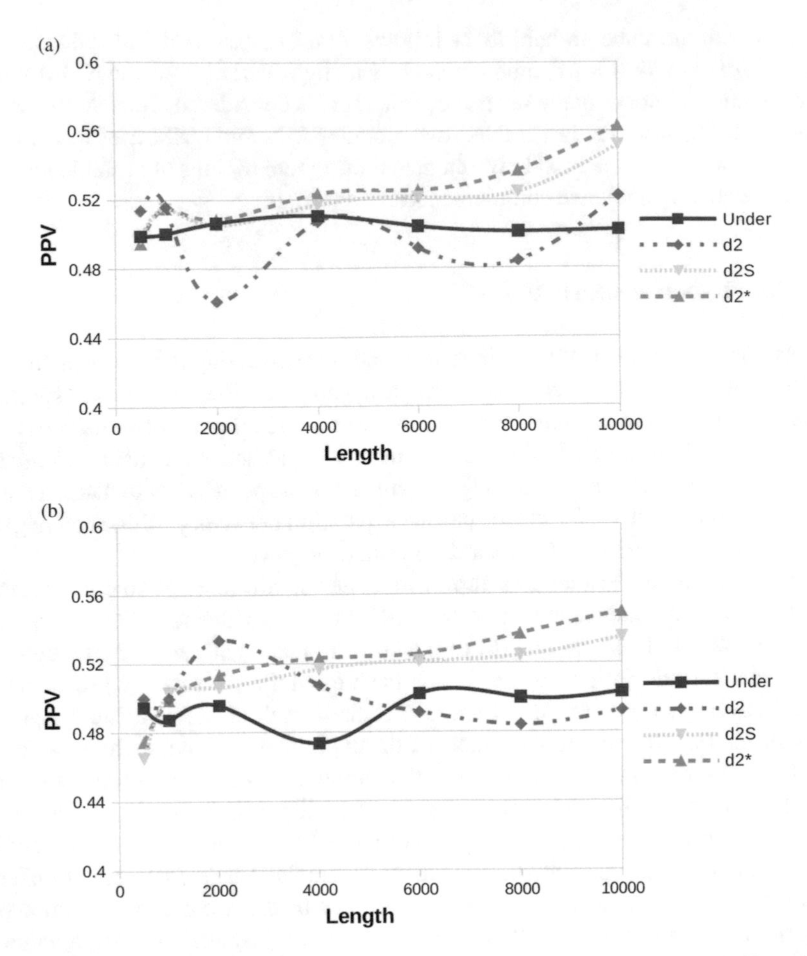

Fig. 14.2 PPV for uniform and GC-rich backgrounds with the CM model. (**a**) Uniform background. (**b**) GC-rich background

$p_T = 1/4$ and GC-rich background with $p_A = p_T = 1/6, p_C = p_G = 1/3$. We measured the PPV of 40 sequences, 20 positive and 20 negative, as the sequence length N varies from 500 to 10,000 bases.

Results for the CM model are shown in Fig. 14.2 with both a uniform background (a) and GC-rich background (b). Using this setup, we observed no significant improvement as N grows (PPV of 0.5 means no statistical power). All measures are almost aligned around a PPV of 0.5, and only for higher values of N (4000 or more) do d_2^* and d_2^s show a slight improvement of their performance, while $\overline{Under_2}$ still behaves badly. This is explained by the fact that the number of patterns inserted grows with the length of the sequence. Thus, longer sequences from the positive set will have more chances to obtain an higher similarity score. However, all methods perform poorly on this data set, as can be seen from the scale of Fig. 14.2. In general,

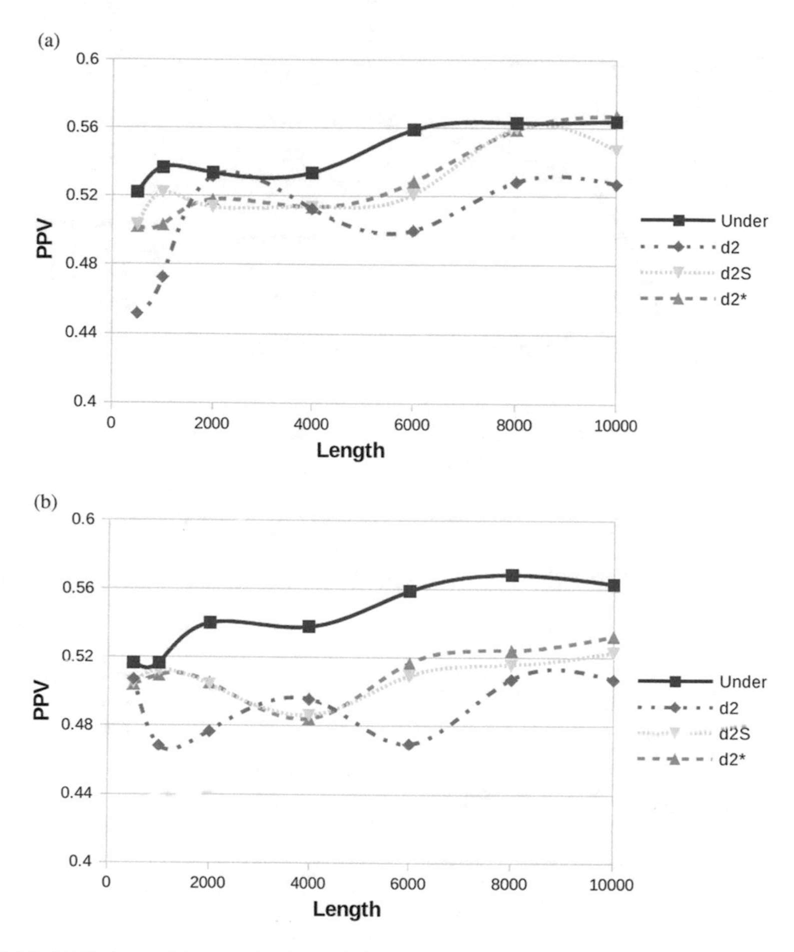

Fig. 14.3 PPV for uniform and GC-rich backgrounds with the SMM model. (**a**) Uniform background. (**b**) GC-rich background

d-type statistics need longer sequences or a higher pattern intensity λ to improve their predictive power (see Fig. 14.6b).

Figures 14.3 and 14.4 show the results for the SMM and FMM models, respectively, with uniform (a) and GC-rich (b) backgrounds. The introduction of multiple motifs does not lead to significant performance improvements for d-type statistics. Even if these statistics also consider the reverse complement, these measures need longer On the other hand, we see a slight improvement of $\overline{Under_2}$ for the SMM model and a significant improvement for the FMM model; this is because introducing the reverse complement (SMM) and also the reverse (FMM) gives better results as the $\overline{Under_2}$ statistic explicitly considers them.

By comparing subfigures (a) and (b) of all Figs. 14.2, 14.3 and 14.4, we can note that changing the background from uniform to GC-rich produces worse PPV values.

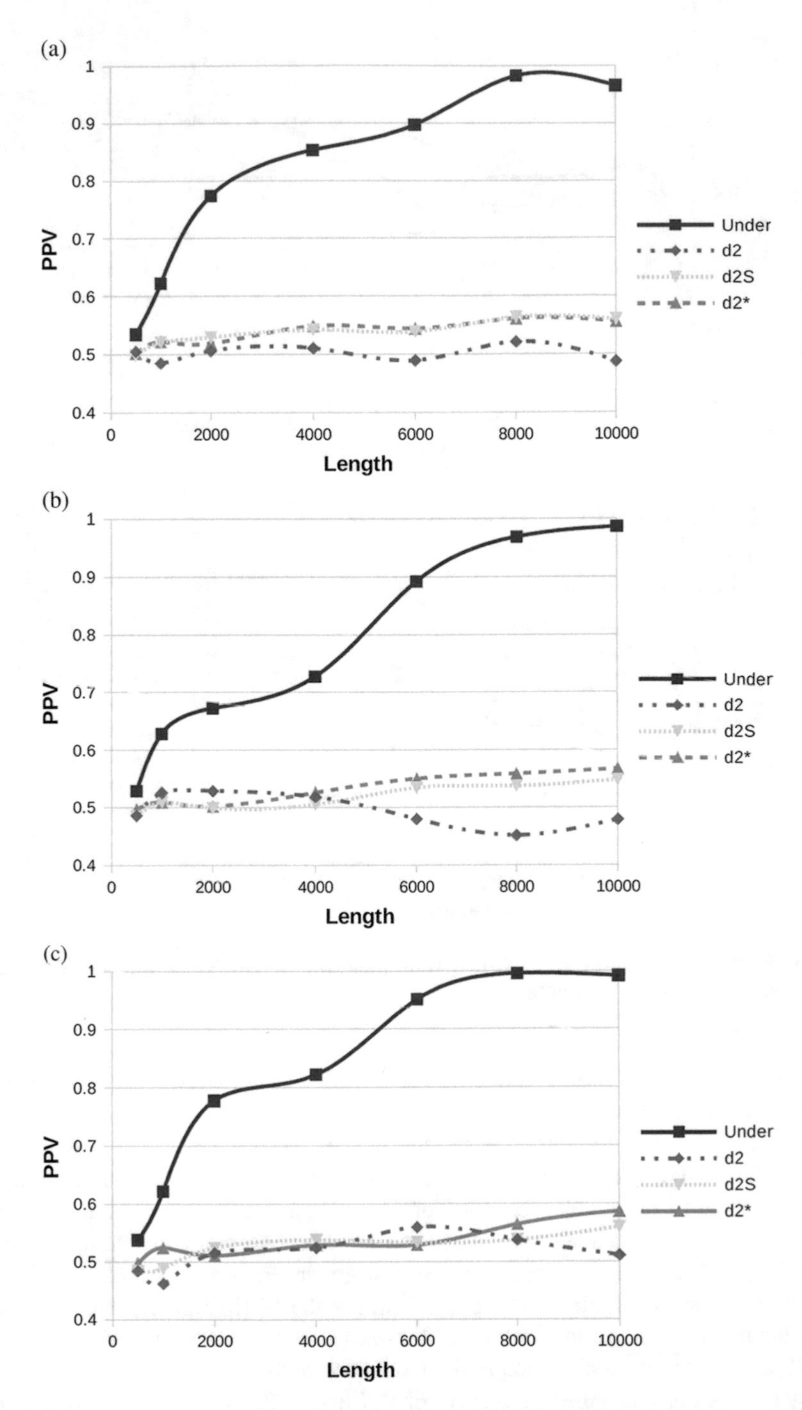

Fig. 14.4 PPV for uniform and GC-rich backgrounds with the FMM model. (**a**) Uniform background. (**b**) GC-rich background. (**c**) Uniform background and double coverage ($\gamma = 10$)

However, this effect becomes significant only for small values of N and when the FMM model is used, while all d-type statistics and all other cases of $\overline{Under_2}$ are almost immune from this effect, probably because performance in these cases is already poor. Finally, in Fig. 14.4c, we double the coverage, $\gamma = 10$. If we compare this plot with Fig. 14.4a, we can note a moderate improvement, especially for longer sequences. Thus, for random backgrounds, increasing the coverage will produce a small performance improvement.

14.3.3.4 Simulations with Drosophila Genome

To assess the performance in a more realistic scenario, in this subsection we use as background real genomic sequences from Drosophila. We first downloaded all the intergenic sequences of the Drosophila genome from *FlyBase*,[4] and then we created the negative backgrounds by picking ten sequences at random for each length varying from 1000 to 10,000. We then generated positive sequences using the foreground models CM and FMM described above. To test the impact of sequencing error, we also performed a set of experiments using the 454 error model provided by *MetaSim* [36] with the FMM foreground; all results are shown in Fig. 14.5.

We observed a consistent trend among all the experiments with $\overline{Under_2}$ always outperforming d-type statistics. Our measure, in fact, always gives better PPVs at all the tested lengths and for all models. As the introduced sequencing error results degrade, however, this effect is more relevant for short sequences where errors become more important and their effects are, therefore, more visible, while at higher lengths the impact of sequencing errors become less significant.

Starting from this latest and more realistic setup, i.e., using the Drosophila genome as background for the FMM model with 454 sequencing errors, we further evaluate how the different parameters affect the performance. Thus, we will compare the next plots with Fig. 14.5c, which was obtained with the following parameters ($\gamma = 5$, $\lambda = 0.001$ and $\beta = 200$). In Fig. 14.6, we report the PPV values while changing only one parameter at a time. If we double the coverage ($\gamma = 10$), subfigure (a), the PPV values do not improve; only with random backgrounds do we see a small improvement (see Fig. 14.4c). If we increase the probability to insert a pattern (λ) in the FMM model, subfigure (b), as expected, all statistics improve and $\overline{Under_2}$ quickly converge to 1. Finally, the use of shorter reads ($\beta = 100$), subfigure (c), does not degrade the PPV of $\overline{Under_2}$, which remains around 0.7.

14.3.3.5 Phylogeny of Genomes Based on NGS Data

In this subsection, we test the ability of alignment-free statistics to reconstruct whole-genome phylogenies of different organisms. We first selected 12 prokaryotic

[4]http://flybase.org, dmel-all-intergenic-r5.49.fasta.

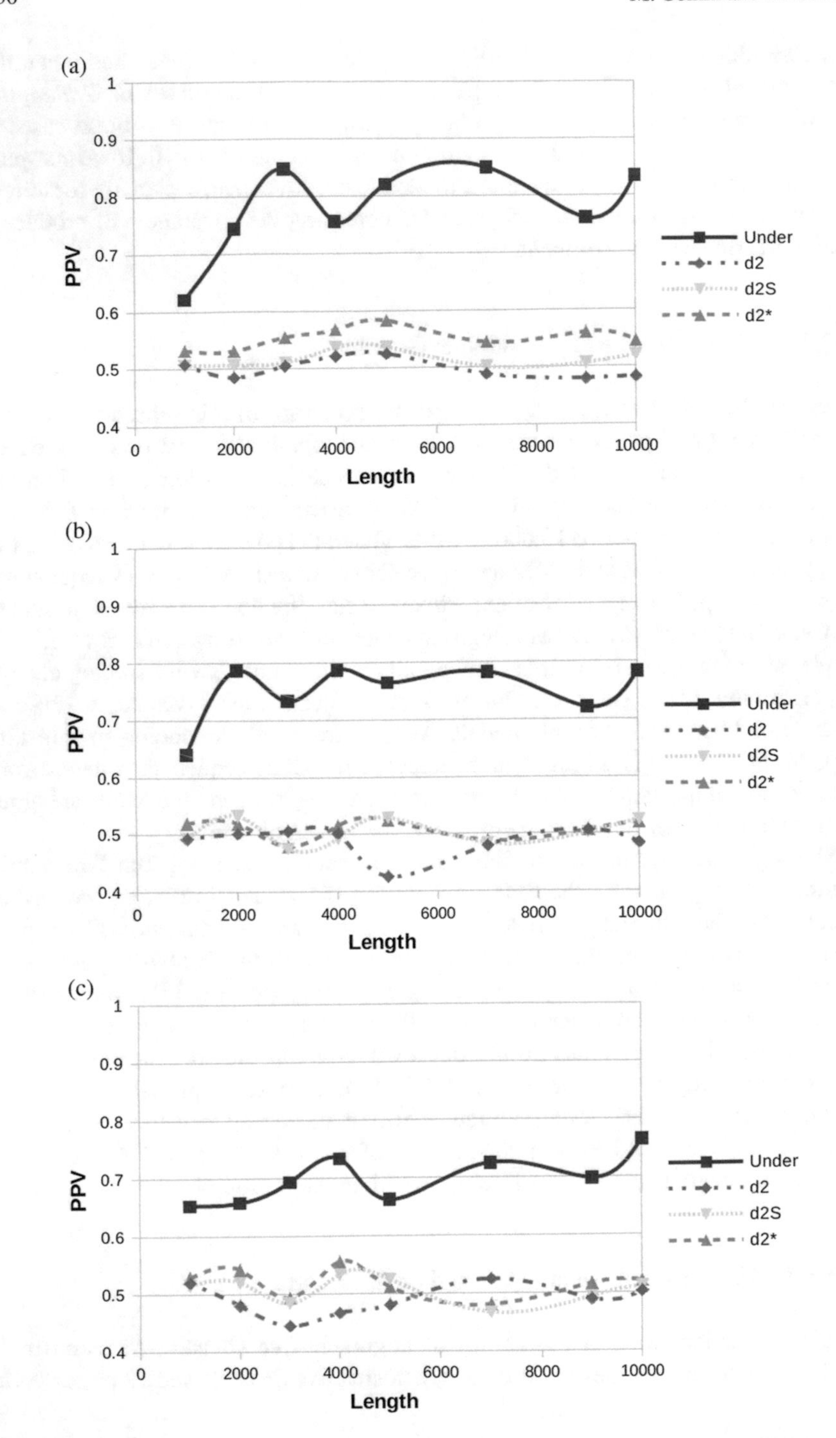

Fig. 14.5 PPV obtained with Drosophila genome as background. (**a**) CM model. (**b**) FMM model. (**c**) FMM model and MetaSim 454 errors model

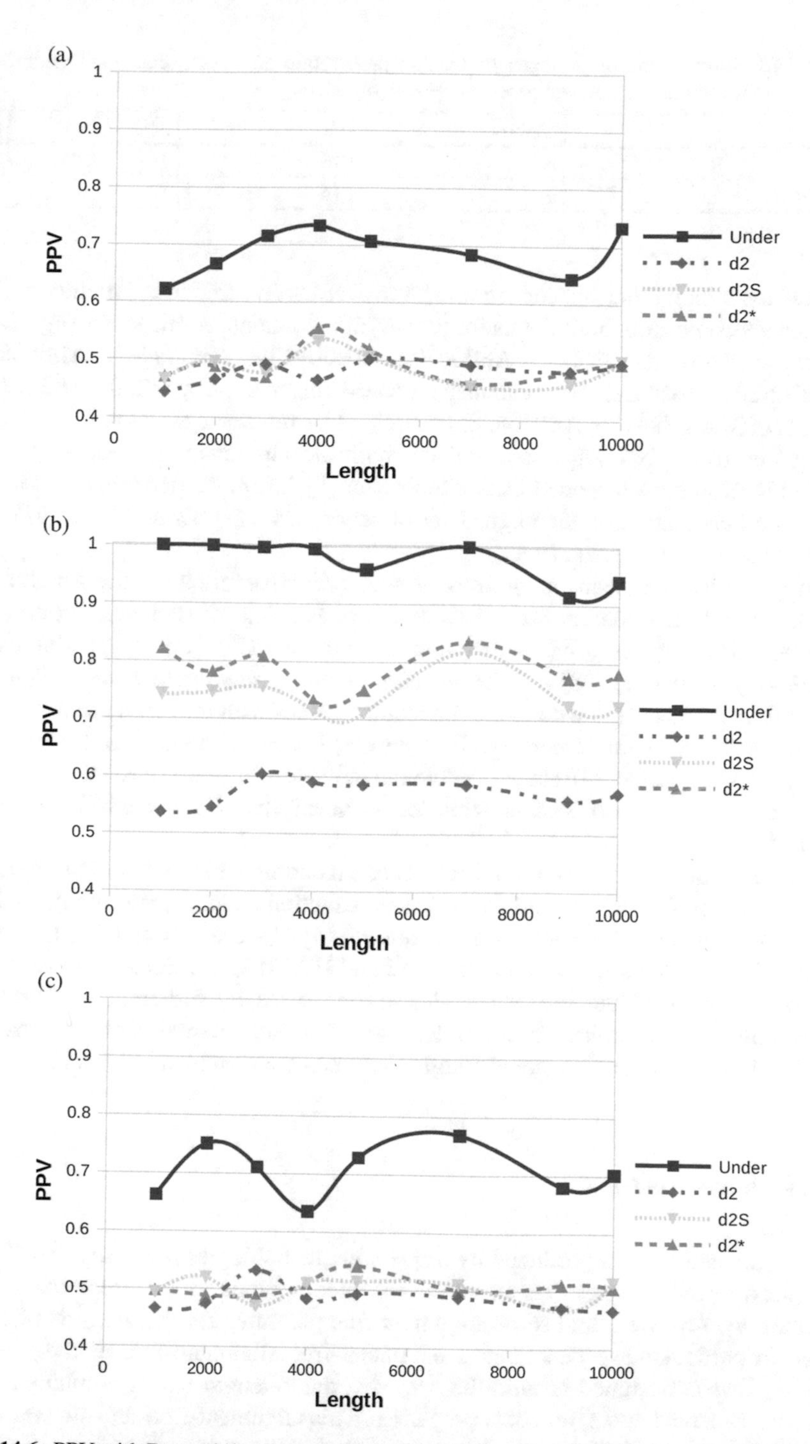

Fig. 14.6 PPV with Drosophila genome as background using the FMM model with 454 sequencing error for various values of parameters (γ, λ and β). (**a**) Double coverage ($\gamma = 10$). (**b**) More implanted patterns ($\lambda = 0.01$). (**c**) Shorter reads ($\beta = 100$)

Table 14.1 Comparison of phylogenetic trees of prokaryotic organisms, computed using NGS data, with the reference taxonomy based on the R–F distance

	$\overline{Under_2}$	d_2	d_2^S	d_2^*
NJ	8	16	14	14
UPGMA	8	16	12	14

organisms among the species in [48] for DNA phylogenomic inference. The organisms come from both the major prokaryotic domains: Archaea, six organisms (accession nos. BA000002, AE000782, AE009439, AE009441, AL096836, AE000520); Bacteria, six organisms (accession nos. AE013218, AL111168, AE002160, AM884176, AE016828, L42023). The reference taxonomy is inferred using the 16S rDNA sequences and the multiple alignment of these sequences available from the Ribosomal Database Project [5]. Then, we perform a maximum likelihood estimation on the aligned set of sequences using *Dnaml* from PHYLIP [19] to compute a reference tree.

We simulate the sequencing process with *MetaSim* following the same setup as above and then compute the distance matrices using all statistics. From these distance matrices, we derive the taxonomies with the PHYLIP [19] software using *neighbor joining* (NJ) [38] and the *unweighted pair group method with arithmetic mean* (UPGMA) [43]. We compare the resulting trees with the reference taxonomy using the *Robinson and Foulds* (R–F) distance [37]. For two unrooted binary trees with $n \geq 3$ leaves, the R–F score is in the range $[0, 2n-6]$. A score equal to 0 means that the two trees are isomorphic, while $2n-6$ means that all non-trivial bipartitions are different.

The R–F distances between the reference taxonomy and the resulting phylogenetic trees, for all statistics and the two reconstruction methods, are summarized in Table 14.1. In general, $\overline{Under_2}$ outperforms all d-type statistics obtaining the lower value with both reconstruction methods NJ and UPGMA. We can also observe that d_2^S and d_2^* obtain comparable results, and, in some cases, the former outperforms the latter, confirming a similar observation in [44]. This latter experiment confirms that $\overline{Under_2}$ is able to detect the genetic signal between unassembled NGS data.

14.4 Conclusions

As the amount of data produced by sequencing technologies increases, the ability to quickly process huge sets of reads becomes a fundamental problem. In this chapter, we reviewed the use of alignment-free measures for the analysis of NGS data. In particular, we described a parameter-free alignment-free method called $\overline{Under_2}$ that is designed around the use of variable-length words combined with specific statistical and syntactical properties. This alignment-free statistic was used to compare sets of NGS reads to detect the evolutionary relationship of unassembled genomes. We evaluated the performance of several alignment-free methods on both synthetic and real data. In almost all simulations, $\overline{Under_2}$ outperformed all other statistics. The performance gain became more evident when real genomes were used.

The use of fast alignment-free statistics can also help to address other important questions that relate to the analysis of reads. One notable area of investigation is to incorporate quality values within current alignment-free statistics. This can play a major role when dealing with noisy reads. Another future direction of investigation is the classification of reads into clusters; this is crucial for modern sequencers producing data from meta-genomes. Also others topics deserve more attention, such as meta-genome assembly and genome diversity estimation, just to name a few.

References

1. Altschul, S.F., Gish, W., Miller, W., Myers, E.W., Lipman, D.J.: Basic local alignment search tool. J. Mol. Biol. **215**(3), 403–410 (1990)
2. Apostolico, A.: Maximal words in sequence comparisons based on subword composition. In: Algorithms and Applications, pp. 34–44. Springer, Berlin/Heidelberg (2010)
3. Blaisdell, B.E.: A measure of the similarity of sets of sequences not requiring sequence alignment. Proc. Natl. Acad. Sci. **83**(14), 5155–5159 (1986)
4. Carneiro, M., Russ, C., Ross, M., Gabriel, S., Nusbaum, C., DePristo, M.: Pacific biosciences sequencing technology for genotyping and variation discovery in human data. BMC Genomics **13**(1), 375 (2012)
5. Cole, J.R., Wang, Q., Cardenas, E., Fish, J., Chai, B., Farris, R.J., Kulam-Syed-Mohideen, A.S., McGarrell, D.M., Marsh, T., Garrity, G.M., Tiedje, J.M.: The ribosomal database project: improved alignments and new tools for rRNA analysis. Nucleic Acids Res. **37**, D141–D145 (2009)
6. Comin, M., Antonello, M.: Fast computation of entropic profiles for the detection of conservation in genomes. In: Proceedings of Pattern Recognition in Bioinformatics PRIB. Lecture Notes in Bioinformatics, vol. 7986, pp. 277–288. Springer, Heidelberg (2013)
7. Comin, M., Antonello, M.: Fast entropic profiler: an information theoretic approach for the discovery of patterns in genomes. IEEE/ACM Trans. Comput. Biol. Bioinform. **11**(3), 500–509 (2014)
8. Comin, M., Schimd, M.: Assembly-free genome comparison based on next-generation sequencing reads and variable length patterns. BMC Bioinform. **15**(Suppl. 9), S1 (2014)
9. Comin, M., Verzotto, D.: Classification of protein sequences by means of irredundant patterns. BMC Bioinform. **11**, S16 (2010)
10. Comin, M., Verzotto, D.: The irredundant class method for remote homology detection of protein sequences. J. Comput. Biol. **18**(12), 1819–1829 (2011)
11. Comin, M., Verzotto, D.: Alignment-free phylogeny of whole genomes using underlying subwords. Algorithms Mol. Biol. **7**(1), 34 (2012)
12. Comin, M., Verzotto, D.: Whole-genome phylogeny by virtue of unic subwords. In: 23rd International Workshop on Database and Expert Systems Applications (DEXA), 2012, pp. 190–194 (2012)
13. Comin, M., Verzotto, D.: Beyond fixed-resolution alignment-free measures for mammalian enhancers sequence comparison. IEEE/ACM Trans. Comput. Biol. Bioinform. **11**(4), 628–637 (2014)

14. Comin, M., Leoni, A., Schimd, M.: Qcluster: extending alignment-free measures with quality values for reads clustering. In: Proceedings of the 14th Workshop on Algorithms in Bioinformatics (WABI). Lecture Notes in BIoinformatics (LNBI), vol. 8701, pp. 1–13 (2014)
15. Dai, Q., Wang, T.: Comparison study on k-word statistical measures for protein: from sequence to 'sequence space'. BMC Bioinform. **9**(1), 1–19 (2008)
16. Djebali, S., Davis, C.A., Merkel, A., Dobin, A., Lassmann, T., Mortazavi, A., Tanzer, A., Lagarde, J., Lin, W., Schlesinger, F., et al.: Landscape of transcription in human cells. Nature **489**(7414), 101–108 (2012)
17. Dohm, J.C., Lottaz, C., Borodina, T., Himmelbauer, H.: Sharcgs, a fast and highly accurate short-read assembly algorithm for de novo genomic sequencing. Genome Res. **17**(11), 1697–1706 (2007)
18. Eid, J., Fehr, A., Gray, J., Luong, K., Lyle, J., Otto, G., Peluso, P., Rank, D., Baybayan, P., Bettman, B., Bibillo, A., Bjornson, K., Chaudhuri, B., Christians, F., Cicero, R., Clark, S., Dalal, R., deWinter, A., Dixon, J., Foquet, M., Gaertner, A., Hardenbol, P., Heiner, C., Hester, K., Holden, D., Kearns, G., Kong, X., Kuse, R., Lacroix, Y., Lin, S., Lundquist, P., Ma, C., Marks, P., Maxham, M., Murphy, D., Park, I., Pham, T., Phillips, M., Roy, J., Sebra, R., Shen, G., Sorenson, J., Tomaney, A., Travers, K., Trulson, M., Vieceli, J., Wegener, J., Wu, D., Yang, A., Zaccarin, D., Zhao, P., Zhong, F., Korlach, J., Turner, S.: Real-time DNA sequencing from single polymerase molecules. Science **323**(5910), 133–138 (2009)
19. Felsenstein, J.: PHYLIP 1984 (Phylogeny Inference Package), Version 3.5c. Department of Genetics, University of Washington, Seattle (1993)
20. Gao, L., Qi, J.: Whole genome molecular phylogeny of large dsdna viruses using composition vector method. BMC Evol. Biol. **7**(1), 1–7 (2007)
21. Göke, J., Schulz, M.H., Lasserre, J., Vingron, M.: Estimation of pairwise sequence similarity of mammalian enhancers with word neighbourhood counts. Bioinformatics **28**, 656–663 (2012)
22. Huang, X., Yang, S.-P.: Generating a genome assembly with PCAP. Curr. Protoc. Bioinformatics **11**(3), 11.3.1–11.3.23 (2005)
23. Jaffe, D.B., Butler, J., Gnerre, S., Mauceli, E., Lindblad-Toh, K., Mesirov, J.P., Zody, M.C., Lander, E.S.: Whole-genome sequence assembly for mammalian genomes: Arachne 2. Genome Res. **13**(1), 91–96 (2003)
24. Kantorovitz, M.R., Robinson, G.E., Sinha, S.: A statistical method for alignment-free comparison of regulatory sequences. Bioinformatics **23**(13), i249–i255 (2007)
25. Koboldt, D.C., Steinberg, K.M., Larson, D.E., Wilson, R.K., Mardis, E.R.: The next-generation sequencing revolution and its impact on genomics. Cell **155**(1), 27–38 (2013)
26. Lander, E.S., et al.: Initial sequencing and analysis of the human genome. Nature **409**(6822), 860–921 (2001)
27. Li, H., Homer, N.: A survey of sequence alignment algorithms for next-generation sequencing. Brief. Bioinform. **11**(5), 473–483 (2010)
28. Lippert, R.A., Huang, H., Waterman, M.S.: Distributional regimes for the number of k-word matches between two random sequences. Proc. Natl. Acad. Sci. **99**(22), 13980–13989 (2002)
29. Liu, X., Wan, L., Li, J., Reinert, G., Waterman, M.S., Sun, F.: New powerful statistics for alignment-free sequence comparison under a pattern transfer model. J. Theor. Biol. **284**(1), 106–116 (2011)
30. Metzker, M.L.: Sequencing technologies – the next generation. Nat. Rev. Genet. **11**(1), 31–46 (2010)
31. Miller, J.R., Koren, S., Sutton, G.: Assembly algorithms for next-generation sequencing data. Genomics **95**(6), 315–327 (2010)
32. Pevzner, P.A., Tang, H., Waterman, M.S.: An Eulerian path approach to dna fragment assembly. Proc. Natl. Acad. Sci. **98**(17), 9748–9753 (2001)
33. Pop, M., Salzberg, S.L.: Bioinformatics challenges of new sequencing technology. Trends Genet. **24**(3), 142–149 (2008)
34. Qi, J., Luo, H., Hao, B.: Cvtree: a phylogenetic tree reconstruction tool based on whole genomes. Nucleic Acids Res. **32**(Suppl. 2), W45–W47 (2004)

35. Reinert, G., Chew, D., Sun, F., Waterman, M.S.: Alignment-free sequence comparison (i): statistics and power. J. Comput. Biol. **16**(12), 1615–1634 (2009)
36. Richter, D.C., Ott, F., Auch, A.F., Schmid, R., Huson, D.H.: Metasim—a sequencing simulator for genomics and metagenomics. PLoS ONE **3**(10), e3373 (2008)
37. Robinson, D.F., Foulds, L.R.: Comparison of phylogenetic trees. Math. Biosci. **53**(1–2), 131–147 (1981)
38. Saitou, N., Nei, M.: The neighbor-joining method: a new method for reconstructing phylogenetic trees. Mol. Biol. Evol. **4**(4), 406–425 (1987)
39. Sanger, F., Nicklen, S., Coulson, A.R.: DNA sequencing with chain-terminating inhibitors. Proc. Natl. Acad. Sci. **74**(12), 5463–5467 (1977)
40. Simpson, J.T., Wong, K., Jackman, S.D., Schein, J.E., Jones, S.J.M., Birol, I.: ABySS: a parallel assembler for short read sequence data. Genome Res. **19**(6), 1117–1123 (2009)
41. Sims, G.E., Jun, S.-R., Wu, G.A., Kim, S.-H.: Alignment-free genome comparison with feature frequency profiles (FFP) and optimal resolutions. Proc. Natl. Acad. Sci. **106**(8), 2677–2682 (2009)
42. Smith, T.F., Waterman, M.S.: Identification of common molecular subsequences. J. Mol. Biol. **147**(1), 195–197 (1981)
43. Sneath, P.H.A., Sokal, R.R.: Unweighted pair group method with arithmetic mean. In: Numerical Taxonomy, pp. 230–234. W. H. Freeman, San Francisco (1973)
44. Song, K., Ren, J., Zhai, Z., Liu, X., Deng, M., Sun, F.: Alignment-free sequence comparison based on next-generation sequencing reads. J. Comput. Biol. **20**(2), 64–79 (2013)
45. Song, K., Ren, J., Reinert, G., Deng, M., Waterman, M.S., Sun, F.: New developments of alignment-free sequence comparison: measures, statistics and next-generation sequencing. Brief. Bioinform. **15**(3), 343–353 (2013). bbt067
46. Staden, R.: A strategy of dna sequencing employing computer programs. Nucleic Acids Res. **6**(7), 2601–2610 (1979)
47. Treangen, T.J., Salzberg, S.L.: Repetitive DNA and next-generation sequencing: computational challenges and solutions. Nat. Rev. Genet. **13**(1), 36–46 (2011)
48. Ulitsky, I., Burstein, D., Tuller, T., Chor, B.: The average common substring approach to phylogenomic reconstruction. J. Comput. Biol. **13**(2), 336–350 (2006)
49. Vinga, S., Almeida, J.: Alignment-free sequence comparison – a review. Bioinformatics **19**(4), 513–523 (2003)
50. Wan, L., Reinert, G., Sun, F., Waterman, M.S.: Alignment-free sequence comparison (II): theoretical power of comparison statistics. J. Comput. Biol. **17**(11), 1467–1490 (2010)
51. Warren, R.L., Sutton, G.G., Jones, S.J.M., Holt, R.A.: Assembling millions of short DNA sequences using SSAKE. Bioinformatics **23**(4), 500–501 (2007)
52. Zerbino, D.R., Birney, E.: Velvet: algorithms for de novo short read assembly using de Bruijn graphs. Genome Res. **18**(5), 821–829 (2008)

Printed in the United States
By Bookmasters